EXOTIC PESTS and DISEASES

Biology and Economics
for BIOSECURITY

EXOTIC PESTS and DISEASES

Biology and Economics for BIOSECURITY

Daniel A. Sumner, Editor

Iowa State Press

A Blackwell Publishing Company

Daniel A. Sumner is Director of the University of California Agricultural Issues Center and the Frank H. Buck, Jr., Professor, Department of Agricultural and Resource Economics, University of California, Davis. He teaches and conducts research and out-reach programs in the area of agricultural economics, policy, and international issues. Dr. Sumner is a former Chair of the International Agricultural Trade Research Consortium, and his research and writing has won American Agricultural Economic Association awards for Quality of Research Discovery, Quality of Communication, and Distinguished Policy Contribution. In recognition of his career contributions, Sumner was named a Fellow of AAEA in 1999.

Iowa State Press
2121 State Avenue, Ames, Iowa 50014

Orders: 1-800-862-6657
Office: 1-515-292-0140
Fax: 1-515-292-3348
Web site: www.iowastatepress.com

First edition, 2003

Library of Congress Cataloging-in-publication data

Exotic pests and diseases: biology and economics for biosecurity/edited by Daniel A. Sumner.—1st ed.
       p. cm.
Includes bibliographical references and index.
       ISBN 0-8138-1966-0 (alk. paper)
    1. Nonindigenous pests—United States. 2. Nonindigenous pests—Control—United States. 3. Agriculture—Economic aspects—United States. 4. Agriculture and state—United States. I. Sumner, Daniel A. (Daniel Alan), 1950–
    SB990.5.U6E95 2003
    632′.9′0973—dc21   2002155232

The last digit is the print number: 9 8 7 6 5 4 3 2 1

Contents

Preface, vii

Contributors, ix

1. Exotic Pests and Public Policy for Biosecurity: 3
   An Introduction and Overview
   Daniel A. Sumner

I. Issues, Principles, Institutions, and History   7

2. Economics of Policy for Exotic Pests and Diseases: Principles and Issues   9
   Daniel A. Sumner
3. Regulatory Framework and Institutional Players   19
   Marcia Kreith and Deborah Golino
4. International Trade Agreements and Sanitary and Phytosanitary Measures   39
   James F. Smith,
5. Historical Perspectives on Exotic Pests and Diseases in California   55
   Susana Iranzo, Alan L. Olmstead, and Paul W. Rhode

II. Exotic Pest and Disease Cases: Examples of Economics and Biology and Policy Evaluation   69

6. Bovine Spongiform Encephalopathy: Lessons from the United Kingdom   71
   José E. Bervejillo and Lovell S. Jarvis
7. Evaluating the Potential Impact of a Foot-and-Mouth Disease Outbreak   85
   Javier Ekboir, Lovell S. Jarvis, and José E. Bervejillo
8. Risk Assessment of Plant-Parasitic Nematodes   99
   Howard Ferris, Karen M. Jetter, Inga A. Zasada, John J. Chitambar,
   Robert C. Venette, Karen M. Klonsky, and J. Ole Becker
9. Ex-Ante Economics of Exotic Disease Policy: Citrus Canker in California   121
   Karen M. Jetter, Edwin L. Civerolo, and Daniel A. Sumner
10. An Insect Pest of Agricultural, Urban, and Wildlife Areas: The Red Imported
    Fire Ant   151
    John H. Klotz, Karen M. Jetter, Les Greenberg, Jay Hamilton,
    John Kabashima, and David F. Williams
11. A Rational Regulatory Policy: The Case of Karnal Bunt   167
    Joseph W. Glauber and Clare Narrod
12. Introduction and Establishment of Exotic Insect and Mite Pests of
    Avocados in California, Changes in Sanitary and Photosanitary Policies,
    and Their Economic and Social Impact   185
    Mark S. Hoddle, Karen M. Jetter, and Joseph Morse

13. Ash Whitefly and Biological Control in the Urban Environment   203
*Timothy D. Paine, Karen M. Jetter, Karen M. Klonsky, Larry G.
Bezark, and Thomas S. Bellows*

14. Economic Consequences of a New Exotic Pest: The Introduction
of Rice Blast Disease in California   215
Jung-Sup Choi, Daniel A. Sumner, Robert K. Webster, and Christopher A. Greer

15. Biological Control of Yellow Starthistle   225
*Karen M. Jetter, Joseph M. DiTomaso, Daniel J. Drake, Karen M. Klonsky,
Michael J. Pitcairn, and Daniel A. Sumner*

Glossary of Terms and Acronyms   243

Index   251

Preface

This book grew out of a large interdisciplinary project at the University of California Agricultural Issues Center. That project resulted in a major public forum and a summary report published in December 1999. A large number of study teams prepared presentations for that forum, and many of those efforts were further developed into the chapters that appear in Part 2 of this book. In addition, at the public forum representatives from the U.S. Department of Agriculture (USDA) and the State of California emphasized the importance of policy measures to protect against the introduction and spread of exotic pests and diseases.

The California Department of Food and Agriculture (CDFA), the USDA and the Division of Agriculture and Natural Resources (DANR) of the University of California supported the original project. Staff members from those organizations joined University faculty and industry representatives as project advisors and on the study teams. Leaders of those institutions were strong advocates for the project. In particular, then Secretary Ann Veneman helped initiate the project and current Secretary William (Bill) Lyons continued the support from the CDFA. At the USDA several leaders of the regional offices of the Animal and Plant Health Inspection Service helped support and guide the project. At DANR W.R. (Reg) Gomes, Henry Vaux, and Joseph Morse provided strong institutional backing.

Finally, turning the results of the larger project into a book manuscript for a broad audience required the help of several staff members at the Agricultural Issues Center. Marcia Kreith, who edited the Conference Summary Report with Ray Coppock, continued to provide guidance on all phases of the project. Karen Jetter, who was a coauthor of several case studies, was indispensable in coordinating the research with biologists. José Bervejillo coauthored the chapters on livestock diseases and helped prepare the manuscript for review. Finally, Gary Beall and Laurie Treacher brought patience, good humor, and professionalism to manuscript preparation and editing.

The threat of exotic pests and diseases cannot be eliminated, but improved analysis of policy alternatives can reduce the costs and increase the benefits of the policies chosen to respond to this threat. The aim of this book is to provide some of the needed analysis and to stimulate additional work on this important topic.

Daniel A. Sumner
July 2002

Contributors

J. Ole Becker, Department of Nematology, University of California, Riverside

Thomas S. Bellows, Department of Entomology, University of California, Riverside

José E. Bervejillo, University of California Agricultural Issues Center

Larry G. Bezark, California Department of Food and Agriculture, Biological Control Division

John J. Chitambar, California Department of Food and Agriculture Plant Pest Diagnostics Center—Nematology Laboratory

Jung-Sup Choi, Korea Rural Economic Institute; San Joaquin Valley Agricultural Sciences Center

Edwin L. Civerolo, USDA-ARS, PWA

Joseph M. DiTomaso, Department of Vegetable Crops, University of California, Davis

Daniel J. Drake, University of California Cooperative Extension, Siskiyou County

Javier Ekboir, CYMMIT, Mexico City

Howard Ferris, Department of Nematology, University of California, Davis

Joseph W. Glauber, United States Department of Agriculture, Office of the Chief Economist

Deborah Golino, Department of Plant Pathology, University of California, Davis

Les Greenberg, Department of Entomology, University of California, Riverside

Christopher A. Greer, Department of Plant Pathology, University of California, Davis

Jay Hamilton, John Jay College of Criminal Justice, City University of New York

Mark S. Hoddle, Department of Entomology, University of California, Riverside

Susana Iranzo, Institute of Governmental Affairs, University of California, Davis

Lovell S. Jarvis, Department of Agricultural and Resource Economics and the Giannini Foundation, University of California, Davis

Karen M. Jetter, University of California Agricultural Issues Center

John Kabashima, University of California Cooperative Extension, Orange County

Karen M. Klonsky, Department of Agricultural and Resource Economics, University of California Davis

John H. Klotz, Department of Entomology, University of California, Riverside

Marcia Kreith, University of California Agricultural Issues Center

Joseph Morse, Department of Entomology, University of California, Riverside and Statewide Program for Agricultural Policy and Pest Management

Clare Narrod, United States Department of Agriculture, Office of Risk Assessment and Cost-Benefit Analysis

Alan L. Olmstead, Department of Economics, University of California, Davis and Institute of Governmental Affairs

Timothy D. Paine, Department of Entomology, University of California, Riverdale

Michael J. Pitcairn, California Department of Food and Agriculture, Biological Control Division

Paul W. Rhode, Department of Economics, University of North Carolina

James F. Smith, University of California, Davis, School of Law

Daniel A. Sumner, University of California Agricultural Issues Center and Department of Agricultural and Resource Economics, University of California, Davis

Robert C. Venette, Department of Entomology, University of Minnesota

Robert K. Webster, Department of Plant Pathology, University of California, Davis

David F. Williams, United States Department of Agriculture, Agricultural Research Service, Center for Medical, Agricultural, and Veterinary Entomology

Inga A. Zasada, Department of Nematology, University of California, Davis

EXOTIC PESTS and DISEASES

Biology and Economics for BIOSECURITY

1

Exotic Pests and Public Policy for Biosecurity: An Introduction and Overview

Daniel A. Sumner

Policies related to the introduction and spread of harmful nonindigenous invasive species (also referred to as exotic pests and diseases) are central to biosecurity.[1] Harmful organisms move across borders readily with the flow of people and commerce and in the context of natural habitats, which themselves cross political boundaries. Most such movements are accidental or incidental to other human or natural activities. However, the purposeful and malicious introduction of such species has long been a concern. Since September 11, 2001, the potential spread of exotic pests and diseases to further the aims of terrorism has had a much higher profile and has heightened interest in a better understanding of biosecurity more generally.

No public issues are more important to agriculture than those related to pests and diseases. The introduction and spread of pests and diseases have the potential to destroy whole industries or to cause massive losses in short periods of time. In some cases, the very existence of crop or livestock production in a location depends on the effective control of plant and animal pests and diseases. Recent experiences with bovine spongiform encephalopathy (BSE) and more recently foot-and-mouth disease (FMD) in Britain reinforce the sense of vulnerability, and historical evidence suggests such occurrences are not unique. (See Chapter 5 for several interesting cases from the history of exotic pests and diseases.)

Introductions of exotic agricultural pests may also have serious implications for the natural environment beyond agriculture. For example, entry and spread of FMD affect deer and other wildlife, and nonindigenous weeds affect

the natural habitat as well as farm and grazing land. Finally, exotic pests can be important for human health. Indeed, some of the most important exotic pest cases, such as BSE, concern human health. Of course, these impacts are linked because anything that affects perceptions of food safety also affects the demand for agricultural products and hence market prices and quantities.

The issues are becoming increasingly complex and urgent. As globalization proceeds, agricultural trade, international travel, and other global connections increase. As a consequence, the probability of spreading exotic pests and diseases increases. Since September 11, 2001, we have been even more aware that policies and programs must cope with the intentional introduction and spread of exotic agricultural pests and diseases.

For decades, agricultural analysts, policymakers, and organizations have focused most of their attention on commodity subsidy programs, such as price and income supports and import barriers. Thousands of scholarly articles, hundreds of books, and many more government reports and other documents have analyzed the effects of these subsidy programs on agricultural markets, farm incomes, and consumer prices. International trade barriers and subsidies have also been studied intensively. These same policies have received similar attention in the popular and trade press. Compared with this huge volume of writing, there has been relatively little analysis of government policies related to exotic agricultural pests and diseases. There has long been much biological research on exotic pests. (See, for example, Chapter 5 and the work summarized in U.S. Congress Office of Technology Assessment 1993.) However, additional research by economists and others on policy evaluation now seems to be forthcoming at a more rapid pace (National Research Council 2000; Anderson, McRae, and Wilson 2001). One reason for this additional attention relates to expanded agricultural trade and new international trade agreements, especially North American Free Trade Agreement (NAFTA) and the creation of the World Trade Organization (WTO). These agreements have raised new issues of national and international relevance. The Sanitary and Phytosanitary (SPS) Agreement in the General Agreement on Tariffs and Trade of 1994 establishes rules for policies concerning introduction and spread of exotic pests.

The goal of this book is to help provide a sound analytical and empirical basis for public policy decisions about exotic pests and diseases and to increase understanding of the interrelated issues. The focus is on agricultural pests and consequences for agriculture, food safety and the environment. The book deals with questions about appropriate roles for government in prevention of the entry and spread of exotic pests in the context of international agreements.

Part I, "Issues, Principles, Institutions, and History," provides background on exotic pest policy[1] issues from several perspectives. In Chapter 2, Sumner reviews the basic notion of and some evidence about the public good aspects of exotic pest policy. He considers the conditions under which economic models suggest collective action rather than leaving the issue to private firms. He concludes that the range over which collective action is economically appropriate is defined by the biology of pest habitats and that additional funding with commodity fees may better link costs to the beneficiaries of the programs. The chapter also reviews briefly methods used to evaluate exotic pest policy.

Chapter 3, by Kreith and Golino, surveys regulations for preventing the movement of exotic pests and diseases across state and national borders and the eradication and control policies applied by the U.S. and state governments. It includes a chronology of the evolution of these policies and regulations. This chapter helps set the stage for the chapters that follow by reviewing the regulatory and institutional framework

that applies to many of the cases that are the focus of the second part of the book.

Smith analyzes the implications of the "Agreement on the Application of Sanitary and Phytosanitary Measures" (SPS Agreement) in Chapter 4. The SPS Agreement is designed to interpret and implement the 1994 provision that WTO members may adopt measures that would otherwise violate WTO provisions, such as national treatment, if the provisions are necessary to protect human, animal, or plant life or health. The chapter analyzes the WTO rules and decisions and focuses on how the rules have been interpreted in the resolution of some important disputes.

In Chapter 5, Iranzo, Olmstead, and Rhode review the early history of exotic pests and diseases in California crop agriculture. When California gained statehood in 1850, the area was relatively free of agricultural pests and diseases. However, by about 1870 a succession of invaders attacked the state's new crops, threatening the commercial survival of many horticultural commodities. Thus, within a few decades, California's farmers went from working in an almost pristine environment to facing an appalling list of enemies in an age when few effective methods had been developed for large-scale pest control. This chapter details campaigns to combat several imported pests and diseases, including powdery mildew, phylloxera, Pierce's disease, San Jose scale, and cottony cushion scale. The efforts to combat these and other problems required the creation of a scientific and institutional infrastructure that still shapes pest and disease control policies.

Part II, "Exotic Pest and Disease Cases: Examples of Combining Economics and Biology for Policy Evaluation," provides 10 interdisciplinary case studies that focus on specific pests or diseases. These cases represent a wide range of threats to U.S. agriculture, wildlands, and the urban landscape and discuss the possible government responses to these threats. Each chapter combines, in an original fashion, biological foundations, economic analysis, and implications for public policy, giving insights to a series of public policy issues of national and international relevance. These chapters measure agricultural impacts broadly in terms of industry costs and returns and market prices that affect consumers and producers.

Chapters 6 and 7 deal with two of the most important diseases facing the livestock industry. BSE, also known as mad cow disease, has affected more than 180,000 bovines in the United Kingdom since 1986. The most transcendental impact of this epidemic was the recognition that the human disease known as the new variant of Creutzfeldt-Jakob disease (vCJD) could have originated in the consumption of BSE-infected meat. No cases of BSE have been detected in the United States, and the Animal and Plant Health Inspection Service (APHIS) has implemented a succession of policy measures to reduce the risk of its entry and spread. Chapter 6 analyzes the epidemiology and the economic aspects of the BSE epidemic, based on the United Kingdom experience, and points toward major policy issues for the United States. In Chapter 7, Ekboir, Jarvis, and Bervejillo consider the economic costs of a potential outbreak of FMD, which is probably the most contagious of all mammalian diseases. This chapter presents an epidemiological model that simulates an FMD outbreak in California's Central Valley. The direct and indirect costs of the outbreak are evaluated under different scenarios. To reduce the costs of an FMD outbreak, proper surveillance mechanisms are essential, operating through government agencies, the livestock industry, and farmers' organizations. An FMD outbreak in California would have major impact on the livestock and related industries throughout the United States through the disruption of domestic and international livestock trade flows.

A large team of scientists and economists led by Ferris and Jetter considers plant-parasitic nematodes in Chapter 8. Five species are selected for their policy and trade implications and their biological and historical significance. Current intervention strategies are described and evaluated. In Chapter 9, Jetter, Civerolo, and Sumner examine the potential impacts of the introduction of citrus canker and review the costs and benefits of eradication or private control. Considerable national and international regulatory efforts are designed to prevent spread of the pathogen to citrus-growing regions around the world where the disease is not endemic but where environmental conditions are conducive to disease development. A model parameterized with market and biological data is used to show how the equilibrium quantities, prices, and other variables respond to introduction of the disease or to eradication. Chapter 10 turns to the red imported fire ant (RIFA), a pest that affects agricultural, urban, and wildlife areas. This chapter evaluates policy options ranging from eradication to letting the RIFA become established combined with private controls and quarantines. The expected costs and benefits of eradication are compared, taking into account uncertainty over the success of the eradication program.

In Chapter 11, Glauber and Narrod combine probabilistic risk assessments with economic analysis. Results show that if the U.S. Department of Agriculture (USDA) had incorporated risk into its benefit-cost analysis of karnal bunt, a disease affecting wheat, it would have reached different conclusions about the impact of its actions. The authors estimate that suboptimal regulatory decisions in the case of karnal bunt cost between $350 million and $390 million per year. They recommend that the USDA incorporate risk assessments into its economic analyses of proposed regulations.

Since 1996, changes in sanitary and phytosanitary regulations and exotic pest introductions have strongly affected the U.S. avocado industry. In 1996, avocado thrips were identified in California avocado groves. In Chapter 12, Hoddle, Jetter, and Morse evaluate the welfare effects of recent trade agreements and the establishment of avocado thrips. Paine, Jetter, Klonsky, Bezark, and Bellows next analyze the program implemented in 1989 to control the urban ash whitefly infestation by introducing a natural enemy that only attacked the exotic pest (biological control). The program ended with the successful control of the ash whitefly by 1992. They assess the costs and benefits of this successful biological control program that required effective collaboration among universities, government, the agricultural industry, and homeowners.

Chapter 14, by Choi, Sumner, Webster, and Greer considers the economic consequences of a new exotic pest that was allowed to become established. Rice blast disease was first found in California in 1996 and has already caused considerable loss to the rice industry. The economic impact of rice blast on the price and quantity of rice production and related economic variables are assessed, as well as the economic benefits and costs of integrated blast control measures. Eradication was deemed biologically difficult and economically questionable and therefore not attempted in this case.

Yellow starthistle, the subject of Chapter 15 by Jetter, DiTomaso and coauthors has been a problem in the western United States for decades. It interferes with grazing and lowers yield and forage quality of rangelands, thus increasing the cost of managing livestock. It can also reduce land value and reduce access to recreational areas. This chapter describes the biology, introduction, and spread of this invasive weed and discusses the feasible intervention and control strategies. The chapter analyzes the economics of biological control approaches to dealing with the weed.

This book adds to the growing literature on the evaluation of policies to deal with exotic pests and diseases. The regulations concerning introduction, eradication, and related measures are evaluated in detail. The case studies that make up the core of the book highlight the importance of combining sound biology with economic analysis to evaluate exotic pest policy.

Notes

[1]This book often uses the shorthand "exotic pest policy" to refer to the host of public decisions related to the entry and spread of nonindigenous harmful invasive species. We use the term "pest" to refer to the host of insects, plant or animal diseases, weeds, etc. that may cause harm to agriculture, the natural environment, or human health.

References

Anderson, Kym, Cheryl McRae, and Davis Wilson Eds., 2001. *The Economics of Quarantine and the SPS Agreement.* CIES-AFFA Biosecurity Australia.

National Research Council. 2000. *Incorporating Science, Economics, and Sociology in Developing Sanitary and Phytosanitary Standards in International Trade.*

U.S. Congress Office of Technology Assessment. 1993. *Harmful Non-Indigenous Species in the U.S.*

PART I

Issues, Principles, Institutions, and History

2

Economics of Policy for Exotic Pests and Diseases: Principles and Issues

Daniel A. Sumner

Government activities regarding exotic pests and diseases are pervasive and important. Such activities include restricting the movements of products and people across internal and external borders, destroying crops and livestock, requiring pesticide treatments on a wide scale, and research and development on the control of harmful species. These activities have large direct budget costs and much larger costs in the markets affected. Research has documented that in many cases large benefits derive from lowered costs of production, improved food quality, reduced human health threat, and improved environmental quality.

Despite their importance, however, appropriate data and analysis have been lacking to document the overall magnitude of the economic threat from exotic pests and diseases and the economic effects of government policy concerning these pests. No comprehensive study has measured the total benefits to agriculture, the environment, or human health regarding efforts to control the entry and spread of exotic plant and animal pests and diseases. Economists have focused much more analysis on explicit government transfer and subsidy programs included under the rubric of farm price and income supports and on agricultural trade protection than on pest policy. Even a cursory check of the academic or government literature in agricultural economics would show hundreds of studies of price and income supports and a relative handful of studies related to pest policy. That situation may be changing with a rapid expansion of the literature related to exotic pests, risk analysis, and the Sanitary and Phytosanitary (SPS) agreement.[1]

Exotic pest policy has much in common with policy related to research and development. Both of these policy areas involve investments that pay off over a period of years, joint public and private activities, and public good rationales for a government role. There has been far more attention paid to agricultural research than to exotic pest policy (Alston et al. 1995), and some of the approaches applied to that topic are useful in studying exotic pest policy.

In this chapter, I use the shorthand *exotic pest policy* to refer to a host of public decisions related to the entry and spread of nonindigenous harmful invasive species. I use the term *pest* to refer to the host of insects, plant or animal diseases, weeds, etc. that may cause harm to agriculture, the natural environment, or human health. Agricultural impacts are observed in industry costs and returns and in market prices that affect consumers and producers. Introduction of exotic agricultural pests may also have serious implications for the natural environment beyond agriculture. For example, entry and spread of foot-and-mouth disease (FMD) affects deer and other wildlife, and nonindigenous weeds affect the natural habitat, as well as farm and grazing land. Finally, exotic pests can be important for human health. Indeed, some of the most important exotic pest cases, such as bovine spongiform encephalopathy (BSE), also known as "mad cow disease," concern human health. These impacts are linked because anything that affects perception of food safety also affects the demand for agricultural products and, hence, market prices and quantities.

First, in this paper I review the public good aspects of control of exotic pests. This discussion is fundamental to the theory of an optimal role for government in prevention of the entry and spread of exotic pests. Under what conditions do standard economic models suggest collective action rather than leaving the issue to private firms? Second, I outline some tools used to assess the economic effects of exotic pest policies. These tools are applied to the individual case studies developed in part 2 of this book. Finally, I review the implementation of exotic pest policy in the United States, including the role of the World Trade Organization (WTO) and the Sanitary and Phytosanitary (SPS) Agreement of 1994 (WTO 1995).

Exotic Pests as Public "Bads" and Pest Exclusion and Eradication as Public "Goods"

The economics of exotic pest policy is based on the economic ideas of public goods and externalities (for a standard textbook treatment see Pindyke and Rubenfeld, 1998). Therefore, as a starting point, we must consider how these economic concepts are defined and how they apply to exotic pests and pest policy.

The Basic Economics of Public Goods

Economists often note an active role for government in markets for goods or services that unaided market forces would fail to provide to a sufficient degree. Public goods are defined as being "nonrival" in consumption, meaning that use by one consumer does not preclude, or make more expensive, consumption by another. Furthermore, excluding the use of public goods or services by those who do not pay is impractical or costly. When these two characteristics describe a market, it is difficult for private producers to profitably provide the good (or service). These market characteristics give rise to the "free rider"[2] problem, the situation under which a user of a service does not voluntarily pay, anticipating that others will pay and the service will continue. Of course, for public goods each consumer has the same incentive, and the service may not be supplied in sufficient quantity or at all.

The benefit of providing an additional unit of a good or service for which consumption by many users can occur without rivalry is calculated as the sum of the individual benefits of consuming that marginal unit. For a "normal" good, the marginal benefit is just the willing-

ness to pay by the marginal user who places the least value on that last unit of the good or service.

Classic examples of public goods include basic scientific research, national defense, and protection from highly communicable human diseases. In the case of basic scientific research, users are other scientists and technology developers who build on the basic science without precluding other scientists from also using the science. For example, when an applied biologist uses the principles of evolution in her research, it obviously does not interfere with the use of evolutionary principles by other scientists. Furthermore, once a principle of basic science is found, it is impractical to exclude other researchers from using that principle. For these reasons, it is difficult to bear the cost of basic scientific research into fundamental scientific principles in anticipation of profitably marketing the results. However, applied research and development that is tied to specific products may be profitably conducted by private firms that market the product innovations themselves. These same ideas apply to protection from communicable diseases. Individuals have clear incentives to purchase typical health care services. However, protecting one person from an epidemic protects everyone in the area, and the cost of protection services is not substantially higher when an additional person is added to the community. Thus underinvestment in such protection is likely without some collective action.

The concept of public goods is related to the notion of "externalities" or external costs and benefits. An external cost occurs when producers or consumers do not incur the full cost of their actions and, thus, do not include some of the costs in their production or consumption decisions. For example, when there is no market incentive or direct charges related to polluting a stream, a profit-maximizing producer does not take into account the cost of the lower-quality water on individuals who also use the stream. External benefits may also occur. A cattle ranch may supply scenic views for which it is often impractical to charge beneficiaries. In that case the rancher may provide less scenery to viewers than if these benefits could be internalized through some fee for viewing. In this example, the external benefit of a scenic view also has public good characteristics because consumption by one viewer does not preclude another viewer from looking, and it costs no more to provide the view for an additional viewer.

These examples should make clear that the existence of public good characteristics provides some rationale for a more active public policy action in an area; however, such characteristics do not preclude private market activity nor do they determine precisely the appropriate form of government action.

Some well-known papers in the economic literature suggest caution about an overeager finding of market failure. Coase (1974) showed that the history of the classic public good example of lighthouses was often inaccurate. Historically, many lighthouses were operated by private firms, which charged shipping companies for their services. They turned on the light to protect paying customers and turned off the light when passing ships had not paid the fee. Cheung (1973) investigated the classic example in which bees owned by honey producers provided an external benefit to fruit growers that needed pollination services. In this case, Cheung did not find an externality, but instead found a well-functioning private market in which orchard owners paid beekeepers for the services provided.

Some services related to exotic pest prevention seem to fall naturally into the category of industry-wide public goods. Furthermore, with respect to exotic pests, one can see a number of potential externalities associated with private behavior. However, the examples described by Coase and Cheung suggest we proceed carefully before presuming specific policy remedies.

Individual farms undertake most agricultural pest protection in the United States. Farmers use products and services bought in relatively competitive private markets to control pests affecting their crops and livestock. They respond to clear private profit incentives to reduce losses and maintain the value of their land and other capital assets. For most pest problems these incentives are deemed sufficient to encourage appropriate responses to pest occurrences. Of course, there are many regulations related to pest management on farms, but most of these are related to concerns about environmental externalities, worker health and safety, or food safety (Antle 1988). The regulations are not directed toward dealing with external costs of pests spreading across farms or the public good

characteristics of pest control. Exotic pests are treated differently. Governments themselves undertake explicit activities related to exclusion or control of exotic pests.

Exotic Pests and Border Measures

We will first consider limiting the spread of exotic pests with border measures to prohibit entry into a nation or a region within a nation. Border measures, such as tariffs, that restrict or tax international trade are often undertaken by national governments and are the prime issue dealt with by the WTO. The WTO-SPS agreement sets international rules for the application of regulations designed to allow countries to protect the health and safety of agriculture, the environment, and a country's residents, while also complying with WTO principles to facilitate trade. (For more details see Chapter 3 and, especially, Chapter 4.)

When an agricultural pest is excluded at the border, benefits accrue to producers and consumers of products whose costs of production would rise (or quality would deteriorate) upon entry and spread of the pest. Exclusion of the pest from a region lowers the marginal costs for a group of producers, and the per-unit cost savings do not depend on the amount of production in the region. The number of other producers that also experience cost savings does not affect the per-unit cost that is saved by any individual producer. That is, the number of direct "users" of exclusion services does not affect the benefit for any single user. Furthermore, the amount of the exclusion service used, the number of farms or acres, does not affect the cost of exclusion services. That is, border control costs about the same to protect an industry of 10,000 acres and 100 farms as for an industry of 100,000 acres and 1,000 farms. From this point of view we may consider the farms as consumers of the border exclusion services and note the nonrivalry in consumption of these services. Alternatively, we may consider the demand for exclusion of exotic pests as a derived demand based on the demand for the final product. Once again, the number of consumers or the amount of consumption of the commodity generally does not affect the cost of providing border exclusion for a pest that may affect the cost of that commodity. From this perspective, too, the criterion of nonrivalry in consumption of pest exclusion services is met. For example, if individual farms do not have to spray for some exotic insect that is kept out of a region, every consumer benefits, and the benefit of one does not preclude or reduce the benefit of others.

Note that the public good nature of exotic pest exclusion is not necessary and hinges fundamentally on the definition of the region. In some cases, expansion of the geographic region protected by an exclusion policy may raise costs of exclusion significantly. Often there are natural barriers to the spread of a pest that provide some natural definition of an area over which the nonrivalry in consumption of pest exclusion services is defined. Consider a pest exclusion area that can be separated into two distinct subregions, N and S, between which it is relatively inexpensive to control pest movements through a pass between N and S. Furthermore, let us assume that both subregions have ports of entry from an outside, infested area. Pest exclusion, then, can proceed separately in the two subregions. For subregion N, exclusion entails controlling entry at its port and monitoring the border with subregion S. If S also controls its port, then the pass may be left open. If S becomes infested with the pest, then subregion N must exclude the pest at the pass as well as its own port. If S and N do not share the same port of entry from the outside area, and if it is cheaper to control the pest at the pass than at the port S, then the principle of nonrivalry is violated between the subregions. Said another way, if we add farms to the exclusion service by adding distinct regions, then protecting those regions may add to the cost of the service. This reasoning hinges on the cost of exclusion at the pass between N and S being lower than the cost of exclusion at the port S. If exclusion at the port is cheaper than exclusion at the pass, it is in the interest of subregion N to control the S port rather than the pass to exclude the pest from N in the most cost-effective manner.

Another way to think about this is in terms of the size of the negative externality between farms in the subregions. If a farm in subregion N becomes infested, then all the farms in subregion N have to deal with the pest. However, if a farm in subregion S becomes infested, the most effective type of control is to close the pass between the regions.

These abstract ideas are applied in practice. For example, rice is grown in two distinct parts of the United States—California and the southern five-state region centering on Arkansas. It is natural to consider excluding a pest from California, even though the pest is established in the rice-growing region of the South. The rice industry in the South does not use the same exclusion services as the rice industry in California. One can model this as a case in which nonrivalry does not hold across regional groups of exclusion services. Holding constant the total effort of rice pest exclusion services, more exclusion services for California reduces the services to growers in the South.

Alternatively, one could model demand for two distinct services, border control for rice pests in the South and border control for rice pests in California, for which nonrivalry does hold in each distinct region. Thus, each separate service provided to a distinct geographic region may be considered a nonrival only within the applicable region. The point here is that nonrivalry may hold only over some subsets of an exotic pest exclusion system and that definitions of region matter.

Next, let us consider the difficulty of excluding nonpayers from consumption of exotic pest border measures. This issue of how investors can capture the returns to an investment often arises in the context of public agricultural research, and the same concerns relate to border measures related to exotic pests. When a pest is kept out of an appropriately defined region, the costs of pest control decrease for all local producers. Furthermore, even if the pest could be allowed to infest the nonpayers initially, that decision would itself damage those that paid for the service because, by definition, the pest would spread within the region to affect nonpayers and payers alike.

The definition of the region is, once again, of prime importance. If we define pest control regions as those over which nonrivalry and nonexcludability apply, then exotic pest border measures are, by definition, public goods within those regions. As with nonrivalry, excludability for nonpayers is technically possible between subregions for which there is some natural barrier. If the pest cannot easily move between regions, then the new region can be offered the border services to keep a pest out, conditional on paying the fee. For example, if it is relatively cheap to control the pass between N and S, then growers in region N have little incentive to pay for border control at port S. But, if the pest moves freely within a region, then pest control regions must be treated as a single unit in terms of the difficulty of excluding nonpayers.

A principle of appropriate financing of any goods or services is to attempt to align the payment for the goods or services with the benefits received. Let us consider how this principle may be applied in the case of exotic pest exclusion services. For exotic pests the producer-beneficiaries are often grouped within natural pest-control regions. However, regions for agricultural production and consumption differ, and both differ from political boundaries. This lack of correspondence raises issues about how to best use fees or taxes to pay for exotic pest exclusion services.

A natural way to raise funds for an industry-wide public good is through producer levies or excise taxes (Alston et al. 1995). When the beneficiaries of the exclusion measures are producers and consumers of production from a well-defined pest control region, then the tax can be applied on production from that region. As with any excise tax, the incidence of the tax is divided between producers and consumers and is based on their own price elasticities of supply and demand. The more that a good produced in the region has close substitutes and, thus, the more elastic is the demand for that good, the less consumers benefit from the exclusion services. The market price would rise only slightly if the pest were to enter, and the market price rises little when the tax is imposed. In this case, the producers would be the main beneficiaries of the exclusion program, and they would bear most of the cost of the tax to pay for the - services. Alternatively, when a region produces a good with few substitutes, consumers benefit from lower costs and would also pay the largest share of the excise tax.

Many agricultural services controlled by governments, such as inspection and grading, research and development, or promotion programs, are paid for from excise taxes or producer levies. However, general tax revenues are most often used to pay the costs of border measures for pest exclusion. One argument is that when consumers or producers of affected products are widespread in the population, an excise tax on the good may have roughly the same incidence as general taxes, and using gen-eral tax funds may simplify tax collection and administration. It is typical for border efforts to exclude exotic pests to be spread across many pests and to apply to many food products simultaneously. Since everyone consumes food, the argument is that everyone benefits from these efforts.

However, this reasoning is not compelling without some further evidence or assumptions about the agricultural supply and demand conditions in the protected region. First, when the tax region and the consumption region are not the same, general taxpayers within the political boundary are, in effect, subsidizing consumers outside the region. Pest exclusion efforts by a single state, say California, lower the product prices for consumers of California produce in both the rest of the United States and in export markets. Second, farmers, farmland owners, and farm workers represent a small share of the whole economy in developed countries. Funding pest exclusion policies from general government revenues allows these beneficiaries to underpay.

There is a more compelling reason that governments undertake border measures related to exotic pests and fund those activities from general revenues. In practice, pest control borders are often defined by political boundaries, and governments have traditionally used many measures to secure their borders. Application of customs duties (import tariffs), immigration control, military defense, internal security, and control of human disease, among other issues, means that nations control their borders. These border control efforts are all considered broad public goods. Government officials operate border controls and mainly use general funds rather than assessments of specific taxes. In that context, it may seem natural for governments to implement and fund border measures with respect to exotic pests. There also may be cost or efficiency gains when implementing exotic pest controls at existing border inspection and enforcement stations. Nonetheless, applying user fees for additional funding for exotic pest border protection at national borders seems appropriate.

Political boundaries are not necessarily the natural boundaries for the flow of biological organisms. For some pests, it is natural to control pest boundaries within a nation. In these cases, the pest control operations are not likely to coincide with other border control measures, and specific financing arrangements may apply. For other pests, joint collective action across national boundaries may be more natural. This applies, for example, to controlling grain and livestock pests and diseases across certain parts of the border between the United States and Canada that do not necessarily correspond to major border crossings. In such cases, we may expect binational funding that draws on industry sources of funds in both countries.

Eradication of Exotic Pests

Eradication means eliminating a pest from a region. Eradication is the extreme case of pest control and, when combined with pest exclusion, can result in a pest-free region. Eradication of pests that spread readily in a habitat region involves collective action because elimination of a pest from one part of the region is naturally short-lived if the pest can simply move back in quickly and easily. Eradication of a pest once it has entered the nation or region is sometimes a "backup" to failed exclusion policies. This was the case, for example, with FMD in Britain. But sometimes eradication is used for pests that have been long established. For example, the boll weevil was eradicated from certain states of the Southeast only in the 1980s. California has considered eradication of yellow starthistle, which has been an important and costly weed for decades (see Chapter 15). The WTO-SPS agreement is relevant to these activities because often eradication is undertaken, in part, to open or reopen export markets.

The same two criteria of nonrivalry and nonexcludability may be considered briefly in the context of eradication programs. Eradication of a pest from a region allows commodity producers to forego private costs of pest management and, perhaps, lower other costs as well. These lower costs decrease market prices to consumers when the region undertaking eradication represents a significant share of the market. The same pest and habitat characteristics that imply that border exclusion measures would lower costs for all producers in a natural pest control region also apply to eradication programs. As long as the region is defined in terms of biological criteria related to the spread of the pest, all those producing commodities affected by the pest in that region (as well as con-

sumers of those commodities) share in the benefits of eradication services. Additional commodity production in the region increases the benefit of eradication services for expanding producers, but does not diminish the benefit of such services for existing producers. Similarly, when a pest is stopped at the border and not allowed to enter a region, eradication for one implies eradication for all in the natural pest habitat region.

These same principles apply to nonexcludability. If some farmers in a natural habitat region (consumer of affected products) refused to pay for eradication services, there would be no way to exclude them from the services and continue to provide the services to their neighbors. That idea is why "eradication" is applied in the first place.

There is an important difference, however, between how costs of eradication and border exclusion programs relate to the number of farms and production quantity. Unlike the costs of border measures, the more an affected commodity is typically grown in a natural pest habitat region, the more costly is eradication. If a crop or livestock enterprise acts as a natural host for a pest, the greater is the extent of the infected commodity and the more difficult or costly is eradication. Often, this implies that the greater the total benefit from eradication, the higher the total costs, and vice versa. Of course, many other factors affect eradication costs, including the existence of multiple hosts, some of which may be wild species, pesticide regulations, and the features of the infested region, such as whether it includes urban as well as rural areas.

Where eradication costs are linked to variables such as the number of livestock or crop acres, a per-unit assessment to fund part of eradication costs would tie the funding of a program to one cost factor and to the benefits. To the extent that per-unit benefits are split between producers and consumers, a user fee or per-unit fee again ties costs to those who benefit.

Institutions and Policies for Exotic Pests

Governments routinely undertake border measures to exclude exotic pests that are not yet established and eradication or other control measures for existing pests. In the United States, both the federal government and individual state governments fund and conduct border protection and eradication. (Chapter 3 contains much detail about regulations in California and the United States.) The WTO provides the forum for developing international rules and for settling disputes related to border protection and eradication.

The Animal and Plant Health Inspection Service (APHIS) is the lead technical agency for exotic pests in the United States. This agency has offices throughout the United States and in many foreign counties. APHIS handles such issues as U.S. border rules for imported farm products, control of smuggled farm goods, and control of accidental entry of farm pests brought into the United States by travelers. In addition, APHIS handles issues that foreign governments raise with respect to export of U.S. farm products. Finally, only partially related to trade, APHIS handles many pest eradication programs in the United States.

APHIS is a large federal agency. Its budget has grown gradually from about $100 million in 1970, to $250 million in 1980, and to between $500 and $600 million in 2001. This budget expansion, while considerable in nominal terms, was slower than that of other regulatory agencies. About 70 percent of this budget is devoted to controlling exotic pests.

APHIS is well-known within agriculture but is less familiar outside agriculture. For example, a leading text in regulatory economics lists APHIS among consumer safety and health agencies rather than among agencies that regulate environmental or industrial practices (Viscusi, Vernon, and Harrington 2000). This categorization misses most of what APHIS does, since most plant and animal health issues have few direct implications for human health.

The APHIS budget comes mainly from general tax revenue rather than user fees or commodity-specific fees or levies. In the United States it has been accepted that the benefits of exotic pest services are broad based and consistent with funding of similar government programs. As with agricultural research, the use of levy funding to supplement the general funds has had only limited success.

The funding for APHIS was increased following September 11, 2001, in response to concerns about bioterrorism. As a matter of national security, the support for increased border surveillance has grown. However, this may cause a net shift of funds from traditional exclusion and eradication programs to protecting against intentional introduction of exotic pests and diseases. The move of APHIS to a new Department of Homeland Security suggests that less attention will be paid to traditional protection of economic interests of agricultural producers and consumers and more attention to broader military or national security concerns.

Some states have active roles in exotic pest exclusion. For example, the California Department of Food and Agriculture (CDFA) takes an active role in exotic pests. Spending some $45 million annually and working closely with APHIS, CDFA handles eradication within the state and also monitors exotic pest outbreaks. CDFA works closely with industry groups on prevention issues and is particularly concerned with movements of pests into California from other parts of the United States. These pest programs are not coincident with national border control. Beneficiaries of the program are producers and consumers of agricultural goods produced in California. This may allow for a larger role of levy funding.

The Uruguay Round Agreement, which created the WTO and began the process of bringing additional disciplines to agricultural trade, also made substantial progress in setting rules related to exotic pests (for details see Chapter 4). The WTO-SPS agreement specified that (a) members may protect themselves from exotic pests, but must use measures that are minimally trade restrictive; (b) members must base rules on scientific principles and evidence; (c) members may use internationally accepted standards or their own standards that then must be science based using risk assessments; and (d) pest control regions may be specified at a subnational level so long as they are distinct in terms of pest control. Note that these rules correspond well to the discussion of public goods listed above, especially the regionalization provisions. The regionalization principle allows for continued exports from clean areas within a country if a pest can be contained within a quarantined area. Importing countries evaluate the effectiveness of the quarantines, but the general idea is that public good aspects of exotic pest exclusion and eradication are defined within biologically determined pest habitat areas.

Putting flesh on these bare-bones principles has been the work of exotic pest regulators and negotiators in each country. As Smith shows in chapter 4, the case law that has been developing around exotic pest disputes has also helped clarify how the principles will be applied. Two issues have received much attention. One issue is how much detailed and quantitative risk assessment is required. The indications so far are that the burden of proof is on those who want to limit trade. They must show that import bans or other measures are truly required to meet certain risk standards and that lesser restrictions would not meet the generally accepted level of risk of entry. Member counties must be careful and systematic about documenting a scientific basis for their border measures with respect to exotic pests and related issues. A second issue is that no special allowances are made for claims of public opinion or political controversy over some pests or related issues. This issue has arisen in the context of European Union (EU) policy where the EU argues that import barriers may lack documented scientific evidence but, nonetheless, reflect the views of either voters or consumers in Europe. In the beef hormone case, which did not deal with an exotic pest but is clearly relevant, this kind of argument was rejected (see Chapter 5).

Evaluation of Exotic Pest Programs

Accepting evidence that exotic pest exclusion and eradication services have public good characteristics does not answer questions about how to design and implement policies. WTO rules do not require that national or regional exotic pest policies balance costs and benefits from the view of producers, consumers, and taxpayers within a nation or subnational jurisdiction. The chapters in Part II of this book develop biological and economic evidence about many individual pests, evaluate policies using standard economic criteria of balancing expected costs with expected benefits, and rank policy alternatives based on the highest net returns to all affected parties.

Economists often measure consumer benefits using the concept of willingness to pay. Any measure that increases the total consumer willingness to pay for a good or service by more than the total expenditure required to purchase the good enhances benefits to consumers. The change in consumer surplus is often a convenient measure of the change in consumer welfare. Producer benefits are measured by the change in producer surplus or the change in total revenue minus costs attributable to variable inputs. Consider a simple border inspection program that kept out an exotic pest and thereby lowered costs of production in a region that produced a large share of the supply of some farm commodity. Under these conditions the returns to land and management on farms in that region would rise and market price would fall. In that way both consumers and producers would gain. If the sum of producer and consumer gains is larger than the cost of operating the program, there is a net welfare gain from the pest exclusion.

Often, however, pest exclusion programs are more complex. Consider, for example, a program that successfully kept out a pest by banning imports from a competitive region that was infested with the pest. In this case producers (or landowners) in the protected regions gain for two reasons. First, they experience lower costs. Second, they experience less competition from the embargoed region. In this case market price may actually rise, and, while producers gain, consumers may lose. If imports from the embargoed region are large without the pest exclusion program, and the cost savings are modest, then the overall societal welfare from successful pest exclusion almost surely falls. The reason is clear. A successful pest exclusion program that restricts international trade reduces gains from trade, and these reductions may be large relative to the savings from keeping out the pest. A program that is biologically successful and compatible with international SPS rules may still harm the economy when consumer and producer interests are both important.

Now, consider an exotic pest that has newly infested an area. Choosing whether to eradicate the pest again implies balancing of costs and benefits of producers, consumers, taxpayers, and perhaps other interests such as environmental quality or wildlife values. The simplest cost to consider is the direct budget cost of agencies undertaking eradication. These costs may be borne by general taxpayers or by industry participants if a levy program were introduced. Often eradication is achieved by limiting production of host crops or livestock in a region. This imposes costs in terms of lost profits that are borne by producers, or by taxpayers if compensation is offered. Of course, higher prices offset some producer losses and are a pure gain to producers who do not have eradication costs but gain from the higher costs to consumers. Assessing these impacts requires careful modeling and data from a variety of sources. Data requirements include biological and agronomic information about the pests, the habitats, and the potential methods of eradication.

Conclusion

This brief overview has considered some exotic pest policy principles. Overall, the exotic pest system in the United States is well developed and works as planned. There is concern, however, about maintaining the system. Funding has remained relatively stagnant compared with the growth of agricultural production and trade in the United States over the past several decades. Funding affects the ability of border measures to keep pests out and the ability of control measures to eradicate introduced pests when entry occurs. The move of exotic pest issues into the Department of Homeland Security raises additional issues.

Continued implementation of the SPS rules in the WTO agreement of 1994 is important globally. To respond to complaints, the nations must be ready to show that border measures are based on sound science and quantitative risk assessments. In addition, data and analysis are required to comply with challenges and to challenge barriers in other countries. All of this requires strong exotic pest infrastructure.

The basic criteria of public goods, nonrivalry, and excludability apply directly to exotic pest border measures and eradication service with three provisos. First, regions over which the criteria apply are defined not by political boundaries, but rather by characteristics of natural pest habitat and spread. Second, for eradication, costs are likely to rise with more production of affected commodities. Third, for some agricultural pests, the public good characteristics may not apply to the general popula-

tion. Only those producers or consumers of the affected products are direct beneficiaries of the services, and these are the ones over which nonrivalry and excludability may apply.[3]

The new WTO rules related to regionalization were a clear recognition of principles of invasion biology and are now well established, although some disputes continue to occur.

Fuller recognition of the industry nature of some of the public good characteristics of exotic pest services may allow better response to the concern over funding. The idea here would be to attempt to supplement state and national budgets for exotic pest services with levies or excise taxes that tie costs of the programs more directly to the beneficiaries. For example, if consumers of citrus in Japan or China were significant beneficiaries of measures to keep citrus canker out of California, or eradication if there were an outbreak, they would pay a share of the costs of these programs through the higher price from an assessment or excise tax. Of course, the initial reaction of industry shifting from using general tax revenue to levies is sure to be negative. But, the current funding system relies on taxpayers who have shown considerable reluctance to fully fund programs demanded by industry.

Notes

[1]This volume represents a large collection of such studies. See also the chapters collected in Orden and Roberts (1997); Anderson, McRae, and Wilson (2001); and the National Research Council (2000), as well as recent academic journal articles by James and Anderson (1998), and Paarlberg and Lee (1998). Also relevant is the work collected in Coppock and Kreith (1999). Citations provided in the case study chapters comprise a large list of studies in this emerging literature.

[2]The term "free rider" is a reference to the example of a bus route in which one additional passenger who boards at a regular stop with other passengers and who takes a seat that would otherwise be empty does not add to the cost of the route. Of course, if all riders were considered the free rider the bus route could not be profitable.

[3]As noted above, some exotic pests affect the natural environment and wild species as well as agriculture. In those cases, consumers of this habitat and services of these species are also beneficiaries. This could be related to specific groups, such as hunters of wild fowl, or the broad public, in the case of national park habitat or endangered species.

References

Alston, Julian M., George W. Norton, and Phillip G. Pardey. 1995. *Science Under Scarcity: Principles and Practice for Agricultural Research Evaluation and Priority Setting.* New York: Cornell University Press.

Anderson, Kym, Cheryl McRae, and David Wilson, Eds. 2001. *The Economics of Quarantine and SPS Agreement.* Centre for International Economic Studies, Adelaide, and AFFA Biosecurity Australia, Canberra, Australia.

Antle, John M. 1988. *Pesticide Policy, Production Risk, and Producer Welfare, An Econometric Approach to Applied Welfare Economics. Resources for the Future.* Washington, D.C.: Johns Hopkins University Press.

Cheung, Steven N.S. 1973. "The Fable of the Bees: An Economic Investigation." *Journal of Law and Economics.* 16:11–33.

Coase, Ronald. 1974. "The Lighthouse in Economics." *Journal of Law and Economics.* 17:357-376.

Coppock, Ray, and Marcia Kreith, Eds. 1999. *Exotic Pests and Diseases: Biology, Economics and Public Policy.* University of California Agricultural Issues Center.

James, Sallie, and Kym Andersen. 1998. "On the Need for More Economic Assessment of Quarantine/SPS Policies." *The Australian Journal of Agricultural and Resource Economics.* 42:425–444.

National Research Council. 2000. *Incorporating Science, Economics and Sociology in Developing Sanitary and Phytosanitary Standards in International Trade.* Washington, D.C.: National Academy Press.

Orden, D., and D. Roberts, Eds. 1997. *Understanding Technical Barriers to Agriculture Trade.* Proceedings of a Conference of the International Agricultural Trade Research Consortium, University of Minnesota, Department of Applied Economics. St. Paul, Minnesota.

Paarlberg, P.L., and J.G. Lee. 1998. "Import Restrictions in the Presence of a Health Risk: An Illustration Using FMD." *American Journal of Agricultural Economics.* 80:175-183.

Pindyck, R.S, and D.L. Rubinfeld. 1998. *Microeconomics,* 5th ed. Upper Saddle River, N.J.: Prentice-Hall Inc. Chapter 18.

Viscusi, W. Kip, John M. Vernon, and Joseph E. Harrington, Jr. 2000. *Economics of Regulation and Antitrust.* Cambridge, Mass.: MIT Press. p. 49.

World Trade Organization (WTO). 1995. "Agreement on the Application of Sanitary and Phytosanitary Measures." In *Results of the Uruguay Round of Multilateral Trade Negotiations: The Legal Texts.* World Trade Organization, Geneva, Switzerland.

3

Regulatory Framework and Institutional Players

Marcia Kreith and Deborah Golino[1]

Introduction

Biological principles, environmental factors, economics, emotion, rhetoric, and politics all have shaped public policies affecting exotic pests and diseases. Sometimes the scientific evidence on which regulation has been based has been solid, at other times shrouded in unknowns. The costs and benefits of specific control measures, and what would happen absent those control measures, are the focus of this book. In this chapter we provide an overview of exotic pest and disease control principles and their regulatory framework. After discussing international treaty obligations, in particular, the Agreement on the Application of Sanitary and Phytosanitary Measures, the chapter concludes with an in-depth discussion on how international obligations may affect U.S. nursery stock phytosanitary regulations. An extensive synopsis of key operational statutes is presented in Appendix 3.1.

Biological Processes and Pest and Disease Control Strategies

Biological processes, in addition to human behavior, determine whether pest and disease control programs will be effective. Although an organism may not survive the trip, reproduce, and thrive in the new environment, successful invaders often go undetected for a long time. Consequences are not always predictable. Successful intervention requires understanding the invasion process, as well as when, where, and how an organism is introduced. Invasion involves several phases: entry of the organism, establishment through at least one reproductive cycle, integration or naturalization into the local environment, and lastly, spread. The modes and pathways of introduction are numerous and may be intentional, unintentional, or natural, and by either legal or illegal means.

Accordingly, five biologically based strategies underlie the extensive and interconnected exotic pest and disease control systems in place in California and the United States:

- Exclusion
- Surveillance and early detection
- Eradication
- Containment
- Suppression

Absent natural constraints to exponential growth, exclusion or early detection and eradication are biologically the most effective control strategies. They have been the preferred control policies of the United States and California since the late 19th century. In addition to inspections both at home and overseas, prohibitions (commonly called embargoes) and, in certain cases, quarantines are used to effect exclusion. When eradication is technically feasible, success depends upon biological, ecological, social, and operational factors, including public support. Where pest or disease population growth is rapid or infestation/infection widespread, eradication may be impossible. Certainly, it will be more costly than for an early small occurrence, whether reckoned in dollars, labor, or potentially adverse consequences of chemical or other treatments. Consequently, eradication programs often result in the elimination of a specific pest or disease within a specific area for a specific duration.

All of these strategies require vigilance, establishment of rational, easily understandable procedures and protocols, and compliance. Scientific data and analyses, risk assessments, and rapid accurate diagnostic tools all hinge on research and education, which can be costly. An informed public is vital to public cooperation and support of sound policies. Well-trained scientists and regulators are crucial to exclusion, detection, and effective response.

Antecedents of Current Institutions and Policy

History is replete with evidence that agricultural pests and diseases have shaped human migrations, wars, and regional diets. In addition, introduced species have transformed natural ecosystems both adversely and beneficially. The westward migration across the Atlantic in the 17th century, and then across North America to California in the 19th century, brought the agricultural crops of Europe to America, including wheat, dairy and beef cattle, potatoes (of South American derivation), tree nuts, and fruits. Soon after arrival, American colonists were faced with developing effective policies to restrict and eradicate the pests and diseases that often arrived with their desired plants, seeds, and livestock. Stem rust fungus of wheat *(Puccinia recondit)* appears to have been the first reported introduction of a plant disease into the American colonies (Wiser 1974). To destroy the local alternate host of the stem rust and thereby suppress the disease, the Connecticut, Massachusetts, and Rhode Island colonies enacted antibarberry *(Berberis vulgaris)* laws, thereby launching a 40-year eradication campaign. Those laws were the first pest abatement laws in the colonies (Ryan 1969). In 1881 the California legislature instituted the nation's first system of plant inspection at points of entry to the state. The state's first successful quarantine interception was in 1891 when a ship cargo of 325,000 orange trees from Tahiti was found to be infested with nine species of insects (Ryan 1969). (See Chapter 5 for a more extensive history of exotic pests and disease of plants in California.)

Not until 1912 did Congress pass the Plant Quarantine Act[2] that authorized federal inspection and quarantine of imported plant material, primarily nursery material. Certification inspections were to be performed by state collaborators. It took 16 more years before the Act was amended in 1928 to seize and destroy plants found to be moving in interstate commerce or brought into the United States (Chock 1983, p. 481). Western states' plant protection officials had by this time created the Western Plant Board to exchange information and prevent the spread of pests and diseases across state lines. Quarantine 37, issued by the U.S. Secretary of Agriculture to control import of seeds, nursery stock, and other plants, set the foundation for present day plant quarantine regulations when it went into effect June 1, 1919[3]—despite objections of U.S. florists, nurserymen, and foreign concerns.

The first response to importation of animal diseases was also at the local level. The first reported introduction of a diseased animal was with the 1843 introduction into New York of a single pleuropneumonia-infected cow from an English ship. When Dutch cattle infected with pleuropneumonia were imported into Massachusetts in 1859 (Wiser 1974), that state established the country's first animal quarantine and destruction program. It took Massachusetts six years to eradicate the outbreak. Alarmed by Massachusetts' mounting costs, in 1865 Congress gave the Secretary of the Treasury authority to prohibit the importation of cattle. Three months later the act was amended to include hides of cattle and provide fines or imprisonment for violators. Nevertheless, it seems to have taken a decade before issuance of the first actual animal quarantine. Finally, after another outbreak of pleuropneumonia in England in 1879, U.S. customs collectors were ordered to quarantine European cattle for 90 days at the importer's expense.

The New World, however, was not the first to enact pest or disease exclusion laws, nor was it solely on the receiving end. Worldwide, the first significant pest legislation was enacted by Germany when it banned import of potatoes in 1875, one year after the Colorado potato beetle's arrival with dirty potatoes. To prevent its entry, England followed with its 1877 Destructive Insects Act.

It is not just individual states or countries that have tried to protect the health and welfare of their citizens and domestic agriculture.[4] International (bilateral and multilateral) agreements and organizations have slowly evolved.[5] Currently, there are roughly 20 binding multilateral agreements at either the global or regional level (FAO 2001). The International

Convention for the Protection of Plants was established in 1929 (League of Nations Treaty). In 1952 these agreements were replaced by the United Nations Food and Agricultural Organization International Plant Protection Convention (IPPC), which is now administered by the IPPC Secretariat in the United Nations Food and Agriculture Organization (UN FAO) Plant Protection Service (Chock 1983, FAO 2002).[6] Today, pertaining to international trade in plants and animals, the IPPC[7] and the regional plant protection organizations operating within the IPPC framework are charged with developing international phytosanitary standards that implement the World Trade Organization Agreement on the Application of Sanitary and Phytosanitary Measures (SPS Agreement).

Spurred by the 1920 introduction into Belgium of rinderpest in zebu cattle transshipped to Brazil from India—the earlier 18th century introduction having been finally eradicated in Europe by the end of the 19th century—in 1924, 28 countries signed the first international agreement to work together as a research and surveillance community to control infectious animal diseases, thereby creating the Office International des Épizooties (OIE). Formation had the support of the Secretary of the League of Nations. By May 2002, 162 countries were members.[8] In 1994 the OIE and the Codex Alimentarius Commission were designated responsible for setting interim World Trade Organization (WTO) animal health and safety standards for international trade.

Government Infrastructure and Jurisdictions: Key Responsible Agencies and Organizations

With their mosaic of legal mandates and authorities, federal, state, and local governments in the United States have created a cooperative system to exclude exotic pests or diseases, or if exclusion fails, to implement containment, eradication, or suppression measures. These domestic agencies in turn collaborate with international governments and organizations in protecting the health and welfare of their territories.

The U.S. Department of Agriculture Animal and Plant Health Inspection Service (USDA—APHIS) is charged with protecting the United States and constituent states against entry—in live, fresh, or processed material—of new pests and diseases primarily of agricultural animals and plants. Live animal imports are regulated by APHIS' Veterinary Services, while plants and plant and animal products are regulated by APHIS' Plant Protection and Quarantine (PPQ). At air, land, and sea ports of entry into the United States, PPQ inspectors have responsibility for inspecting luggage and cargo to intercept illegal plant and animal products and disease vectors. In addition, PPQ and Veterinary Services also certify the health and pest- and disease-free status of interstate movements of federally quarantined items, foreign imports, and exports. The U.S. Secretary of Agriculture is responsible for promptly reporting the outbreak of an OIE "List A" animal disease to OIE.

Pests and diseases of wildlife and habitat are primarily concerns of the USDA Forestry Service, APHIS Wildlife Service, Department of Interior's Fish and Wildlife Service and Bureau of Land Management, and the National Marine Fisheries Service. At entry ports, Fish and Wildlife Service is the lead agency inspecting for internationally or domestically prohibited fish and wildlife imports.

Human health is the major emphasis of the federal Centers for Disease Control and Prevention, the Office of Public Health and Science, and the Food and Drug Administration, all in the Department of Health and Human Services. Also, the USDA Food Safety Inspection Service conducts slaughter epidemiology surveys. These agencies provide surveillance data, scientific analysis, and advice to APHIS. The U.S. Postal Service and Customs Service are also major cooperators on exclusion efforts. The Department of Defense cooperates with APHIS to assure that returning cargo and troops do not bring contaminated soil and pests. Aspects of bioterrorism are under the purview of a wider gamut of law enforcement agencies, including the FBI and local police, and recently the president has proposed creating the Department of Homeland Security.[9]

In addition to its jurisdiction over movement of harmful pests and diseases and host materials across international borders, the federal government regulates interstate movements when there are federal domestic quarantines. Moreover, it is the contracting party for international agreements that are between countries or within the framework of quasi-governmental international organizations such as WTO or the

United Nations. Individual U.S. states do not make binding agreements with foreign countries without congressional approval, although enforcement measures may be delegated to the state. This may be accomplished by regulation or cooperative agreement,[10] or state personnel may be designated as "federal collaborators" for enforcement. Commercial trade, especially in horticulture, often depends upon overseas inspection by U.S. personnel, or under U.S. oversight, by officials of the exporting country's national plant protection organization.

While federal foreign and domestic regulations preempt state laws and regulations, states are responsible for within-state movements and control measures. In the absence of either federal domestic regulations or conflicts with such regulations, states may impose their own interstate quarantine regulations to protect against and respond to pest and disease introductions. Federal-state-county collaboration and cooperation in pest prevention is common. Quarantine enforcement activities at air, sea, and land ports of entry can be a federal-state partnership arrangement with each agency fulfilling their respective roles and responsibilities and reporting pest findings to each other to ensure the highest possible level of joint exclusion effectiveness. The federal and state governments also work together in diagnostics and eradication of federally regulated pests. Smuggling interdiction is another area of collaboration, as with the Closing the Los Angeles Area Market Pathway Project in California, known as the CLAMP Project, to identify smuggling routes and intercept agricultural contraband already in U.S. commerce.[11]

In California, in tandem with APHIS responsibilities with foreign commerce, the Department of Food and Agriculture (CDFA) and the county departments of agriculture have joint responsibilities for exclusion, detection, and eradication of new invaders, as well as control and containment of established agricultural pests and diseases. CDFA's scope includes pests and diseases that affect urban and natural areas, although except for its major responsibility in ensuring a safe food and fiber supply, human diseases are not CDFA's primary concern. CDFA has 16 border agricultural inspection stations in addition to county agricultural commissioners to enforce California Food and Agricultural

Code and federal and state quarantines. The USDA cooperates with CDFA in many endeavors. The CDFA, however, additionally enforces state requirements on a number of nonindigenous organisms that are not regulated by the federal government, such as plum curculio, Caribbean fruit fly, and European shoot moth.

Other California agencies with pest and disease responsibilities include the Department of Fish and Game, Department of Forestry and Fire Protection, Water Resources Control Board, and the Governor's Office of Emergency Services. Also a major player through its regulation of pesticides and pesticide use is the Department of Pesticide Regulation.

Locally, county agricultural commissioners enforce California's regulatory structure concerning exclusion and control of exotic pests and plant and animal diseases, make inspections, and under delegated authority, provide pest- and disease-free certifications for movement and export of plants, animals, and related products. With the Department of Pesticide Regulation, they also have joint responsibilities to ensure safe local use of pesticides.

A number of prominent associations and boards also affect the creation and application of U.S. public policies. Although not formed pursuant to federal or state law, their membership consists of government officials, private sector representatives, or a mix of public and private organization representatives. Among the more prominent are such national organizations as the U.S. Animal Health Association and the National Plant Board, an organization comprised of regulatory officials of each state and the Commonwealth of Puerto Rico formed in 1925 to promote (1) greater uniformity and efficiency in plant pest regulations, (2) enforcement of quarantines and inspections among the states, and (3) to maintain states' contacts with federal agencies.[12] In addition, the National Invasive Species Council was created by executive order in 1999 to coordinate federal activities concerning invasive species, including on wildlands.

The major international standard-setting organizations are (1) the Codex Alimentarius Commission of the UN FAO and WHO, (2) the OIE, and (3) the FAO International Plant Protection Convention[13] and its regional plant protection organizations (RPPOs).[14] Representative members from the national plant protection organizations of Canada, the United States, and Mexico serve on the North American Plant Protection Organization, which has primary responsibility for developing regional plant protection standards to protect its member states from the entry and establishment of pests while facilitating trade.

Members of the WTO have agreed to take their international trade disputes relating to plant and animal health and safety measures to the WTO Dispute Settlement Body.

Regulatory Tools

Embargoes, certification, confiscation, destruction of pests or infected hosts, regulated lists, permits, surveillance, reports of detection, hold orders, and quarantines are the principal tools used to implement policies of pest and disease exclusion. Pesticides, heat treatments, quarantines, release of biological control agents and sterile insects, vaccinations, and follow-up monitoring are used with eradication, containment, or suppression activities.

Embargoes (Prohibitions)

Two divergent strategies underlie U.S. entrance regulations aimed at excluding introduction of injurious plant and animal diseases and pests. One requires that plants and animals and related products be pest and disease free (import permits and pest-free/disease-free certification may be required to ensure success of implementation). The other lists prohibited animal species, invasive noxious plants, and plants that are known hosts of named exotic pests from specified countries, i.e., quarantine pests. This second approach uses what is sometimes described as the "dirty list" or "black list." This is the policy used by the 1981 amended Lacey Act of 1900. Under the Lacey Act the organism must first be proven detrimental before it is dirty listed. Reacting and listing after proof of harm has greater risk than a forward-looking approach. For this reason, a third policy has been suggested by those who advocate that plant species not be introduced unless they have been shown not to be harmful to the environment—only species on the "clean" or "white" list would be allowed. Prevention of negative

impacts to native habitat by nonindigenous plants is the primary concern of proponents of this clean list approach.

Embargoes may be highly selective against importation of a particular species or a product from a particular country where the pest or disease occurs, or they may be broadly based. For example, in general but with a few exceptions, Title 7 CFR 319.37—commonly called Quarantine 37—broadly prohibits commercial importation of nursery stock and other propagative materials of grapes, *Citrus,* strawberries, *Prunus* (peaches, cherries, almonds, etc.), apples, pears, and sweet potatoes primarily because of various diseases. In addition, Quarantine 37 (7 CFR 319.37-2) has an extensive list of plants and plant material ("prohibited articles") whose import is prohibited from explicitly listed countries. Nevertheless, prohibited plant articles may be imported or offered for entry into the United States if they meet conditions in 7 CFR 319.37-2(c). Similarly, lists of wild animals, birds, reptiles, crustaceans, etc. whose import for release is prohibited by the Lacey Act—such as the Java sparrow or any species of mongoose—are enumerated in 18 USC 42 and in Title 50 CFR 16.11-16.15.

Some plants and plant products are absolutely prohibited, while others are restricted, requiring treatment as a condition of entry to prevent the introduction of plant pests. In addition, plant material may not be imported with sand, soil, or earth, only with certain specified growing media. If pests or diseases are found, the plants are treated, destroyed, or refused entry.

In contrast, import of all animals or animal products is regulated according to the disease status of the exporting country. In general, entrance of live animals is prohibited from countries with a disease not in the United States, while quarantines may be required from countries[15] without the disease, but at some risk. Allowable animal products from countries with the foreign animal disease must be cooked. Furthermore, there are no treatments for animal products upon reaching the entry port. They are either refused entry or destroyed. Foot and mouth disease, rinderpest, classical swine fever, bovine spongiform encephalopathy, exotic Newcastle disease, and many other OIE List A diseases are on the USDA list of quarantine diseases.

Quarantines, Hold Orders Controlling Movements, and Confiscation or Destruction

Quarantines are legal instruments to impose prohibitions and restrictions aimed at exclusion (or containment and eradication) of harmful pests. Typically, they are enforced by inspections. Federal Quarantine 37 and Quarantine 56 are the most extensive among a number of federal foreign plant quarantines. Quarantine 56 (7 CFR 319.56 et seq.) regulates the entrance from various countries of fruits and vegetables intended for consumption, whereas Quarantine 37 pertains to propagative materials. In addition, there are federal domestic quarantines that pertain to interstate movements. APHIS' Regulated Pest List[16] provides a comprehensive listing of most of the external and domestic quarantine plant pests found in Title 7, Code of Federal Regulations, Parts 300-399. Since the nature of pest status is dynamic, the list does not include all pests against which APHIS might take action at any given point in time. The USDA describes those plant pests for which they may take quarantine action—refuse entry, treat, or destroy— as "actionable plant pests."

The California Department of Food and Agriculture's Plant Quarantine Manual[17] contains the state's quarantines in detail and the pests they cover. Many are federally actionable pests, but the CDFA also takes action on pests of state concern. In April 2002, the online manual links to 26 state exterior quarantines, 16 state interior quarantines, 20 federal domestic quarantines, and federal Hawaiian and territorial quarantines and conditions—movement is prohibited or regulated for over 100 fruits, flowers, herbs, and vegetables from Hawaii and the territories.

In addition, California takes regulatory action under statute against shipments that, although not in violation of a state exterior or federal quarantine, are infested with pests that are not known to occur in California or that are under official control in the state. Also, shipments infested with pests not under official control in the state may be rejected by county departments of agriculture when such pests are widely but not generally distributed in the state, or the commissioner determines that the pest is not present in the county or area where the material is to be planted.

All states have authority to quarantine premises with animals that are disease positive, display suspect clinical signs, or that may have been exposed to a disease foreign to the country or state. Animals (other than birds) intended for entry into the United States are held under quarantine surveillance at one of three APHIS animal import centers—Newburg, New York; Miami, Florida; or Los Angeles, California. Birds and poultry are held either at Newburg, Miami, or Otay Mesa, California.

Many plant articles (including nuts and seeds) are allowed entry pursuant to Quarantine 37 only under a permit that provides for planting and inspection under prescribed post-entry quarantine conditions.

Import Permits

Permits, usually written, are required for importing fresh fruits and vegetables, animals and animal products, and logs and lumber from foreign countries. Permits are similarly required prior to import of live plant pests and noxious weeds. They are issued only to U.S. residents, and applicants must show a home or business street address. Permits are free, with the exception of the General Permit to Engage in the Business of Importing, Exporting, or Re-Exporting Terrestrial Plants listed on the Convention on International Trade in Endangered Species of Wild Fauna and Flora (CITES), for which the fee is $70.[18]

Generally, a USDA veterinary permit is needed to import animals or materials derived from animals or exposed to animal-source materials to ensure that foreign animal diseases are not introduced. Dairy products (except butter and cheese) and meat products (e.g., meat pies, prepared foods) from countries with livestock diseases exotic to the United States require an import permit. Other animal materials that require a permit include animal tissues, blood, cells or cell lines of livestock or poultry origin, RNA/DNA extracts, hormones, enzymes, monoclonal antibodies for in vivo use in nonhuman species, certain polyclonal antibodies, antisera, bulk shipments of test kit reagents, and microorganisms, including bacteria, viruses, protozoa, and fungi.

To import fruits and vegetables admissible under Quarantine 56 for the food market, a separate "56 permit" must be obtained for shipments from each country and for each port of first arrival in the United States. Valid for five years, the 56 permit is free.

In conformance with Quarantine 37 more than 400 plant genera may be imported for propagation or nursery sale without post-entry quarantine, but they require a permit.[19] In addition, some materials such as seeds and bulbs, require pretreatments, and others require clearance at a specially equipped Plant Protection and Quarantine Inspection Station listed on the permit. Cut flowers, except those with berries attached or regulated under CITES, do not require a permit.

Prior to September 11, 2001, for research purposes it was possible to import into the United States "prohibited plants and plant products," including live plant pests or noxious weeds—or to move them across state borders—but only after obtaining a permit from APHIS. Although there were unannounced inspection provisions and a requirement for destruction at the completion of the intended use or permit expiration date, the system was based largely on trust. To reduce the threat of bioterrorism, a USDA moratorium was instituted on issuing import permits for plant pests, plant pathogens, genetically modified organisms, and biological control organisms while import policies and regulations on these organisms are under review and revision. At the time of this writing, applications are being approved for interstate movement and for environmental release of an organism already imported under an APHIS permit (November 7, 2001 moratorium). In addition, California requires a permit to move noxious weeds within or into the state.

Certification of Pest-Free Status Prior to Movement of Plants, Animals, and Animal Products

To import plant propagative materials, Quarantine 37 requires a phytosanitary certificate issued by a plant protection official in the country where the material was grown that certifies it has been inspected and is free from injurious plant diseases, insects, or other pests and meets other U.S. requirements for import. The certificate must be issued no more than 15 days prior to shipment. Even with certification, in all circumstances treatment or rejection is required if pests are found upon inspection at the port of entry or as a result of preclearance inspections overseas.

For most classes of livestock and poultry, entry into California of live agricultural animals from other states requires a veterinarian's health certificate showing that the animals described meet California entry requirements, which may be quite specific as to species, age, and sex. Even so, additional pre-entry tests or inspections may be required. California's certification requirements supplement basic requirements of the USDA. In 2001, CDFA issued 5,236 permits for interstate livestock movement into the state, with shipments from nearly every state (Ashcraft 2002).

In the event of an interior quarantine—as currently with the glassy-winged sharpshooter and the red imported fire ant—CDFA requires certification for within-state movement of pest-free or disease-free plants, animals or related products and equipment.

Although the United States does not require a phytosanitary certificate for export of commodities, many importing countries' national plant protection organizations have imposed certification requirements for entrance of commodities they regulate. APHIS certification is provided as a service to exporters who must comply with these requirements.

In addition to government-mandated certification of imports, a voluntary certification system attesting to the disease-free nature of domestic nursery stock has evolved in a number of states. This will be discussed later.

Surveillance, Monitoring, Detection, and Reporting of Listed Pests or Diseases

The diagnosis or the occurrence of undiagnosed or unusual animal disease conditions that are possibly reportable and/or of foreign origin must be reported to the state and federal animal health officials. For California veterinarians, there are three sets of lists of animal diseases of significance to animal health and well-being, animal trade and commerce, and/or of public health concern. (To a large extent the lists overlap.) They are as follows:

1. California Reportable Disease List a and b.[20] List (a) Emergency Animal Diseases—must be reported by telephone within 24 hours of dis-

covery; List (b) Domestic Diseases of Regulatory Importance—report by mail within three days of discovery.

2. USDA Reportable National Program Diseases. These diseases are the target of state/federal cooperative eradication programs.

3. OIE Reportable Diseases List A and B.[21]

Recently, CDFA consolidated and organized these lists by affected species (Whiteford 2002). The list is presented in Appendix 3.2. Both bovine spongiform encephalopathy and foot and mouth disease, which are discussed in Chapters 6 and 7, must be reported within 24 hours of discovery.

In the area of plant protection, the California Food and Agricultural Code mandates that CDFA "shall prevent the introduction and spread of injurious insect or animal pests, plant diseases, and noxious weeds." CDFA Plant Health and Pest Prevention Service rates pests and diseases as A, B, C, D, or Q. These ratings reflect CDFA's assessment of the statewide importance of the pest, the likelihood that eradication or control efforts would be successful, and the present distribution, if any, of the pest within the state. Serving as policy guidelines they indicate the most appropriate action to take against a pest under general circumstances. Local conditions may dictate more stringent actions at the discretion of the county agricultural commissioner, and the rating may change as circumstances change or more information becomes available.

An A-rated pest is an organism of known economic importance and is subject to action by CDFA, including eradication, quarantine, containment, rejection of shipments, or other holding actions. Entrance of seeds and articles or commodities containing these pests is prohibited. A Q-rated pest is a pest that is intercepted in a shipment entering the state or a newly detected organism that seems likely to be of economic importance, but information on it is limited. Q-rated pests are treated as A-rated pests, pending a full evaluation.[22]

Enforcement of the health certificate and requirements of movement permits also is performed by staff of California's 16 border agricultural inspection stations. Using data collected at the border stations in 2001, the CDFA Animal Health Branch recorded 30,970 incoming live animal shipments, mostly commercial, which represented 3.5 million dairy cattle, beef cattle, swine, sheep, goats, and horses, plus additional shipments of 9.5 million poultry birds and 293.4 million fertile eggs (Ashcraft 2002). In 1999, CDFA Plant Health and Pest Prevention Services reported monitoring 364,752 commercial plant shipments at its border inspection stations. Of these, 1,803 shipments were rejected, and another 27,052 were sent under "Warning-Hold Inspection Notices" to the final destination county agricultural commissioners for final disposition. That same year CDFA intercepted 70,835 commercial lots of plant materials that were infested or not properly certified for entry into the state.

A joint state-federal program, CLAMP, to identify smuggling routes and intercept contraband agricultural material already in U.S. commerce, seized 41,252 pounds of illegally imported plant and animal materials from retail and wholesale markets, nurseries, swap meets, warehouses, etc. in 1999 (California Department of Food and Agriculture Plant Health and Pest Prevention Services 2000). These materials seized in the Los Angeles area arrived from locations around the world, including other U.S. states. Very serious pests were found, either as contraband noxious weeds or "hitchhiking" pests and diseases. Eleven were rated "actionable" by USDA and another 22 were rated by CDFA as either A (9 species found), Q (13 species), or B (1 species). Among the actionable pests found were giant salvinia, citrus blackspot, citrus canker, water spinach seed, and smut fungus.

International port and border entries, too, must be monitored for pest and disease arrival. The quantities of incoming produce, cut flowers, and propagative material are large. For example, during an average month in the winter of 2001-2002, 118.3 million kilograms of fruits and vegetables entered California in 8,325 separate shipments. In 1999, at the Los Angeles International Airport alone, the USDA made 3,000 reportable entomology interceptions and 16,037 interceptions of meat and poultry, seizing 29,351 pounds of meat and poultry (APHIS 1999).

Sanitary and Phytosanitary Treatments

Vaccinations For some time the OIE has recognized a category called "free with vaccination." However, APHIS does not recognize a country as free of certain diseases if that country carried out vaccination for those diseases. This is because widespread use of vaccines in an affected population can suppress the symptoms of the disease, giving the impression that the disease has been eradicated when it has not. In addition, with some vaccines, vaccinated animals are able to pass on the disease even though they themselves will not manifest clinical disease. Nevertheless, certain animal products from such countries, if they have been properly treated, may be allowed entrance by USDA.

Pesticides   The detection of exotic pests may trigger eradication campaigns that use various pesticides. Similarly, fumigation before export may be a condition imposed by overseas markets. Although only pest-free plant materials may be imported to the United States, protocols often allow for chemical treatments should pests be found upon arrival. Required plant quarantine program treatments are described in the *Federal Plant Protection and Quarantine Program Treatment Manual,* which is incorporated by reference into the quarantine regulations.

Incineration or Heat   Waste from ships and airplanes arriving from foreign countries must be incinerated or heat treated to reduce disease risk. (California regulations are more stringent than those of USDA and require longer heat treatment.) Facilities handling foreign garbage are licensed and routinely inspected. During 2001, CDFA personnel conducted 600 inspections of 121 licensed facilities and 24 military facilities (Whiteford 2002).

Differing agency mandates, objectives, regulations, and acceptable risk levels can lead to what appears to be regulatory incongruence. For example, many plant materials being imported for food are subject to less stringent requirements than similar articles being imported for nursery propagation. Traditional agricultural crops are usually subject to higher standards than house plants and cut flowers. Certain aquatic plants also have been declared noxious weeds, yet for hobby aquariums, requirements for their interstate movement is largely unenforced. Animal health laws, some dating back to the 1880s, have been scattered throughout the U.S. Code. In May 2002, legislation[23] was

signed modernizing and consolidating the body of laws protecting the health of agricultural animals. No doubt, other measures to increase transparency, scientific bases, and global harmonization will further reduce inconsistencies.

Legal and Regulatory Authority and Concepts in the United States

Within the federal-state system, legal authority for agency activity derives from federal and state laws passed by Congress and state legislatures. The trend has been for legislatures to pass framework laws that delegate subsequent development of the detailed specifics of regulations or implementation measures to the implementing agency in the executive branch. Thus, pest and disease control regulation is governed not only by authorizing *statutes* but by agency *regulations* that interpret and apply those statutes. In actuality, *case law* also may affect implementation, as do annual appropriations.

The Tenth Amendment of the U.S. Constitution specifies, "The powers not delegated to the United States by the Constitution, nor prohibited by it to the States, are reserved to the States respectively, or to the People." Under many circumstances this means individual states may enact statutes and regulations that are more rigorous than their federal counterparts. While some federal laws, such as the Lacey Act, Federal Noxious Weed Act, Federal Plant Quarantine Act, Plant Pest Act,[24] and Federal Insecticide, Fungicide, and Rodenticide Act assert predominance over state law, they then explicitly permit more restrictive state law in certain areas, as for example in pesticide registration. Also germane to exotic pests and disease regulation, the Constitution grants Congress the power "to regulate Commerce with foreign Nations, and among the several States, and with the Indian tribes." With the consent of Congress, however, states may enter into agreement or compact with another state or foreign power. Congressional approval has not been necessary for California's contracts with foreign universities for basic and applied research or with the government of Mexico to supply sterile Mexican fruit flies. Similarly, following the Memorandum of Understanding to cooperate on trade, tourism, public safety and health, and environmental and coastal quality concerns that was signed by the governors of California and the adjacent Mexican state of Baja California in December 2001, the respective secretaries of agriculture signed an agreement to cooperate on pest detection and eradication, and to share information on plant and animal disease.

Enactment of statutes (the laws passed by Congress or state legislatures and approved by the president or governors) and promulgation of regulations by federal and state agencies are neither a capricious nor speedy process in the United States. While the Congress and state legislative bodies each have their own rules governing how legislation is created, in general there are formal opportunities for public review and input.

The process by which agencies promulgate regulations is much more highly codified to provide opportunity for public input and administrative review prior to adoption. In California, every unit or individual in the executive branch, unless expressly exempted by statute, must[25] follow the rule-making requirements spelled out in the California Administrative Procedure Act (APA)[26] and in regulations adopted by the Office of Administrative Law (OAL).[27] A 45-day opportunity to submit written, faxed, or e-mail comments on all or any part of a proposed rule-making action starts when notice of the proposed rule-making is published in the *California Regulatory Notice Register* (http://ww.oal.ca.gov/notice.htm); hearings may be held, and opportunity must be provided to comment on proposed modifications.

According to OAL, "APA requirements are designed to provide the public with a meaningful opportunity to participate in the adoption of regulations by California state agencies and to ensure the creation of an adequate record for the public and for OAL and judicial review" (Office of Administrative Law 2001). Furthermore, "a regulation must be easily understandable, have a rationale, and be the least burdensome, effective alternative. A regulation may not alter, amend, enlarge, or restrict a statute, or be inconsistent or in conflict with a statute." Upon approval by OAL and adoption by the rule-making agency, California regulations are printed in the California Code of Regulations (CCR) now also available on the Web at http://www.calregs.com. Title 3 of the CCR concerns Food and Agriculture; Title 27 concerns Environmental Protection. Not to be

confused with the CCR, are the 29 California Codes, which contain the statutes passed by the California legislature (accessible at http://www.leginfo.ca.gov/calaw.html).

Similarly, at the federal level, in recent years the rule-making process of agencies has become more transparent and open to public input. The federal Administrative Procedure Act[28] first enacted in 1946, and the Administrative Procedure Technical Amendments Act of 1991[29] prescribe the steps required prior to adoption of any agency regulation. Proposed regulations must be posted in the *Federal Register* (http://www.access.gpo.gov/su_docs/aces/aces1 40. html), which is published daily, seeking public input during a legally specified period. Public hearings may be held and numerous iterations may be required, and all require notice in the *Federal Register*. The regulation, if adopted, becomes part of the Code of Federal Regulation (CFR) (http://www.gpo.gov/nara/cfr/index.html). Title 7 of the CFR is Agriculture, Title 9 relates to Animal and Animal Products, and Title 50 is Wildlife and Fisheries. Laws passed by Congress are published in the U.S. Code (USC) (United States Code 2002). For example, Agriculture is located in Title 7 of the USC (there are 50 Titles).

As will be discussed below, the SPS Agreement requires transparency. It would seem that regulations issued by CDFA and USDA pursuant to the respective APA procedures should be in conformance with the transparency requirement.

WTO SPS Principles

A whole new set of international trade requirements and terminology—regionalization, equivalence, harmonization, transparency—came into being on January 1, 1995, when the WTO replaced the General Agreement on Tariffs and Trade (GATT). Recognizing the potential that countries may increase use of sanitary or phytosanitary measures as a new form of trade protectionism, the Uruguay Round Agreements included the SPS Agreement.

As a result of the SPS and the North American Free Trade Agreement (NAFTA), plant quarantine authorities must be able to demonstrate the threat of a particular disease or pest that makes necessary a particular SPS requirement, for example outright prohibition, postharvest treatments, or growing season requirements. Accordingly, after undertaking risk assessments and many years of bilateral discussions, the United States now allows several previously prohibited commodities to enter the country under specified conditions. For example, since January 1997, and of great concern to California and Florida avocado growers, the broad 1914 U.S. prohibition on import of Mexican avocados—first imposed to exclude seed weevils—no longer exists. Now, Haas avocados grown in approved orchards in the state of Michoacán, Mexico, may be imported through specified ports and routes to 19 northeastern states and the District of Columbia during the months of November through February, provided APHIS inspectors determine the avocados, orchards, packinghouses, and shipping procedures meet pests-free safeguards enumerated in the rule—including field surveys, trapping and field bait treatments, field sanitation practices, host resistance, post-harvest safeguards, winter shipping, packinghouse inspections, port-of-arrival inspections, and limited U.S. distribution. The avocado rule (7 CFR 319.56-2ff) is based on a "systems approach" with overlapping safeguards expected to operate sequentially to progressively reduce risk of introduction of several species to an insignificant level.

In a move toward implementing regionalization, and in a spirit of transparency, APHIS has posted 11 factors (Federal Register 1997) they will consider in evaluating requests to export animals or animal products to the United States from distinct or definable regions. Among the factors are: status of the disease in the region and in adjacent regions, vaccination status, infrastructure of veterinary services organizations in the region, emergency response capacity in the region, and the degree to which the region is separated from regions of higher risk through physical or other barriers.

U.S. implementation of other details of the SPS Agreement continues to evolve, and additional risk assessments and policy decisions are ongoing. These too are likely to be controversial. Globalization of trade and travel presents a compelling reason to harmonize with international agreements, priorities, and standards. At the same time, there is a concern USDA priorities will shift away from plant and animal protection as the USDA attempts to facilitate fair and open markets. In this regard, an overview of

the exotic pest and disease issues facing the U.S. nursery industry serves to illustrate some of the policy issues currently under discussion.

A Perspective on International Trade Rules: Implications for Horticultural Nursery Crops in the United States

An umbrella of strict federal quarantine regulations in combination with voluntary state certification programs has served to protect the health of nursery stock in the United States. At the same time, it has heavily restricted foreign imports. This mix of voluntary and mandatory regulations will probably need to evolve to become more consistent with the WTO phytosanitary measures, which require scientific justification, equivalency in application among member states, and eventually, conformance with IPPC standards or acceptance of comparable standards that achieve the receiving country's phytosanitary criteria.

The introduction of certain important horticultural crops into the United States is restricted by Federal Quarantine 37 in an effort to avoid the importation of injurious plant diseases and pests. Plants for crops such as grapes, *Citrus*, strawberries, *Prunus* (peaches, cherries, almonds, etc.), apples, pears, and sweet potatoes are known as "prohibited articles" and, with few exceptions, cannot be imported commercially as nursery stock. Even when dormant bare root plants are imported, nursery stock of these valuable genera can harbor damaging diseases caused by viruses, viroids, and phytoplasms. This broad prohibition has safeguarded U.S. producers of these crops for many years, with a minimum of government expense. APHIS enforces these regulations and supervises importation of germplasm of these genera.

Current U.S. Scheme for Importing Prohibited Horticultural Nursery Stock

In the United States today, prohibited crops enter under APHIS permits through a system of post-entry quarantine facilities supervised by scientists familiar with the particular crop and its pests. Some of these facilities, such as the USDA's Plant Germplasm Quarantine Office, do this work as part of the federal germplasm system.[30] There are also state-supported facilities that help with this work, such as the National Research Support Project 5 in Prosser, Washington;[31] the National Grape Importation and Clean Stock Program at Foundation Plant Materials Service, the University of California, Davis;[32] and the Citrus Clonal Protection Program, University of California, Riverside.[33] In addition, a number of research scientists throughout the country hold APHIS permits to allow germplasm of prohibited genera to enter the country. Before determining the terms of the permit, APHIS consults with regulatory officials of the state where post-entry quarantine will occur.

The quarantine process varies for each crop, depending on the biology of the plant type, the diseases of concern, and the testing procedures required for each disease. Generally, plant material from single plants is imported in the smallest quantity possible without soil, roots, or leaves. For grapes, for example, dormant cuttings are imported and treated with insecticides and fungicides before propagation; they are then subject to two years of laboratory and biological testing. *Citrus* is not a deciduous crop; therefore green bud sticks are imported. Strawberries are imported as dormant plants or tissue-culture plants. The actual strawberry plant that enters the country is heat treated, runners are propagated, and the original plant is subsequently destroyed to prevent introduction of red steele disease. For all these crops, the testing procedures require months to years to complete.

With the protection that is afforded by these strict quarantine regulations, U.S. growers have been protected from diseased nursery stock imported from abroad, whether it is infected with exotic pathogens or economically important pathogens that are already in the country. From a disease control perspective, this has been an efficient way to manage the plant health issues that can affect these valuable perennial crops. Since there is no cure or treatment for viruses in these crops, planting infected nursery stock material can result in decreased yields and quality, which occur over decades in fields, vineyards, and orchards.

Complementary to the federal quarantine regulations, a series of *voluntary* state certification programs have arisen in those states with significant nursery industries for these horticultural crops. These voluntary programs are well respected. Although all nurseries in the United States do not necessarily participate in the programs, the programs themselves set the standard for nursery stock and provide the initial propagating material for most commercial nursery stock. As a result of these programs and dedicated extension work by regulators and researchers over the last two generations, the disease- and pest-free standard for U.S. nursery stock is world class.

International Nursery Crop Phytosanitary Standards

The IPPC Secretariat is currently establishing guidelines and definitions and coordinating the efforts of RPPOs to establish consistent regional standards. The RPPOs are being asked to make the first efforts at harmonized standards because geographically contiguous areas often share common exotic pest and pathogen concerns and often work in concert to establish both internal and external control programs.

The North American Plant Protection Organization (NAPPO) is engaged in the process of creating regional standards for trade in a number of important nursery crops. A potato standard has recently been approved. Another panel has been working for several years in an effort to develop a grape standard. In 1999, panels began meeting to develop standards for *Citrus, Malus* (apple and crab apple), and *Prunus* fruit trees. Panels for additional crops are planned for the near future.

As the NAPPO panels have worked to develop a standard for a number of important nursery crops, a common problem has arisen for U.S. panel members attempting to follow new global standards while protecting U.S. growers. U.S. clean stock programs depend heavily on the umbrella of our current quarantine regulations. However, our certification programs are largely a state-by-state patchwork of voluntary programs. Most participants are doubtful that voluntary control programs, lacking government mandates, will be viewed as nondiscriminatory and will constitute sufficient control for the international community to allow the programs to generate a U.S. list of "regulated nonquarantine pests." Consequently, it has been difficult for NAPPO working groups to develop standards that both satisfy the IPPC and provide U.S. growers with the level of protection they now enjoy against disease.

Possible Scenarios For the Future

Discussions have commenced about possible solutions to this dilemma. As work continues on NAPPO standards for these crops, and as nonquarantine injurious diseases are removed from the regional lists and ultimately from national quarantine lists, more open trade should result between NAPPO countries and with the rest of the world. If no other action is taken by the United States in this arena, this could result not only in more open competition for the nursery industries but also possible importation of damaging pests and diseases and degradation of quality and a loss of farm productivity. Many nursery growers, regulators, and researchers find this prospect unacceptable.

Phytosanitary quality under the current voluntary system for horticultural nursery stock is very high; U.S. nursery products have ranked at the top when evaluated by independent testing agencies. Although nursery stock for these crops does not enter the country directly from foreign countries without going through a quarantine process, many foreign nurseries have invested in the United States and brought new plant materials, techniques, and ideas to the U.S. industry. Because very little stock has entered the United States, a large regulatory infrastructure to supervise imports has not been needed, and the current system is inexpensive.

A federal program of regulation, either mandatory certification programs or official control programs for target diseases for each commodity, could allow IPPC classification of these economically important diseases as regulated nonquarantine pests. Formal state or domestic regional regulations might also serve this purpose. By establishing formal domestic regulations, only imported nursery stock meeting high standards of freedom from specific domestic diseases could enter the country. The idea of a federal mandatory certification program, however, has no existing model in the United States. Many nurserymen and growers find the idea intrusive and contrary to American ideals of free choice, trade, and competition.

Furthermore, any program would require funding to enforce. This could come from industry, state, or federal funds, but is likely to be far more expensive than our current exclusionary system. The citrus nursery industry already has some mandatory state control programs for Tristeza virus and is exploring the concept of mandatory state or federal certification programs. Discussions are just beginning among grape growers, scientists, and regulators about similar possible actions.

It is unlikely that the international pressures on the U.S. nursery industry to clarify and harmonize standards will subside. Furthermore, in the international community the United States is most often a vocal force for more open markets. This issue is unlikely to attract the necessary support to make a change in IPPC regulations or WTO policies. Although it might be a number of years before a change in our current practices is forced by either a WTO challenge or changes in U.S. regulations as a result of international agreements, many think it wise to discuss issues, solutions, and implementation before that time.

This chapter has summarized the domestic and international regulatory principles and institutions for exclusion or control of exotic pests and diseases in the United States and California. It provided an overview of historical experiences with undesirable introductions and sketched out the evolution of relevant government regulations and organizations. After relegating a synopsis of key statutes to an appendix, it described regulatory tools available to implement exclusion policies, such as prohibitions, surveillance activities, and quarantines, and other activities used to implement policies of eradication, containment, or suppression—for example, pesticides, heat treatment, release of biological control agents, or quarantines. Noting that Chapter 4 presents a more complete discussion of the WTO SPS Agreement, this chapter highlighted examples of recent U.S. regulatory changes that would appear to be consistent with the NAFTA and WTO SPS Agreement. Illustrative of the challenges ahead as the world community continues to develop international sanitary and phytosanitary standards, the chapter concluded with a discussion of some of the issues facing the U.S. nursery industry.

Notes

[1] The authors wish to thank Richard Breitmeyer, DVM, Animal Health and Food Safety Services, California Department of Food and Agriculture; Bill Callison and Dorothea Zadig, Plant Health and Pest Prevention Services, California Department of Food and Agriculture; Paul O. Ugstad, DVM, Veterinary Services, USDA Animal and Plant Health Inspection Service; and Helene Wright, Plant Protection and Quarantine, USDA Animal and Plant Health Inspection Service, for their review and valuable comments on a previous draft of this chapter. Final responsibility for content rests with the authors, however.

[2] 37 Stat.L., 315.

[3] Quarantine 37 was issued November 18, 1918, by the Federal Horticultural Board. It is found in 7 CFR 319.37.

[4] The 1942 Mexican Border Act authorized the Secretary of Agriculture to work with Mexican authorities to prevent transborder pests, and the Organic Act of 1944 gave the Secretary authority to cooperate with farmers' associations, individuals, and Mexico to detect and control plant pests (Chock, 1983).

[5] European countries signed the first international plant protection agreement in 1881, the *Phylloxera vastrix* Convention.

[6] IPPC Secretariat created in 1992 in anticipation of the SPS Agreement; IPPC last amended in 1997. (See Glossary.)

[7] 117 members as of June 24, 2002.

[8] http://www.oie.int/eng/oie/en_oie.htm (May 2001 October 24, 2000 download).

[9] The proposal for the new department would shift APHIS from USDA to the new department.

[10] Phytosanitary certification for export is performed in California under a memorandum of understanding between the USDA and CDFA and between CDFA and the county agricultural commissioners, who actually issue the certificates.

[11] In the first three months of 2000, CLAMP intercepted materials with serious pests originating from the state of Georgia (found in 14,900 pounds of pecans), China (found in 88 pounds of dried citrus peel), and Mexico (found in 1,040 pounds of mangos). Among the 1998 seizures were 871 wooden crates infested with long-horn beetle larvae and flat-headed borers.

[12] In 2000, The National Plant Board prepared a Model Nursery Law and a Model Plant Pest Law for use by the states.

[13] The IPPC Secretariat in the United Nations Food and Agricultural Organization Plant Protection Service now administers the multilateral IPPC treaty.

[14] The WTO-SPS measures are the sanitary and phytosanitary measures of GATT 1994, also known as the Uruguay Round of Multilateral Trade Negotiations, which established the World Trade Organization. The SPS standards have or are being developed by OIE, Codex Alimentarius, and IPPOIPPC. (See

Chapter 4 for more detail.)

[15] With the exception of birds and poultry from Mexico, animals from Canada and Mexico are inspected prior tobefore leaving and are not held in quarantine.

[16] The "regulated pest list" is found at http://www.aphis.usda.gov/ppq/regpestlist/ (January 25, 2002 download).

[17] This California manual is available on the Web at http://www.cdfa.ca.gov/phpps/pe/pqm.htm (February 14, 2002 download).

[18] Information about federal permits and downloadable forms is available from http://www.aphis.usda.gov/ppq/permits/ (January 25, 2002 download).

[19] As part of APHIS participation in the international movement toward greater regulatory transparency, the list of allowable plant materials can be found on the Internet at http://www.aphis.usda.gov/ppq/permits/nursery.htm. (January 25, 2002 download).

[20] http://www.cdfa.ca.gov/ahfss/ah/emergency_management.htm (January 28, 1998 download).

[21] OIE classification lists are available at http://www.oie.int/eng/maladies/en_classification.htm (October 22, 2002 download).

[22] "B"-rated pests are subject to action by CDFA only when found in a nursery, and otherwise are subject to eradication, containment, control, or holding action at the discretion of the individual agriculture commissioner. A "C"-rated pest is not subject to state action except to provide for general pest cleanliness in nurseries. Individual agricultural commissioners may elect to take additional action against the pest within their counties. A "D" rating indicates that an organism is of little or no economic importance and no action is taken against it. In all cases, the law governs what regulatory action is taken. There are times when, for example, laws governing nursery stock standard of pest cleanliness may require that it be free of "C"-rated pests.

[23] Animal Health Protection Act, P.L. 107–171.

[24] The Plant Protection Act, which consolidated 10 statutes, includes a specific provision for states to petition the Secretary for permission to enforce more restrictive measures on interstate movements of regulated pests and articles.

[25] CA Government Code §11346.

[26] CA Government Code §11340; available at http://www.leginfo.ca.gov/calaw.htm/.

[27] California Code of Regulations, Title 1, §§11-90; available at http://ccr.oal.ca.gov/.

[28] 5 USC §551 et seq.

[29] 5 USC §§561nt.

[30] http://www.barc.usda.gov/psi/fl/gfqo.html (May 25, 1999 URL).

[31] http://.nrsp5.wsu.edu (May 25, 1999 URL).

[32] http://fpms.ucdavis.edu (May 25, 1999 URL).

[33] http://www.cc.pp.uct.edu (October 25, 2002 download).

At the IPPC, the International Plant Protection

Convention, a treaty was signed by 117 contracting countries; http://www.fao.org/WAICENT/FaoInfo/Agricult/AGP/AGPP/PQ.

References

APHIS. 1999. FY 1999 Report.

Ashcraft, Mark. 2002. California Department of Food and Agriculture Animal Health and Food Safety Services. Personal communication to M. Kreith, June 26, 2002.

California Department of Food and Agriculture Plant Health and Pest Prevention Services. 2000. *1999 Report to the Western Plant Board, Annual Report. Sacramento.*

Chock, Alvin Keali'i. 1983. "International Cooperation on Controlling Exotic Pests." In Charles L. Wilson and Charles L. Graham, Eds., *Exotic Plant Pests and North American Agriculture.* New York: Academic Press.

FAO. 2001. *The State of Food and Agriculture 2001*; January 12, 2002 online at http://www.fao.org/docrep/003/x9800e/x9800e18.htm.

FAO. 2002. FAO website accessed January 12, 2002 http://www.fao.org/WAICENT/FAOINFO/AGRICULT/AGP/AGPP/PQ/En/Conven/evolut.htm/.

Federal Register. 1997. 62(208):56028–56029.

League of Nations Treaty. 1929. *International Convention for the Protection of Plants of 1929,* Apr.

16, 1929. 126 League of Nations Treaty Series 305. Geneva: League of Nations.

Office of Administrative Law, State of California. 2001. *How to Participate in the Rulemaking Process.* Available June 20, 2002 at http://www.oal.ca.gov/document/howtoparticipate.pdf. Sacramento: Office of Administrative Law.

Ryan, Harold J. 1969. *Plant Quarantines in California. A Committee Report.* Berkeley: University of California Division of Agricultural Sciences.

United States Code. 2002. Published by Office of the Law Revision Counsel, the U.S. House of Representatives. October 22, 2002 access at URL: http://uscode.house.gov/usc.htm and Government Printing Office, URL:http://www.access.gpo.gov/congress/cong013.html/.

United States. 2002. *Federal Register.* October 22, 2002 available at http://www.access.gpo.gov/su_docs/aces140.html. Government Printing Office.

Whiteford, Annette. 2002. California Department of Food and Agriculture Animal Health and Food Safety Services. Personal communication to M. Kreith, July 2, 2002.

Wiser, Vivian. 1974. *Protecting American Agriculture. Inspection and Quarantine of Imported Plants and Animals.* Agricultural Economic Report No. 266. ERS, USDA. Washington: Economic Research Service, U.S. Department of Agriculture.

Appendix 3.1

Statutes of the United States and California Important to the Control of Exotic Pests and Diseases

Marcia Kreith

Key Federal Statutes

Prominent statutes operational at the federal level that pertain to control of exotic pests and diseases are found in the *U.S. Code Title 7. Agriculture*, and also in *Title 16. Agriculture*.

U.S. Code Title 7. Agriculture

Chapter 7. Insect Pests Generally
Chapter 7B. Plant Pests
Chapter 8. Nursery Stock and Other Plants and Plant Products

U.S. Code Title 16. Conservation

Chapter 35. Endangered Species. This chapter includes prohibitions on the import of endangered species into the United States and export or damage to endangered species.

Additional Federal Laws Listed by Popular Name

Animal Health Protection Act was signed into law on May 13, 2002, as Title X, Subtitles E, §10401 et seq. and F §10501 et seq.[1] of P.L. 107-171, the Farm Security and Rural Investment Act of 2002.

Plant Protection Act (PPA). 2000. (Title IV of the Federal Crop Insurance and Agricultural Risk Protection Act of 2000, P.L. 106-224.) 7 USC §§7701, 7702, 7711-7718, 7731-7736, 7751-7758. It amends and replaces the Federal Plant Pest Act, Plant Quarantine Act of 1912, the Federal Noxious Weed Act, and seven other plant health laws. It requires that the processes used in developing regulations that govern import requests be based on sound science and be transparent and accessible.

National Invasive Species Act of 1996 (NISA). 16 USC §4701 nt. Chapter 67 deals with aquatic nuisance prevention control. NISA reauthorized and made changes to the *Nonindigenous Aquatic Nuisance Prevention and Control Act of 1990 (NANPOA).*

Federal Noxious Weed Act of 1974. 7 USC §§2801-2814. Amended 1990, 1994, 1997. Sections 2801-1813 repealed by P.L. 106-224, Title IV of the Agricultural Risk Protection Act, June 20, 2000. Section 2814 provides for nonindigenous species control plans on federal lands.

Lacey Act. 1900. 18 USC §§42 to 44. Substantially amended in 1981 (16 USC 3371-3378), 1984, 1988, and 1994. The act prohibited interstate trade of endangered wildlife killed in violation of states' laws, such as the passenger pigeon, and banned the importation of mongooses, fruit bats, English sparrows, starlings, and others species of threat to U.S. crops. Prohibitions were subsequently expanded to cover international trade, treaties, and foreign laws, and also, plants.

Lacey Act Amendments of 1981. 16 USC §3371 et seq. 2002. Repealed 18 USC §43 and §44 of original 1900 Lacey Act. These amendments regulate introduction of certain nonindigenous species, including illegally taken fish and wildlife and rare plant species listed by the Convention on International Trade in Endangered Species of Wild Flora and Fauna (CITES).

Migratory Bird Treaty Act. 1918. Amended 1960, 1969, 1974, 1978, 1998. 16 USC §§703-708, 709a, 710, 711, 668aa, 668bb, 668cc-1. This act implements the U.S. commitment to international conventions for the protection of shared migratory bird resources. Each convention protects selected species common to at least two of the signatory countries.

Endangered Species Act of 1973 (ESA). Amended 1977, 1978, 1979, 1980, 1982, 1984, 1986, 1988. 16 USC, many sections. 1988 amendment §§1536 and 1538.

Endangered Species Act Amendments of 1978. 16 USC, §1531 et seq. has secondary effects on control of exotic pests and diseases, such as through regulation of pesticide registration and use.

National Environmental Policy Act of 1969 (NEPA). 42 USC §4321 et seq. has secondary effects on control of exotic pests. It established the Council on Environmental Quality.

Federal Seed Act. 1939 and subsequent amendments. 7 USC §§1551 to 1611. Its primary purpose is variety certification, but it prohibits importation of agricultural or vegetable seeds if they contain noxious weed seeds as defined in Subchapter III.

Federal Environmental Pesticide Control Act. 1972. 7 USC §§136 to 136y; 15 USC §§1261 and 1471; 21 USC §§321 and 346a. This act has secondary effects on control of exotic pests and diseases. It incorporates and supersedes the Insecticide Act of 1910.

Federal Insecticide, Fungicide, Rodenticide Act Amendments of 1947 (FIFRA). 7 USC 136 et seq. 2002. This act has secondary effects on control of exotic pests and diseases. By amending the 1910 Federal Insecticide Act, FIFRA shifted emphasis to protection of health and the environment and required pesticide re-registration.

Key California Statutes

Statutory authority for California pest control (including for agency regulations) derives from federal statutes and state laws, primarily, the *California Food and Agricultural Code.*

California Food and Agricultural Code

Division 4. Plant Quarantine and Pest Control. Sections 5321-5323 provide authority for investigation and regulation.

Division 5. Animal and Poultry Quarantine and Pest Control are regulated pursuant to §§9501-9702[1], and other sections of the code. They provide authority for actions of the State Veterinarian.

Division 6. Pest Control Operations.

Division 7. Agricultural Chemicals, Livestock Remedies, and Commercial Feeds.

Division 8. Vessel and Aircraft Garbage.

Additional California Laws Listed by Popular Name

Plant Quarantine Inspection Act. 1967. Cal. Food and Agricultural Code §5341 et seq.

California Airport and Maritime Plant Quarantine, Inspection, and Plant Protection Act. 1990. Cal. Food and Agricultural Code §§5350-5353.

---

[1]§ is the symbol used in the legal profession to represent "section," §§ designates "sections"; et seq. is the abbreviation for et sequitur, meaning "and following" section(s).

Appendix 3.2

California List of Reportable Conditions for Animals and Animal Products

Pursuant to Section 9101 of the California Food and Agricultural Code and Title 9 Code of Federal Regulations Section 161.3(f)

Reportable conditions pose or may pose significant threats to public health, animal health, the environment, or the food supply. Any licensed veterinarian, any person operating a diagnostic laboratory, or any person who has been informed, recognizes or should recognize, by virtue of education, experience, or occupation, that any animal or animal product is, or may be affected by, has been exposed to, or may be transmitting or carrying any of the following conditions, must report that information.

Any diseases or conditions caused by exposure to pesticides, toxins, heavy metals, or other toxicants, any animal disease not known to exist in the United States, any disease for which a control program exists, or an unexplained increase in the number of diseased animals or deaths must be reported. Conditions that are, or have the potential to be a public health, animal health, or food safety threat must be reported within 24 hours.

These conditions must be reported either to your closest Department of Food and Agriculture, Animal Health Branch (AHB) District Office: Redding 530-225-2140, Modesto 209-491-9350, Tulare 559-685-3500, Ontario 909-947-4462, the AHB Headquarters at 1220 N Street, Room A-107, Sacramento, California 95814, telephone 916-654-1447, facsimile 916-653-2215, or the USDA Animal and Plant Health Services, Veterinary Services (VS) office at 916-857-6170 or toll free at 877-741-3690.

List effective July 2, 2002

Emergency Conditions. Report within 24 hours to CDFA Animal Health Branch or APHIS Veterinary Service.

Multiple Species
Anthrax (*Bacillus anthracis*)
Screw worms (*Cochliomyia hominivorax* or *Chrysomya bezziana*)

Bovine
African trypanosomiasis (Tsetse fly diseases)
Bovine babesiosis (piroplasmosis)
Bovine spongiform encephalopathy (Mad Cow)
Contagious bovine pleuropneumonia (*Mycoplasma mycoides mycoides small colony*)
Foot-and-mouth disease (Hoof-and-mouth)
Heartwater (*Cowdria ruminantium*)
Hemorrhagic septicemia (*Pasteurella multocida serotypes B:2 or E:2* not known to occur in US)
Lumpy skin disease
Malignant catarrhal fever (African type)
Rift Valley fever
Rinderpest (Cattle plague)
Theileriosis (Corridor disease, East Coast fever)
Vesicular stomatitis

Caprine/Ovine
Contagious agalactia (*Mycoplasma* species)
Contagious caprine pleuropneumonia (*Mycoplasma capricolum capripneumoniae*)
Foot-and-mouth disease (Hoof-and-mouth)
Nairobi sheep disease
Peste des petits ruminants (Goat plague)

Pulmonary adenomatosis (Viral neoplastic pneumonia)
Rift Valley fever
Salmonella abortus ovis
Sheep and goat pox

Porcine
African swine fever
Foot-and-mouth disease (Hoof-and-mouth)
Hog Cholera (Classical swine fever)
Japanese encephalitis
Nipah virus
Swine vesicular disease
Teschen (Enterovirus encephalomyelitis)
Vesicular exanthema
Vesicular stomatitis

Commercial Poultry
Exotic Newcastle disease (Viscerotrophic velogenic Newcastle disease)
Highly pathogenic avian influenza (Fowl plague)

Equine
African horse sickness
Dourine (*Trypanosoma equiperdum*)
Epizootic lymphangitis (equine blastomycosis, equine histoplasmosis)
Equine piroplasmosis (*Babesia equi, B. caballi*)
Glanders (Farcy) (*Pseudomonas mallei*)
Hendra virus (Equine Morbillivirus)
Horse pox
Japanese encephalitis
Surra (*Trypanosoma evansi*)
Venezuelan equine encephalomyelitis
Vesicular stomatitis
West Nile Virus)

Other Species
Chronic Wasting Disease in cervids
Viral hemorrhagic disease of rabbits (calicivirus)

Conditions of Regulatory Importance. Report within two days of discovery to CDFA Animal Health Branch or APHIS Veterinary Service.

Multiple Species
Rabies of livestock

Bovine
Bovine brucellosis (*Brucella abortus*)
Bovine tuberculosis (*Mycobacterium bovis*)
Cattle scabies (*Sarcoptes scabiei* var. *bovis*)
Trichomoniasis (*Tritrichomonas fetus*)

Caprine/Ovine
Caprine and ovine brucellosis (excluding *Brucella ovis*)
Scrapie
Sheep scabies (Body mange) (*Psoroptes ovis*)

Porcine
Porcine brucellosis (*Brucella suis*)
Pseudorabies (Aujeszky's disease)

Commercial Poultry
Ornithosis (Psittacosis or avian chlamydiosis) (*Chlamydia psittaci*)
Pullorum disease (Fowl typhoid) (*Salmonella gallinarum* and *pullorum*)

Equine
Contagious equine metritis (*Taylorella [Haemophilus] equigenitalis*)
Equine encephalomyelitis (Eastern and Western equine encephalitis)
Equine infectious anemia (Swamp fever)

Other Species
Brucellosis and tuberculosis in cervids
Duck viral enteritis (Duck plague) (all ducks)

Monitored Conditions. Report by telephone, mail or electronic methods on a monthly or as detected basis.

Multiple Species
Avian tuberculosis (*Mycobacterium avium*)
Bluetongue
Echinococcosis/Hydatidosis (*Echinococcus granulosus* or *E. multiloculans*)
Johne's disease (paratuberculosis) (*Mycobacterium avium paratuberculosis*)
Leptospirosis
Q Fever (*Coxiella burnetii*)
Trichinellosis (*Trichinella spiralis*)

Bovine
Anaplasmosis (*Anaplasma marginale* or *A. centrale*)
Bovine cysticercosis (*Taenia saginata* in man)
Bovine genital campylobacteriosis (*Campylobacter fetus venerealis*)
Dermatophilosis (Streptothricosis, mycotic dermatitis) (*Dermatophilus congolensis*)
Enzootic bovine leukosis (Bovine leukemia)
Infectious bovine rhinotracheitis (Bovine herpesvirus-1)
Malignant catarrhal fever (North American)

Caprine/Ovine
Brucella ovis (Ovine epididymitis)
Caprine (contagious) arthritis/encephalitis
Enzootic abortion of ewes (Ovine Chlamydiosis) (*Chlamydia psittaci*)
Maedi-Visna (Ovine progressive pneumonia)

Porcine
Atrophic rhinitis (*Bordetella bronchiseptica, Pasteurella multocida*)
Porcine cysticercosis (*Taenia solium* in man)
Porcine reproductive and respiratory syndrome
Transmissible gastroenteritis

Commercial Poultry
Avian infectious bronchitis
Avian infectious laryngotracheitis
Duck viral hepatitis
Fowl cholera (*Pasteurella multocida*)
Fowl pox
Infectious bursal disease (Gumboro disease)
Marek's disease
Mycoplasmosis (*Mycoplasma gallisepticum*)

Equine
Equine influenza
Equine rhinopneumonitis
Equine viral arteritis
Horse mange (multiple types)

Commercial Fish for Human Consumption
Epizootic hematopoietic necrosis
Infectious hematopoietic necrosis
Onchorynchus masou virus disease
Spring viremia of carp
Viral hemorrhagic septicemia

Other Species
Hemorrhagic diseases of deer (Bluetongue, Adenovirus, and Epizootic hemorrhagic disease)
Tularemia and Myxomatosis in commercial rabbits

4

International Trade Agreements and Sanitary and Phytosanitary Measures

James F. Smith[1]

Introduction

Following World War II, the United States and Western Europe built a new world economic order based on the principles of free trade and decentralized market economies. The architects of the new era met in Bretton Woods, New Hampshire, in 1944. They created the International Monetary Fund (IMF) to oversee currency exchange policies, the World Bank to grant loans to developing countries, and the International Trade Organization (ITO) to establish and enforce multilateral trade rules. The IMF and the World Bank survive to this day. The ITO was stillborn due to the opposition of the United States, which perceived it as a threat to U.S. sovereignty. Instead, the General Agreement on Tariff and Trade (GATT), the provisional agreement of the ITO, became the international trade code. Its secretariat was the "institution" for enforcement of these rules. The GATT was remarkably successful in reducing tariff barriers to trade through eight rounds of multilateral trade negotiations (1948-1994). Reduction of nontariff barriers to trade was a more intractable problem. These included agricultural subsidies and sanitary and phytosanitary (SPS) measures.

The SPS Provisions

Agriculture and SPS measures have been a "special case" in the GATT from the outset (Stewart 1993). Prior to the Eighth GATT round, the Uruguay round (1986–1994), member nations enjoyed considerable latitude in restricting imports to protect human, animal, or plant life or health under GATT, Article XX(b).

GATT Article XX(b) provides an exception to GATT obligations (McNiel 1998). The GATT Article XX(b) exception prohibits members from using such measures as disguised trade barriers or from arbitrarily discriminating against imports (McNiel 1998). However, the GATT provides little guidance on the limits of the importing countries' sovereign prerogative in restricting imports on health and safety grounds (Johanson and Bryant 1996).

Thus, GATT process was unpredictable in dealing with SPS measures claimed to violate the treaty (Cromer 1995). If the exporting country challenged SPS measures, in a GATT dispute settlement process the panel would first address whether the SPS measures violated the GATT's national treatment (discrimination against imports) or import quota prohibitions (ban of allegedly harmful products). If the GATT Panel found such a violation, the second issue was whether the SPS measure qualified as an exception to the GATT obligation under Article XX(b) because it was "necessary to protect human, animal or plant life or health." When Thailand banned imported cigarettes, purportedly as a health measure, the GATT Panel ruled that, despite the laudable goal of reducing cigarette consumption, banning imported cigarettes was not "necessary" because domestic production was not restricted.

When negotiators used the term "necessary" in the North American Free Trade Agreement's (NAFTA's) SPS provisions, environmentalist and consumer groups argued that this would unduly restrict a sovereign nation's prerogative to protect itself from health hazards to people, animals, or plants (Gaines 1993).

[1] Acknowledgements to Jaime Raba, Lysle Buchbinder, Lily Chen, Joe Cruz, Paul Moncrief, Jonathan Warner, Erin Wester-Main.

When the Uruguay Round commenced, developing nations called for the revision of SPS laws (Seilhamer 1998). They emphasized the inability of developing countries to comply with SPS measures (Stewart 1993). When food or drugs are produced in developing countries that do not meet the level required by a developed country, a trade barrier could exist without adequate assurance that the measure is necessary (Cromer 1995).

The NAFTA/SPS (Josling and Barichello 1993) and World Trade Organization (WTO) (Stewart 1993) SPS agreements attempt to balance a sovereign prerogative to protect itself from external health threats that may be introduced through foreign imports, and the multilateral goal of eliminating disguised protectionist restrictions. U.S. law, which implements these international agreements, reflects the tension between these goals (Walker 1998).

The negotiators, including the U.S. delegation, emphasized the overarching importance of conditioning SPS restrictions on sound science (Millimet 1995; Daniel 1994). Indeed, Maruyama has termed the principle of "sound science" a "new pillar" of the WTO comparable to the GATT precepts of "most favored nation" and "national treatment" (Maruyama 1998; Walker 1998).

The WTO members agreed to replace the former GATT Article XX(b) defense, for health and safety measures, with a detailed SPS code (Franzen 1998; Schaefer 1998). The NAFTA/SPS and WTO/SPS agreements that were drafted at the same time are quite similar (Steinberg 1995; Gaines 1993). However, there are differences that may prove decisive (Steinberg 1995; Wirth 1994). This article focuses on the WTO/SPS rather than the NAFTA/SPS for several reasons. The WTO/SPS applies to over 130 WTO members, including the NAFTA parties. The WTO Appellate Body has interpreted the critical provisions in three separate cases (Hudec 1999).

Finally, because the WTO/SPS and the NAFTA/SPS are quite similar, and the latter has not yet been interpreted by a NAFTA dispute settlement panel, it is likely that the WTO/SPS interpretations will be highly persuasive should there be a formal NAFTA dispute.

Instead of creating SPS standards, the WTO/SPS agreement provides rules for the adoption of such measures (Johanson and Bryant 1996; Maruyama 1998). While members may take measures to protect health and life within their territories, they may do so only if such measures are not inconsistent with the provisions of the SPS Agreement. Article 2.3 of the SPS provides:

> Members shall ensure that their sanitary and phytosanitary measures do not arbitrarily or unjustifiably discriminate between Members where identical or similar conditions prevail, including between their own territory and that of other Members. Sanitary and phytosanitary measures shall not be applied in a manner which would constitute a disguised restriction on international trade.

To achieve the dual objective of protecting the member's sovereign prerogative to protect human, animal, and plant life and health, but to refrain from arbitrary, unjustifiably discriminatory or disguised restrictions on trade, the WTO/SPS encourages, but does not require, adherence to international standards. While a member is free to choose a higher level of protection, it must justify such a measure through sound science (risk assessment).

The only exception to the requirement of scientific justification or risk assessment is a temporary SPS measure, which a member may adopt when "relevant scientific evidence is insufficient" and certain other requirements are fulfilled. Additionally, "[m]embers shall avoid arbitrary or unjustifiable distinctions in the levels it considers to be appropriate in different situations."

Furthermore, "Members shall ensure that such measures are not more trade-restrictive than required to achieve their appropriate level of sanitary and phytosanitary protection, taking into account technical and economic feasibility." The WTO/SPS also require that importing members accept SPS measures of exporting members who seek the same level of protection, albeit by differing means (equivalency), and that regional differences be taken into account. Moreover, a member is to provide information to other WTO members on their SPS measures (transparency). Accordingly, to comply with the WTO/SPS a member's SPS measure

1. must conform to the international standard, if any, or be able to scientifically justify the measure through a risk assessment (WTO/SPS, arts. 2.2, 3.3 and 5.1);

2. may be temporarily adopted, although it would not qualify as a permanent measure, only as provided by the WTO/SPS (article 5.7);

3. must avoid arbitrary or unjustified distinctions in levels of protection (WTO/SPS article 5.5;

4. be no more trade restrictive than necessary to achieve the appropriate level of protection (WTO/SPS article 5.6);

5. accept an exporting member's SPS measures as equivalent if that member objectively demonstrates that its measures achieve the same level of protection (WTO/SPS article 4);

6. take into account the level of prevalence of specific diseases or pests, existence of eradication or control programs (WTO/SPS article 5.6);

7. notify other members of changes in their SPS measures (WTO/SPS article 7); and

8. may not be applied through the control, inspection, and approval procedures to limit arbitrarily or unjustifiably the importation of foreign products (WTO/SPS article 8).

These eight requirements of the WTO/SPS and the WTO Appellate Body's interpretation of the requirements follow.

The SPS Compliance Requirements

Conform or Justify by Risk Assessment
(WTO/SPS articles 2.2, 3.1, 3.2, 3.3, 5.1, Annex A(4))

The first requirement is closely related to the goal of harmonization in that members are encouraged to conform to international standards or to scientifically justify the SPS measure through a risk assessment. The multiple references to international standards in the WTO/SPS and the WTO Appellate Body interpretation of the WTO/SPS illustrate the fundamental importance of this first requirement, namely to conform or justify through risk assessment.

In 1994, before the WTO/SPS became effective, Barcello characterized the agreement as containing "only a weak effort to harmonize S&P [SPS] standards" and the agreement effort to harmonize S&P standards as "weak" and "hortatory" (1994). Harmonization removes inconsistent worldwide standards but is controversial if it is seen as compelling the least common denominator of protection. Barcello noted that the major incentive for harmonization is the

presumption that an SPS measure "based on" an international standard is "deemed to be necessary . . . and presumed to be consistent with the relevant provisions of the S&P Agreement and GATT, 1994" (WTO/SPS, article 3.2). However, the WTO Appellate Body interpretation of the WTO/SPS has made this view less tenable.

In the beef hormone case, the Appellate Body ruled that once the complaining body establishes a "prima facie [apparent] case of inconsistency with a particular provision of the SPS . . . the burden of proof moves to the defending party." A complaining party would have been hard-pressed to attack a conforming SPS measure.

Moreover, a complaining party may request that the trade restricting country provide full information about the risk assessment. Clearly an importing country may choose to conform its measures to the international standard, adopt the international standard with adjustments, adopt an "equivalent measure" (Walker 1998), or establish sanitary measures that provide "a higher level of sanitary . . . protection" that are "based on" the international standards. In the later instance, the Agreement requires "scientific justification," namely that the sanitary measure is selected in accordance with the risk assessment provisions of SPS Article 5 (WTO/SPS, article 3.3). In cases where there are no international standards or the importing country adopts only some of the international standard, the restricting import country may well be called on to justify the measure with a pertinent risk assessment.

The WTO/SPS specifically defines international standards by reference to three international organizations, the Codex Alimentarius Commission (Codex), the International Plant Protection Convention (IPPC), and the International Office of Épizooties (OIE). Terence P. Stewart and David S. Johanson have cogently summarized the cumulative effect of the numerous explicit and implicit references to these international bodies (Stewart and Johanson 1998). Indeed, a member who does not use an international standard must explain why (Stewart and Johanson 1998). As one U.S. official succinctly put it "what we have here are some norms which are guiding sovereignty, and the thing to do if you are in the sovereignty maintaining business, is to go out and screw around with the norms if you can" (Schaefer 1998).

Of the first three WTO/SPS disputes, the United States was the complaining party in two of them and the defending parties were the European Community (*EC Beef Hormones*) and Japan (*Japanese Agricultural*). In the remaining case, *Australian Salmon*, Canada complained against restrictions. In these cases the importing countries sought a level of protection that was higher than the international standard, or there was no international standard. In each case the WTO panel and the WTO Appellate Body found that the importing party failed to scientifically justify the measure or that its risk assessment was inadequate.

It may be that these cases were relatively easy ones based on the facts that the risk factors were not especially compelling from either a scientific or common sense view. Nonetheless, the language of the opinions suggests that the WTO Appellate Body is notifying the world trading community of its interpretation of the WTO/SPS.

EC Beef Hormones Case In the late 1980s, the European Communities (EC) prohibited growth hormones in beef production and banned hormone-treated meat imports. The EC's justification for these measures was that the hormones are carcinogenic, and using them for growth promotion adds to the risk already faced by consumers from background levels of hormones. The United States argued that the measure violated the European Communities' obligations under Article 3.1 of the SPS Agreement in that it was not based on the international standards.

The Codex maintained standards for five of the six hormones under dispute, which provided that these five hormones, when used according to sound veterinary practices for purposes of growth promotion in beef cattle, do not pose risks to human health. The Appellate Body found that the SPS Agreement does not require a WTO member to base its SPS measures upon international standards but such measures must be based upon risk assessments as described in article 5.1.

The Appellate Body held that a risk assessment need not establish a minimum quantifiable risk, nor exclude factors that are not susceptible of quantitative analysis by the empirical or experimental laboratory methods. It further held

that the importing party need not demonstrate that "it actually took into account a risk assessment when it enacted or maintained the measure. . . ." Despite the stated flexibility, the Appellate Body ruled against the EC, finding that they "did not assess risks arising from the failure of observance of good veterinary practice combined with problems of control of the use of hormones for growth promotion purposes." The EC produced one expert's opinion who stated that of 110,000 women who would get breast cancer, several thousand would do so from "the total intake of exogenous estrogens from every source" and "one of those 110,000 would come from eating meat containing estrogens as a growth promoter, if used as prescribed."

The Appellate Body rejected the opinion as a risk assessment because it "does not purport to be the result of scientific studies carried out by him or under his supervision focusing specifically on residues of hormones in meat from cattle fattened with such hormones" and "that the single divergent opinion expressed by Dr. Lucier is not reasonably sufficient to overturn the contrary conclusions reached in the scientific studies."

The Appellate Body emphasized that the "opinions of individual scientists have not evaluated the carcinogenic potential of those hormones when used specifically for growth promotion purposes" nor "the specific potential for carcinogenic effects arising from the presence in food," more specifically, "meat or meat products" of residues of the hormones in dispute." The EC argued that the SPS Agreement contemplated the sovereign prerogative to ban possibly dangerous substances despite the absence of scientific evidence under the "precautionary principle."

The Appellate Body disagreed, ruling "the precautionary principle has not been written into the SPS Agreement as a ground for justifying SPS measures that are otherwise inconsistent with the Agreement. This suggests that science is exalted over parochial concerns. However, the science itself appears to have been politicized in that the 1995 Session of the Codex Alimentarius approved the growth hormones in question, at the request of the United States, on a secret vote in which 33 delegates voted for the standard, 29 opposed, and 7 abstained (Stewart and Johanson 1998).

Australian Salmon Dispute   In 1995 Canada challenged Australia's ban on fresh, chilled, and frozen salmon from Canada. Australia argued that the ban was necessary to protect Australian fish from diseases that could have damaging economic and biological consequences for Australia's fisheries. OIE standards did not exist for all of the 24 diseases from which Australia was seeking protection, and the OIE had no guidelines for salmon as a specific product. The Appellate Body, like the panel, found that Australia's policy as applied to ocean-caught salmon was not based upon a risk assessment and therefore violated Articles 2.2 and 5.1. The Appellate Body articulated a three-pronged test for risk assessment under WTO/SPS, Article 5.1 as follows:

1. identify the diseases whose entry, establishment or spread a Member wants to prevent within its territory, as well as the potential biological and economic consequences associated with the entry, establishment or spread of these diseases;

2. evaluate the likelihood of entry, establishment or spread of these diseases, as well as the associated potential biological and economic consequences; and

3. evaluate the likelihood of entry, establishment or spread of these diseases according to the SPS measures which might be applied.

The Appellate Body emphasized that "likelihood" means "probability," and that under the definition of "risk" and "risk assessment" of the OIE Guidelines for Risk Assessment a possibility is not sufficient. Thus, for the Appellate Body, the "risk" must be an ascertainable risk. A theoretical uncertainty is "not the kind of risk which, under Article 5.1, is to be assessed." The Appellate Body found Australia's risk assessment inadequate because it did not include the "evaluation of the likelihood of entry, establishment or spread" of the diseases of concern "and of the associated potential biological and economic consequences" or "evaluate or assess the SPS' measure's relative effectiveness in reducing the overall disease risk."

Japanese Agricultural Products   In 1997, the United States challenged Japan's import approval process for certain agricultural prod-

ucts. Japan prohibited the importation of individual varieties of the same product to control codling moth until each variety had been tested by the required quarantine treatment (methyl bromide fumigation and cold storage). Although Japan had approved the importation of red delicious apples because the United States had proven that this variety could be effectively treated for the codling moth, Japan banned other apple varieties from the United States.

The Appellate Body ruled that Japan's varietal testing measure was not based upon scientific principles, in violation of Article 2.2. With respect to the varietal testing for apricots, the Appellate Body noted that Japan's claimed risk assessment did not discuss or even refer to the varietal testing requirement or to any other phytosanitary measure that might be taken to reduce the risk. The Appellate Body concluded, ". . . [t]herefore . . . the risk assessment does not . . . evaluate the likelihood of the entry, establishment or spread of codling moth . . . within the meaning of Article 5.1." The codling moth dispute with Japan lingered for a decade before the WTO/SPS dispute settlement procedure required Japan to present a transparent description of its quarantine concerns. Once they had done so, in the case of apples, or simply failed to do so, in the case of apricots, the absence of sound science became apparent (Johanson and Bryant, 1996).

Does this mean that all U.S. SPS measures must conform to an international standard or have a risk assessment prepared? Such strategic thinking is in order to protect one's import protocols. However, because of the enormous market, political, and technological resources of the United States it is not likely that many U.S. trading partners will challenge its SPS restrictions, or insist on maintaining theirs, in the event of an SPS dispute. Most SPS trade disputes are resolved by negotiations, and the relative economic power of the parties in the dispute greatly influences the outcome of the negotiations. However in the event of a dispute between one of the four "Quad" members of the WTO (United States, Canada, Japan, and the European Union), where power disparity is less of an issue, a formal dispute process is more likely in cases of a substantial conflict of economic interest or culture (Echols 1998).

Temporary Measures Pending Further Scientific Investigation
(WTO/SPS article 5.7)

Members may adopt provisional SPS measures "[i]n cases where relevant scientific evidence is insufficient" and certain other requirements are fulfilled (WTO/SPS article 5.7). In *EC Hormones*, the Appellate Body stated that the precautionary principle found reflection in Article 5.7. In the Japanese Agricultural varietal testing case, the WTO Appellate Body sets out four requirements of Article 5.7 that must be met in order to adopt and maintain a provisional SPS measure. A member may provisionally adopt an SPS measure if this measure is (1) imposed in respect of a situation where "relevant scientific information is insufficient," and (2) adopted "on the basis of available pertinent information."

But such a provisional measure may not be maintained unless the member that adopted the measure (1) "seek[s] to obtain the additional information necessary for a more objective assessment of risk," and (2) "review[s] the measure accordingly within a reasonable period of time." The Appellate Body held that "[w]henever one of these four requirements is not met, the measure at issue is inconsistent with Article 5.7."

The member's obligation is to "seek to obtain" additional information to allow the member to conduct "a more objective assessment of risk." Accordingly, the information sought must be germane to conducting such a risk assessment, i.e., the evaluation of the likelihood of entry, establishment, or spread of a pest or pests, according to the SPS measures that might be applied. The Appellate Body ruled that Japan could not rely on this provision because it had not undertaken to "examine the appropriateness" of the SPS measure at issue.

Finally, the member adopting a provisional SPS measure is to "review the measure accordingly within a reasonable period of time." The Appellate Body held that what constitutes a "reasonable period of time . . . has to be established on a case-by-case basis and depends on the specific circumstances of each case, including the difficulty of obtaining the additional information necessary for the review and the characteristics of the provisional SPS measure."

The Appellate Body noted that the SPS measure in issue had been in effect in January 1995, when the WTO/SPS became effective, four years earlier, and in this particular case "collecting the necessary additional information would be relatively easy."

Avoid Arbitrary or Unjustified Distinctions in Levels of Protection
(WTO/SPS article 5.5)

In *EC Beef Hormones*, the Appellate Body disagreed with the panel's conclusion that the evidence showed a trade-restricting *purpose* and reversed the panel's finding that the hormones regulation violated Article 5.5. As Robert E. Hudec observed, "the Appellate Body offered no criticism of the purpose analysis" and "threw itself into a detailed analysis of the evidence relating to the issue of purpose, giving every indication that it thought the purpose analysis was a proper issue to be considered under [A]rticle 5.5." (Hudec 1998). However, the absence of a trade restrictive purpose will not insulate the measure from an adverse Appellate Body ruling, (Hudec 1998) as occurred in *EC Beef Hormones*. It does appear unlikely that a party who has presented an adequate risk assessment would be found in violation of Article 5.5. Rather, as in the *Australian Salmon*, an inadequate risk assessment combined with other factors persuaded the Appellate Body that Australia's measure violated Article 5.5.

In *Australian Salmon*, the Appellate Body held that in order to find a violation of WTO/SPS under Article 5.5

1. the member concerned must have adopted different appropriate levels of sanitary protection in "different" but comparable situations";

2. those levels of protection must exhibit differences which are "arbitrary or unjustifiable"; and

3. the measure embodying those differences results in "discrimination or a disguised restriction on international trade."

Under the first prong the situations must have in common a risk of entry, establishment, or spread of one disease of concern. The WTO panel found this to be the case here in that "two categories of non-salmonids [herring used as bait and live ornamental finfish], for which more lenient sanitary measures apply, can be presumed to represent at least as high a risk— if not a higher risk—than the risk associated with . . . [ocean-caught Pacific salmon]." On this basis the Appellate Body found the first two prongs of their WTO/SPS test to have been met. Article 5.5 is fulfilled. In applying the third prong of Article 5.5, the panel considered that "the arbitrary character of the differences in levels of protection is a 'warning signal'" with respect to the rather substantial difference in levels of protection between an import prohibition on ocean-caught Pacific salmon. Another warning signal the panel considered was the insufficient risk assessment. Finally the Appellate Body approved the panel's consideration of the protectionist role that the domestic industry may have played in securing the adoption of the measure. Thus the legislative history of the measure's "aim and effect" is relevant under Article 5.5.

Is the Measure More Trade Restrictive Than "Necessary"?
(WTO/SPS, article 5.6)

In *Australian Salmon*, the Appellate Body established the following three-pronged test to establish whether the SPS measure violated Article 5.6. The SPS measure

1. is reasonably available taking into account technical and economic feasibility;
2. achieves the member's appropriate level of sanitary or phytosanitary protection; and
3. is significantly less restrictive to trade than the SPS measure contested.

Each prong is required in order to find a violation. As to the first prong, the WTO panel or Appellate Body may find that there are alternative SPS measures that are reasonably available, taking into account technical and economic feasibility. Or, the complaining party may suggest an alternative measure. As to the second prong, the determination of "the level of protection deemed appropriate by the Member" is a prerogative of the member concerned and not of a panel or of the Appellate Body. The Appellate Body emphasized that the "appropriate level of protection" and the "SPS measure" are not the same thing. The first is an objective, the second is an instrument chosen to attain or implement that objective. Article 5.6 requires an examination of whether alternative SPS measures would meet the appropriate level of protection as de-

termined by the member concerned. Thus, a member is obliged to determine the appropriate level of protection. If the member does not determine its appropriate level of protection, or does so with insufficient precision, panels may determine the appropriate level of protection on the basis of the level of protection reflected in the SPS measure actually applied.

The Appellate Body has not made adverse findings regarding Article 5.6 in the cases it has considered. In *Australian Salmon* the Appellate Body found that the panel did not evaluate or assess the alternative measures' relative effectiveness in reducing the overall disease risk. For that reason it was not in a position to complete the examination of whether there was another measure that achieved the appropriate level of sanitary protection. In *Japanese Agricultural* the Appellate Body deferred to the panel's fact-finding that the complaining party had failed to demonstrate that the alternative measure would achieve the chosen level of protection.

Accept an Exporting Member's SPS Measures as Equivalent If It Achieves the Same Level of Protection
(WTO/SPS article 4)

The SPS Agreement mandates members to accept SPS measures of other members as equivalent. SPS measures of the exporting member that achieve the same level of protection as the importing member should be accepted by that member even if the method to reach that level of protection differs (Johanson and Byrant 1996). This requirement remedies the refusal of some members to import produce because of inconsequential differences in inspection or food safety standards (Maruyama 1998). To conform with this requirement, it must be objectively determined what methods achieve equivalent levels of protection (Johanson and Byrant 1996). This permits WTO members to use different means to ensure the same level of protection. Thus, this is an alternative de facto route to harmonization. Stewart and Johanson report that the European Union (EU) and New Zealand have accepted an animal product equivalency agreement including dairy goods. The EU and the United States are in similar negotiations (1999).

Consider Prevalence of Specific Diseases or Pests, Existence of Eradication or Control Programs in Specific Regions of the Exporting Country
(WTO/SPS article 5.6)

Countries may have different producing regions, and certain pests and diseases may not be found in all of these regions. The WTO/SPS required members to recognize disease-free or pest-free areas and areas with low prevalence for certain pests and diseases. Exporting countries that claim to have pest- or disease-free areas must provide evidence supporting these claims to importing countries. Indeed, overly broad restrictions of exports when a more confined area would suffice may violate WTO/SPS Article 6 (Stewart and Johanson 1998).

Notify Other Members of SPS Measures (Transparency)
(WTO/SPS article 7, Annex B(1))

The WTO and the NAFTA emphasize *transparency*, namely the requirement that measures that affect trade be prepublished, subject to comments from the private sector, and that the government provide appropriate administrative and judicial review of domestic action that might affect trade. The WTO/SPS requires that members promptly publish their SPS measures and provide for a point of central inquiry concerning such measures. When there is no pertinent international standard, a member is to publish at an "early stage" and to advise members through the WTO Secretariat (Johanson and Bryant 1996). Moreover a member is to publish the notice of an intended measure before it is enacted (Stewart and Johanson 1999).

In *Japanese Agricultural*, the Appellate Body ruled that even though the varietal testing requirement was not mandatory, it was a "phytosanitary regulation" subject to the publication requirement in Annex B. Moreover, the Appellate Body was not impressed by Japan's claim that the measure was set out in the Experimental Guide, which was not legally enforceable. Rather the critical point was generally applicable to the nature of the varietal testing requirement and its actual impact. Thus, its essential character was similar to laws, decrees, and ordinances.

Control, Inspection, and Approval Procedures May Not Arbitrarily or Unjustifiably Limit the Importation of Foreign Products
(WTO/SPS article 8)

The WTO/SPS Article 8 prohibits members from using control, inspection, and approval procedures to arbitrarily or unjustifiably limit the importation of foreign products. Timely inspection of some agricultural products, especially perishable ones, is critical. The guiding principles of Article 8 are timeliness, reasonableness, equity, necessity, and nondiscriminatory treatment. Stewart and Johanson report that in 1998 Argentina, Australia, and India protested the EU's proposed aflatoxins regulation as costly, overburdensome, trade distorting, and unnecessary to protect human health. The EU relaxed its sampling requirements at a meeting of the WTO's SPS Committee (Stewart and Johanson 1999).

The WTO, Federal and State Law

When Congress so legislates, federal law can preempt state law, although such law is a traditional state police power like food safety (Mertz 1992; Schaefer 1998). However, unless such federal "preemption" occurs and there is, in the words of the United States Supreme Court, "no inevitable collision between the two schemes of regulation, the states may continue to regulate in the area despite the dissimilarity of the standards." Congress often authorizes the states to establish their own SPS standards that do not conflict with federal law. The Congress of the United States recognized these principles of federalism in the implementing legislation of the WTO and the NAFTA. But in terms of international trade, it is of critical importance that federal and state SPS import and export regulations are well coordinated to avoid confusion and problems of transparency. More importantly, under the WTO and the NAFTA, it is solely the national government that is responsible for state or provincial measures. This means that the federal government would have to defend challenged state SPS measures. The Uruguay Round Agreements Act contemplates close cooperation between the U.S. Trade Representative (USTR) and the states in trade matters,

WTO dispute settlements, and in challenges to state SPS measures.

The Agreement on Technical Barriers to Trade and Labeling

The Agreement on Technical Barriers to Trade (TBT) covers consumer and environmental concerns not covered by SPS. Some examples of these concerns include biotechnology or nutritional labeling for consumer preferences unrelated to a demonstrated threat to human health. The TBT relies on a test of whether a measure discriminates against imported products (including products of WTO members and nonmembers).

The TBT addresses "standards," "technical regulations," and "conformity assessment procedures." It employs the term *standards* to refer to voluntary product standards. *Technical regulations*, however, is the term designating mandatory product standards. *Conformity assessment procedures* are methods used to determine whether a product is safe.

Under the TBT, each WTO member is to ensure that imported products are treated no less favorably than domestic or other imported "like products." Moreover, technical regulations are not to be more trade restrictive than necessary to fulfill a "legitimate objective." Some of these legitimate objectives include national security and preventing consumer deception. This also includes the protection of the environment, human health and safety, and animal and plant life or health.

Europeans have faced multiple food safety and animal health disasters. Recent problems include mad cow disease, foot-and-mouth disease, and numerous cases of microbial contamination. Largely because of this, the EU has imposed a de facto moratorium on the approval of new, genetically modified varieties of agricultural products. Additionally, the EU Commission has announced new labeling and tracing rules. Other countries, such as Japan, Australia, and New Zealand, have measures addressing biotechnology as well.

The EU labeling rules require that all food and feed derived from biotechnology bear a biotech label. This labeling is mandatory regardless of whether the genetic alteration is de-

tectable or not. The tracing rules require extensive documentation of the biotechnological history in a commodity chain as well. The EU's approval process for biotechnology-based food products is slow and nontransparent.

There are tenable arguments that the EU biotechnology labeling regulations violate both the WTO/SPS and the TBT. The labeling rules do not appear to be scientifically based. They are also arguably more trade restrictive than necessary. The United States, however, has yet to challenge them. Clearly, the SPS requires that labeling measures be based on scientific research. However, it is not clear whether the TBT precludes labeling based on considerations such as religious or ethical convictions or even lack of trust in particular science.

As actions shift from the SPS to the TBT, one can imagine a myriad of issues being raised by labeling: Are chocolates eco-friendly? Are eggs from free-range hens? Is leather from placid cows? Are products from family farms? Also, the question to be asked is whether these consumer preferences are legitimate objectives.

The question of what is unjustifiable or arbitrary discrimination looms large. The United States recently won a huge victory before the WTO Appellate Body in its ban of shrimp harvested by methods that threatened sea turtles in *U.S. Shrimp-Turtle II*. The decision focused on the good faith efforts to negotiate with the exporters. However, given the concerns within the EU and elsewhere about biotechnology-based food products, labeling may be a political necessity. However, one continues to question whether these labeling requirements are legal under the WTO. Its legality may ultimately depend on how burdensome it is for exporting countries to comply, as well as its political necessity.

Implications of the WTO/SPS Agreement

The WTO/SPS Agreement has changed the regulatory environment for agricultural industries. This is equally true for industries that wish to protect and expand their export market or protect their product from exotic pests and diseases. The exporter who suspects that another country's SPS measures are politically driven nontariff barriers or simply not founded on sound science has several international remedies. The WTO/SPS provides procedures to pin down the importing country's rationale for the SPS measure in question. For example, the exporter may

1. compare the measure with the international standard, if any;
2. request that the WTO/SPS committee provide any information on file with the committee regarding the SPS measure; and
3. request that the United States Department of Agriculture (USDA) or the United States Trade Representative (USTR) seek further information from the importing country regarding any scientific justification and risk assessment that the importing country may have.

Once the exporter obtains this information the exporter may undertake a scientific, economic, and legal analysis of whether the SPS measure in question is vulnerable to challenge under the WTO/SPS principles described above. This information may suggest additional steps or modifications that the exporter may make to strengthen its case or provide further documentation or test results that the exporter may provide to the importing country. These informal discussions alone may persuade the importing country to modify or eliminate their SPS barriers.

For example, the USDA or USTR may be in a position to advise the importing country that sufficient information was available to make a scientifically based decision. During such negotiations with the Japanese over California tomatoes concerning blue mold, Japan agreed to reconsider previous positions and tentatively accepted U.S. research and testing data. If informal discussions fail, the exporter may then make the case with U.S. authorities for initiating a WTO or NAFTA consultation that may lead to a dispute settlement procedure.

With respect to import protocols, the mere fact that certain restrictions have been employed in the past does not obviate the necessity of testing such restrictions under the principles of the WTO/SPS. This is essentially a process whereby the federal or state authorities adopt the international standards or be prepared to document that a risk assessment justifies a higher standard. Additionally, the authorities and affected industries may attempt to persuade the international body to modify their standard along the lines of the U.S. or California protocols. These remedies underscore the importance of being informed on the standards, recommendations, and guidelines of the international body designated in the WTO/SPS.

With these principles in mind, a discussion of possible strategies and problems of certain exotic pests follows. The case studies are foot-and-mouth disease, exotic Newcastle disease, and the Mediterranean fruit fly.

Foot-and-Mouth Disease

Foot-and-mouth disease (FMD) is a highly contagious viral disease that affects cloven-hoofed animals such as cattle, swine, sheep, goats, and deer. Animals, people, or materials that bring the virus into physical contact with susceptible animals can spread FMD (see Chapter 7). Recent outbreaks have occurred in Greece (1996), Taiwan (1997), the Philippines (1998), India (1999), Great Britain (2001), and South Korea (2001 and 2002).

The international standard for FMD was developed under the auspices of OIE. The OIE seeks to ensure that scientifically justified standards govern international trade in animals and animal products, including the development of diagnostic tests and vaccines. The OIE's Epizootics Commission, one of the specialist commissions, assists in identifying the most appropriate strategies and measures for FMD prevention and control. The commission convenes groups of experts and can recognize without further consultation that a member country or a zone within its territory is FMD free if outbreaks are eradicated in accordance with OIE standards.

OIE classified FMD as a List A disease. List A diseases are transmissible diseases that have the potential for very serious and rapid spread and are of major importance in the international trade of animals and animal products. List A diseases require more stringent notification and reporting requirements for member countries. The International Animal Health Code provides regulatory standards for international trade. It includes provisions for notifications and epizootiological information, certification for international trade, import risk analysis, import and export procedures, and risk analysis for biologicals for veterinary use.

Because it is a country that has previously eradicated FMD, the risk analysis of the International Animal Health Code (IAHC) of the OIE provides that the United States is an FMD-free region. The IAHC encourages a country to design its own methodology for carrying out risk analysis. Risk analysis may involve risk assessment, evaluation of veterinary services, and zoning and regionalization of countries. Import risk assessment should be transparent so that the exporting country may be provided with a clear and documented decision on the conditions imposed for importation or refusal of importation.

If a region is declared FMD free for the first time, USDA regulations require it to change its recognition of disease status. The process includes publishing a proposed rule based on full technical and scientific information, allowing public comment, and issuing a final rule. For example, the Veterinary Agreement of the United States and the EU authorizes a declaration process. The initial presumption is that the regionalization decision taken by the other party will be accepted, allowing for exceptional cases in which, for justifiable reasons, the party feels the need to take recourse under the safeguard provision. For example, where the EU takes a decision to restrict an area that has previously been recognized as disease free, the United States would accept the EU's regionalization decision without having to take further actions. After appropriate measures and after the EU lifts restrictions on that area, the United States would accept that decision.

OIE recognized Uruguay as a FMD-free country in 1994. The disease was eliminated through an intensive vaccination program of the cattle population and through movement restrictions. In 1994, Animal and Plant Health Inspection Service (APHIS) officials conducted an on-site evaluation of Uruguay's animal health program with regard to FMD. The evaluation consisted of a review of Uruguay's veterinary services, diagnostic procedures, vaccination practices, and administration of laws and regulations. APHIS officials evaluated all border-crossing points and determined that the country's veterinary infrastructure was sufficient to maintain them. The regional sanitary situation also reduced the risk of FMD spreading into Uruguay from Argentina or Paraguay. For example, until recently, Argentina had not reported a focus of FMD since April of 1994

and Paraguay had been FMD free for one full year in all its territory. Considering these factors, APHIS officials concluded that Uruguay was FMD free and that the country's veterinary infrastructure was outstanding. Federal regulations require that a health certificate signed by a veterinary official of Uruguay accompany meat and other animal products imported into the United States from Uruguay to confirm that they have not been commingled, directly or indirectly, with meat or animal products from a country where FMD exists. A department-approved foreign meat inspection certificate must also accompany meat and other animal products consigned by Uruguay. These required certifications verify that meat and other animal products from Uruguay meet the conditions of U.S. regulations. These certification procedures are in compliance with OIE standards.

However, beginning in August 2000, Argentina sustained a major FMD epidemic that worsened during the early months of 2001 and spread to Uruguay. Before this outbreak, FMD was believed to be eradicated in Argentina. The OIE had granted Argentina the status of "FMD-free without vaccination," since it had ended its vaccination program in 1999. Argentina maintains that the 2000 problem originated from 10 cattle illegally imported from Paraguay. The EU and others, however, have suggested that Argentina has never been completely free of FMD. In response, it promised a massive overhaul of its testing system and is vaccinating all cattle north of the 42nd parallel, roughly 85% of its cattle. The recent experience of Argentina suggests that eradication can be short-lived when internal controls and monitoring along borders are compromised. Work must be done in Argentina (and perhaps also Paraguay, Brazil, and Uruguay) to eradicate FMD and keep it out permanently. The whole problem indicates clearly how important regional and global cooperation is in eliminating the threat of FMD. Also of great importance is the well-planned and sustained implementation of control measures.

Pest-Free Regions U.S. regulations recognize certain pest-free regions. For example, the government of Brazil may request that the APHIS administrator recognize Rio de Janeiro as a region. The Brazilian officials must provide necessary and valid information for risk assessment before the United States will conduct the risk assessment of importation. If the risk is of negligible level, the United States will determine the import conditions for the region.

Equivalency The WTO/SPS, Article 4 concept of equivalency is set forth in the regulation. Based on information from its trading partners, the United States may identify mutually agreeable risk management measures to reduce risk to a negligible level. The exporting region has the burden of proof to demonstrate to the United States (the importing country) that the region meets the standards equivalent to the United States' standards or to the acceptable standards.

Risk Assessment APHIS recognizes the identifiable and measurable gradations of risk presented by animals and animal products and that these gradations are often tied more to facts such as geography, ecosystems, epidemiological surveillance, and the effectiveness of disease control programs than to national political boundaries. APHIS policy is to assess risk along a continuum and to determine on a case-by-case basis what import condition will reduce the risk of disease introduction to a negligible level. APHIS will use risk categories as a benchmark to assist regions in evaluating where the regions can expect to fall on a risk level spectrum and what general import conditions may apply. Factors such as proximity between regions will not be given a predetermined weight in the assessment process because of the varying climatic and other ecological factors for each region. For the veterinary infrastructure factor, APHIS concedes that the evaluation will be somewhat subjective. However, until the OIE develops an objective measure of infrastructure, APHIS will consider the veterinary infrastructure of the region on a case-by-case basis and may include on-site visits.

Transparency If the importation is allowed, APHIS will publish within the *Federal Register* notice of the proposed importation and the conditions under which the importation would be allowed. Public comments on the proposal can be made during a period of time, usually within 90 days. During the comment period, the public will have access, both in hard copy and electronically, to the information upon which APHIS based its risk analysis, as well as to the methodology used in conducting the analysis. If

APHIS allows the importation, the importation conditions will be published in a final rule in the *Federal Register*.

Incorporation of the OIE Standards Commentators for proposed rules of regionalization recommended that APHIS review internationally accepted guidelines for regionalization, risk analysis, and risk assessment. They specifically cited various OIE reports on technical items presented to the international committee or regional commission and the OIE's International Health Code (Cane 1994; Moreley, Acree, and Williams 1990-1991). APHIS responded that it has incorporated concepts from these references into the policy on regionalization and risk assessment. However, APHIS has not incorporated OIE's diagnostic test for use on animals being imported. APHIS agrees that tests approved by the OIE would generally meet the scientific validity requirements of an equivalent approved test. But APHIS administrators have retained the flexibility to not use any test if evidence shows that it is not valid, even though it is included in the OIE approved tests list. APHIS administrators feel they must have the flexibility to use new tests when deemed appropriate, even if they are not on the OIE list.

It is difficult to compare U.S. and OIE standards because of considerable differences in transparency. While the OIE standards are very specific and easily accessible, the details of the U.S. standards are less so. The United States has adopted the case-by-case methodology for risk analysis and does not have a standardized risk analysis. The information regarding risk assessment appears to be available only by request during the public comment period of the proposed declaration. Even with the published declaration, such as for Uruguay, the publication lacks precise details on risk assessment. Because of the lack of accessible and uniform information of U.S. risk assessment, the exporting country bears the burden of unpredictable risk assessment by the U.S.

However, despite the transparency problem, U.S. standards can still comply with the OIE standards. The United States is allowed to design its own methodology of risk assessment and it has incorporated some of the OIE components of risk analysis, such as evaluation of the veterinary services and zoning of countries, as factors in assessing risk of importation. Arguably, U.S. standards are transparent since they do list all the information needed from an exporting country to conduct its risk assessment. The unpredictability only arises from the uncertain weight of each factor for the risk assessment of a particular country or region.

Exotic Newcastle Disease

Export   U.S. law requires that animals offered for export be accompanied by a health certificate. This includes birds or poultry that may have been exposed to exotic Newcastle disease (END). In the case of birds or poultry, the certificate must state that the animals were inspected within 30 days of the date of movement and were healthy and free of any communicable disease. All animals must also be inspected within 24 hours of export by an APHIS veterinarian at an export inspection facility at an approved port. The APHIS veterinarian will issue the export certificate if the animals are found to be healthy and free from communicable disease.

Import   Live birds and poultry offered for importation must be accompanied by a certificate by an authorized veterinarian from the country of export stating that they are free of communicable diseases. The birds or poultry, the premises where they were kept, and adjoining premises must not have been exposed to END within 90 days of importation or the birds will be quarantined. Before birds and poultry may be imported, they are subject to quarantine for not less than 30 days, subject to extensions, at an approved facility. They are inspected and tested for communicable diseases. If birds show signs of, or are found to have been exposed to, communicable diseases during the quarantine, they are refused entry or destroyed.

The United States has designated countries as being free of END. These nations are Australia, Canada, Chile, Costa Rica, Denmark, Fiji, Finland, France, Great Britain, Greece, Iceland, Luxembourg, New Zealand, Republic of Ireland, Sweden, and Switzerland. Products, including eggs, that originate or move through an infected area may be imported, but are subject to special provisions. There must be no evidence that the flock was exposed to END. These birds must be tested at 10-day intervals by a veterinarian of the national government at an approved laboratory. Finally, in the 60-day

period before importation, dead birds must be removed from the flock weekly and tested for END. In that same period, a sample of at least 10 live birds must be randomly tested for the disease.

Internal Controls   When END is found in an area within the United States, that area is quarantined. Currently, no regions of the United States are quarantined because of END. Infected birds or eggs may not be moved interstate from a quarantined area. Other birds or poultry may be moved from a quarantined area by permit, only under strict conditions. While eggs not infected with END may be moved from a quarantined area, hatching eggs must be held for 30 days after hatching to ensure they are not infected with END.

International Standard (OIE)   The OIE publishes the IAHC, which provides regulations concerning END. The OIE considers a country free from the disease when it is shown that it has not been present for at least three years. Because the incubation period of END is 21 days, the OIE standard considers a country to be infected with the disease until at least 21 days after the last confirmed case and the completion of eradication and disinfection procedures.

The United States requires a 30-day quarantine for birds and poultry products offered for import. The OIE guidelines contemplate that the certificate will be issued if the birds have been quarantined for at least 21 days in the nation of export. The U.S. 30-day quarantine is arguably a higher level of protection than called for in the OIE standards. If challenged, this difference could call for scientific justification, namely a risk assessment (WTO/SPS, articles 2.2, 3.3, 5.1, Annex A(4)). A complaining party may argue that the U.S. regulation does not conform to the OIE standard, while the United States could respond that it, at the very least, is based on it. But in general, the U.S. restrictions on internal movement of birds and poultry are sufficiently similar to those imposed on imports that there would be little room to argue that the regulations were discriminatory or arbitrary (WTO/SPS article 2.3, 5.5).

A second problem is transparency. The U.S. regulations, like those on END, are not easily understood. To some extent, this is a product of a high level of detail (WTO/SPS, Annex B). A possible solution to these problems is for the United States to adopt the OIE standards. This creates a presumption of validity that would probably be invulnerable to attack (WTO/SPS article 3.2). Would this reduce the level of protection currently provided? Alternatively, the United States may publish the differences between the OIE and U.S. regulations and the risk assessments that justify those distinctions (WTO/SPS article 12(4)). Also, the United States could modify the structure of its regulatory regime to be more parallel to that of the OIE.

Mediterranean Fruit Fly

In 1995 California enacted legislation requiring the California Secretary of Food and Agriculture to adopt quarantine regulations established by the USDA. These regulations identify areas in which the Mediterranean fruit fly (Medfly) currently exists. They authorize administrators to establish a quarantine zone of less than the entire state. The treatment methods include vapor heat, cold treatment, fumigation with methyl bromide, fumigation plus refrigeration, and irradiation and are determined by the host fruit. The purpose of these regulations is to prevent the spread of the Medfly into or throughout the United States. The regulations authorize port of entry inspectors to restrict entry and the USDA to control movement of the Medfly or infected plants. For export, USDA provides for certification of the phytosanitary conditions of plants and plant products. An inspector assesses the compliance of the exports with the regulations of the receiving country.

The WTO/SPS identifies the secretariat of the IPPC as the organization providing international standards for SPS measures to protect plant resources from harmful pests (phytosanitary measures). The IPPC has promulgated International Standards for Phytosanitary Measures (ISPMs). However, the IPPC does not have a specific standard for the Medfly. Accordingly, federal and California SPS measures on the Medfly must be supported by a risk assessment. While the identification of the pest and its potential economic and biological consequences would appear to be relatively easy to document, the likelihood of entry and relevance of the import controls to restricting entry would

also have to be documented. Because California is a net exporter of host fruits for the Medfly, it would appear likely that the SPS measures of importing countries could prove quite costly to California industry. For example, before joining the WTO, China prohibited imports of all citrus, apples, table grapes, and cherries from the United States because of detection of the Medfly in the Los Angeles area. A likely cause of future disputes with importing countries is the size of the area covered by the quarantine boundaries. In the United States, if fertilized female fruit flies are detected, the free area must be canceled in an 8-km (4.5-mile) radius around the area where they are captured. Other countries require the free area to be considerably larger. Australia requires an 80-km (50-mile) radius quarantine region. Korea follows the U.S. quarantine protocols unless Medflys are found outside a 2-km area. In that case Korea insists that the entire political division (county) be quarantined. Taiwan demands the quarantine area be an additional 20-km radius beyond the county boundary. These differences suggest a potential equivalency dispute. The United States may contend that the chosen level of protection of Australia, Korea, and Taiwan is achieved by the U.S. protocol (WTO/SPS article 6). Should these larger quarantine areas prove to be significantly trade restricting, the United States could follow the discovery and possible dispute settlement strategies outlined above.

Implications of WTO/SPS

Of the three case studies, the FMD and END controls of the U.S. standards are substantially equivalent to OIE standards. The U.S. Medfly import and export protocols have been adopted without benefit of an IPPC standard. This suggests the importance of industry and government working with the IPPC so that their Medfly standards are congruent with those of the United States. The END dissimilarity of quarantine time may prove contentious. With regard to FMD, there appears to be a concerted effort among nations to harmonize the declaration process for the importing countries and the standards to eradicate the disease for exporting countries. Because U.S. standards for both END and FMD are not as transparent as the OIE standards, redrafting and restructuring them may avoid future disputes.

Conclusion

The WTO/SPS and the WTO/TBT are complimentary agreements that address the regulation of imported products, attempting to balance the interests of state autonomy and the objectives of WTO/GATT. However, because they are essentially mutually exclusive, their differences are also rather pronounced. While the SPS is rather narrow in scope, the TBT is far more expansive. The SPS replaces the earlier regime of the general exception in Article XX(b) of GATT, setting out a number of strict requirements for imposition of SPS restrictions, whereas the TBT purports to expound upon the GATT obligations of members when they impose technical standards, regulations, etc. Under the SPS, case law has developed stating a defending nation must satisfy the eight requirements discussed above for a protective measure to be valid. However, under the TBT, the primary consideration is that a measure is "not more trade restrictive than necessary to fulfill a legitimate objective." While this analysis is actually more sophisticated than it may at first appear, it is nonetheless a far less ambitious legal standard to overcome and may be more friendly to defending states.

At first blush, the WTO/SPS appears to be extremely pro-exporter and insensitive—perhaps even antagonistic—toward environmental protection and the health and safety interests of importing nations. Given the major cases, *EC Beef Hormones*, *Australian Salmon*, and *Japanese Agricultural*, a reasonably tenable interpretation is that relatively unimpeded trade shall occur at the expense of legitimate environmental and health and safety measures. However, the case law is still developing, and its ultimate trajectory may not actually be as disadvantageous to importing nations as it seems. Another viable interpretation is that, based on the facts of those cases, to varying extents the defending states really were engaging in veiled protectionism as evidenced by a certain degree of bad faith. However, the good faith evidenced in negotiations with the exporters in *U.S. Shrimp-Turtle II* was central to the Appellate Body's analysis, which upheld a ban on shrimp-harvesting methods that threatened sea turtles. The SPS may prove less mechanical in operation and more like the TBT. The ostensibly pro-exporter cases may instead stand for the proposition that absent a showing of good faith, the defending party will have a difficult time overcoming its eight requirements. This will become more certain only as case law develops more fully. Although the Appellate Body is not bound or necessarily driven by precedent, the three cases may help to establish a perimeter of nonviable environmental and health and safety restrictions, rather than stand for a general policy favoring exporters.

References

Barcello, John J. III. 1994. "Product Standards To Protect the Local Environment—the GATT and the Uruguay Round Sanitary and Phytosanitary Agreement." *Cornell Int'l. L. J.* 27:755.

Cane, B.G. 1994. *The Concept of Regionalization in Establishing Disease-Free Areas.* Cited at 62 Fed. Reg. 56000, 56004 (1997).

Cromer, Julie. 1995. "Sanitary and Phytosanitary Measures: What They Could Mean for Health and Safety Regulations under GATT." *Harv. Int'l. L. J.* 36:557.

Daniel, Jr., Al J. 1994. "Agriculture Reform: The European Community, The Uruguay Round, and International Dispute." *Ark. L. Rev.* 46:873, 893.

Echols, Marsha A. 1998. "Food Safety Regulation in the European Union and the United States." *Colum. J. Eur. L.* 4:525.

Franzen, Rick. 1998. "GATT Takes a Bite out of the Organic Food Production Act of 1990." *Minn. J. Global Trade.* 7:399, 409.

Gaines, Sanford E. 1993. "Environmental Laws and Regulations After the NAFTA." 1 *U.S. Mex. L.J.* 20:4-1.

Hudec, Robert E. 1998. "The New WTO Dispute Settlement Procedure: An Overview of the First Three Years." *Minn. J. Global Trade.* 8:1.

Johanson, David S., and William L. Bryant. 1996. "Eliminating Phytosanitary and Sanitary Trade Barriers: The Effects of the Uruguay Round Agreement on California Agricultural Exports." *San Joaquin Ag. L. Rev.* 6:1, 4.

Josling, Tim, and Rick Barichello. 1993. "Agriculture in the NAFTA: A Preliminary Assessment." *Commentary.* 43.

Kellar, J.A. 1992. *The Application of Risk Analysis to International Trade in Animals and Animal Products.*

Maruyama, Warren H. 1998. "A New Pillar of the WTO: Sound Science." *Int'l. Law.* 32:651.

McNiel, Dale E. 1998. "The First Case Under the WTO's Sanitary and Phytosanitary Agreement: The European Union's Hormone Ban." *Va. J. Int'l. L.* 39:89, 93-94.

Mertz, Gregory J. 1993. "Dead but Not Forgotten: California's Big Green Initiative and the Need to

Restrict State Regulation of Pesticides." 60 *Geo. Wash. L. Rev.* 506, 531:4-14.

Millimet, Robert M. 1995. "New Agreements on Sanitary and Phytosanitary Measures: An Analysis of the U.S. Ban on DDT." *Transnat'l. L. & Contemp. Probs.* S:443, 466.

Moreley, R.S., J. Acree, and S. Williams. 1990–1991. *Animal Import Risk Analysis (AIRA): Harmonizing Our Approach.*

Schaefer, Matthew. 1998. "Sovereignty Revisited: Discussion After the Speeches of Paul Martin and Matthew Schaefer." *Can.-U.S. L. J.* 24:385, 388–389.

Seilhamer, Lisa K. 1998. "The Sanitary and Phytosanitary Agreement Applied: The WTO Hormone Beef Case." *Envtl. Law.* 4:537.

Steinberg, Richard H. 1995. "Trade-Environment Negotiations in the EU, NAFTA, and GATT/WTO." Berkeley Roundtable on the International Economy, Working Paper 75. Berkeley, CA.

Stewart, Terence P., Ed. 1993. *The GATT Uruguay Round: A Negotiating History (1986-1992).*

Stewart, Terence P., and David S. Johanson. 1998. "The SPS Agreement of the World Trade Organization: The Roles of the Codex Alimentarius Commission, the International Plant Protection Convention, and the International Office of Epizootics." *Syracuse J. Int'l. L. & Com.* 26:27.

Stewart, Terence P., and David S. Johanson. 1999. "The SPS Agreement of the World Trade Organization and the International Trade of Dairy Products." *Food Drug L. J.* 54:55.

Walker, Vern R. 1998. "Keeping the WTO from Becoming the 'World Trans-Science Organization': Scientific Uncertainty, Science Policy, and Fact Finding in the Growth Hormones Dispute." *Cornell Int'l. L. J.* 31:251, 273–277.

Wirth, David A. 1994. "The Role of Science in the Uruguay Round and NAFTA Trade Disciplines." *Cornell Int'l. L. J.* 27:817.

Cases

EC Beef Hormones

EC Measures Concerning Meat and Meat Imports (Hormones) Report of the Appellate Body of the World Trade Organization (United States), WT/DS48/13, AB-1994-4, 29 May 1998 (Appellate Body).

Australian Salmon

GATT Dispute Appellate Body Report on Canadian Complaint Concerning Australian Measures Affecting the Importation of Salmon, WT/DS18/AB/R, AB-1998-5, 7 Oct. 1998 (Appellate Body).

Japanese Agricultural

Measures Affecting Agricultural Products, WT/DS76/AB/R, AB-1998-8, 22 Feb. 1990 (Appellate Body).

U.S. Shrimp-Turtle II

WTO Appellate Body Report, United States — Import Prohibition of Certain Shrimp and Shrimp Products, Recourse to Article 21.5 by Malaysia, WT/DS58/AB/RW, paras. 122-34, 22 Oct. 2001 (Appellate Body).

5

Historical Perspectives on Exotic Pests and Diseases in California

Susana Iranzo, Alan L. Olmstead, and Paul W. Rhode

Introduction

Pests and diseases have been destroying livestock and crops since the dawn of agriculture. The biblical accounts of plagues of locust and frogs, whether or not apocryphal, offer a hint that such problems existed in antiquity. This chapter picks up the story of pests and diseases at the beginning of modern agriculture in California in the mid-19th century. From the 1850s on, vast quantities of nursery stock and scores of new varieties of plants and animals were introduced into the state. In addition, the organization and density of agricultural production along with the supporting transportation, financial, and scientific infrastructures evolved rapidly. This created an ideal setting for all sorts of noxious plant pests and diseases to flourish.

California offers an unusually fertile ground for studying the impact of diseases and pests and for examining individual and collective control and eradication efforts. Given the remarkable array of crops grown in the state, California could host a large number of plant enemies. Moreover, the rapid introduction of new crops over the 19th century created what can be considered an enormous natural experiment. When the waves of farmers arrived following the Gold Rush, California was largely free of harmful insects and diseases. The growth of agriculture based on nonnative plants required importing nursery stock from other states and countries. Accompanying the new plants were pests and diseases that within a few decades were ravaging the state's crops. Their destructive power in some cases was so severe that they marked the end of the prosperity in leading producing areas. But perhaps the most interesting aspect of this history is the organized responses by the state's agricultural community to these new challenges. Just as the state was

largely pristine territory before the surge in development, it was also largely devoid of the political, scientific, legal, and commercial infrastructures needed to combat the new threats. The spread of diseases and pests prompted collective action and research efforts that led to the eradication or at least the containment of the pest problems.

This chapter offers a brief historical account of a few key diseases and pests that had a significant impact on California horticulture in its formative years. This examination sheds light on the unusually successful, innovative, and productive research and outreach programs that emerged in the public and private sectors.[1] For crop after crop, the creative efforts of leading farmers, scientists, and government agencies overcame the "free rider" problem to literally save large-scale commercial agriculture. Table 5.1 provides a summary account of many of the significant institutional changes enacted to help protect agriculture. We do not attempt to measure the economic rates of return on these investments, but by any reasonable accounting they must have been enormous. The following accounts of the early campaigns against exotic pests and diseases will help illustrate some of the generic problems associated with pest control and eradication. Invariably, these campaigns were complicated because of the problems of imperfect information, of capital constraints, of externalities, and the need to lower the transaction costs associated with collective action.

Threats to the State's Vineyards

We start by examining three diseases that attacked what has become the state's leading crop—grapes. In the 19th century the vines of

Table 5.1 Partial list of U.S. and California efforts in plant protection (California efforts are in bold)

Year	Law/Institution	Purpose
1870	**First California plant pest control legislation**	Various statues empowered counties to pay bounties for gophers and squirrels. Later, in 1883, the California Political Code gave county boards of supervisors power to destroy gophers, squirrels, other wild animals, noxious weeds, and insects injurious to fruit or fruit trees, or vines, or vegetable or plant life.
1880	**Creation of the Board of State Viticultural Commissioners**	Supplement the university's work in controlling grape pests and diseases with special emphasis on phylloxera. Remedy oriented rather than research oriented—the university was responsible for experimental and research work.
1881	**California passes the first American law granting plant quarantine authority**	The Act enlarges the duties and powers of the Board of Viticultural Commissioners and authorizes the appointment of a state viticultural health officer, who is empowered to restrain the importations into the state of vines or other material that might be diseased.
1881	**Creation of County Boards of Horticultural Commissioners by County Boards of Supervisors**	Eradicate specific scale bugs, codling moth and other insects. The county boards were empowered to inspect properties upon complaint and to require treatment of insect infestations. By 1882 county boards had been appointed in 21 counties.
1882	**University of California offers its first course in economic entomology**	
1883	**Creation of the State Board of Horticultural Commissioners**	Empowered with authority to issue regulations to prevent the spread of orchard pests and to appoint an "inspector of fruit pests" and "quarantine guardians" as enforcement officers.
1885	**First explicit legislative authority to inspect incoming interstate and foreign shipments**	Besides the local inspections, now the state inspector of fruit pests or quarantine guardian was authorized to inspect fruit packages, trees, etc., brought into the state from other states or from a foreign country.
1886	**First county plant quarantine ordinance**	Ventura county was the first county prohibiting transportation within the county of anything infected with scales, bugs, or other injurious insects. Other counties followed, and by 1912 at least 20 counties had enacted several ordinances against the entry of pests.
1890	**Initiation of maritime inspection of cargoes of foreign vessels**	
1899	**California State Quarantine Law**	The Act required the holding and inspection of incoming shipments of potential pest carriers and disposal of infestations to the satisfaction of a state quarantine officer or quarantine guardian of the district or county. Labeling of shipments was required, hosts of certain peach diseases were embargoed from infested areas, and importation of certain pest mammals was prohibited.
1903	**The State Board of Horticulture is replaced by the State Commissioner of Horticulture**	New body empowered to promulgate interstate and intrastate quarantines.
1905 (March 3)	Insect Pest Act	Prohibited the importation and transportation, interstate, of live insects that are injurious to plants.
1905	**First California Quarantine Order**	Issued because of the citrus whitefly of Florida.

Table 5.1. (continued)

Year	Law/Institution	Purpose
1907	**Establishment of the Southern California Pathological Laboratory at Whittier**	**Do research studies on plant diseases and insect problems in Southern California.**
1910 (April 26)	**National Insecticide Act**	
1912 (August 12)	Federal Plant Quarantine Act	Prevent the importation of infested and diseased plants.
1912	Creation of the Federal Horticultural Board	Enforce the Plant Quarantine Act
1912	**Establishment of the Citrus Experiment station and Graduate School of Tropical Agriculture at Riverside**	**Superseded the Southern California Pathological Laboratory. Strong divisions in entomology and plant pathology.**
1912	**Work started at the University Farm at Davis**	**Carry out entomology and plant pathology research for the university.**
1912	**Development of the Agricultural Extension's County Farm Advisor Service**	
1915	Terminal inspection of plants in the U.S. post offices begins	
1919	Creation of the Western Plant Quarantine Board	
1919	**Creation of the State Department of Agriculture**	**Take over some of the duties of the State Commissioner of Horticulture.**
1919	Federal Quarantine Law No. 37	Regulate the movement of plants and plant products
1920	Federal Quarantine Law No. 43	Quarantine against the European corn borer.
1921	**Initiation of California border inspection of incoming motor traffic**	**Stations established on the roads coming from Nevada and Arizona. The original purpose was to prevent the introduction of alfalfa weevil. By 1963, 18 stations were in operation on all major highways entering from Oregon, Nevada and Arizona.**
1924	Quarantine on grapes from Spain	Prevent the introduction of Mediterranean fruit fly.
1925	Organization of the National Plant Quarantine Board	
1926	Federal Bulb Quarantine	
1928	Creation of the Plant Quarantine and Control Administration	Supersede the Federal Horticultural Board in its task of inspection of imports of nursery stock and other plants and prevention of plant pests.

Sources: Compiled from Weber 1930, pp. 1–90; Essig 1940, p. 40; Smith et al. 1946, pp. 239–315; Ryan et al. 1969, pp. 4–11.

California, and those in most of the world, were seriously threatened and at least once faced commercial extinction. The villains—powdery mildew, phylloxera, and Pierce's disease—still scourge the world's vineyards.

Powdery Mildew

California was largely spared the destructive impacts of powdery mildew (*Uncinula necator*) because the state's wine grape industry did not really take off until after reasonably effective control measures were developed in Europe.

This represents a case in which California farmers were able to borrow a technology developed mostly in France and England. Powdery mildew (also known as oidium) was almost certainly indigenous to native vines found in the eastern states of the United States, and until the mid 19th century the disease was probably unknown in California and Europe. It was but one of a number of American diseases that doomed every effort to establish commercial wine grape production in the eastern and midwestern states. Over the ages native American vines evolved to coexist with this and other diseases.

But the vines of Europe (*Vitis vinifera*), which were to become the mainstay of the California grape and wine industries, had no prior exposure to this disease and lacked the defenses to ward off its effects (Pinney 1989).

The first serious attacks of powdery mildew outside of its native habitat occurred in England in 1845. According to E.C. Large (1940, p. 44):

> The disease appeared on the young shoots, tendrils and leaves, like a dusting of white and pulverulent meal; it spread rapidly on to the grapes themselves, withering the bunches when they were small and green, or causing the grapes to crack and expose their seeds when they were attacked later. The disease was accompanied by an unpleasant mouldy smell, and it ended in the total decay of the fruit.

By the late 1840s, oidium was ravaging vines across France, and by the early 1850s it was endemic throughout much of Europe, Asia Minor, and North Africa. The results were devastating, with losses often ranging between 50 and 90 percent of the crop. The area hardest hit was Madeira, where most of the population depended on the vines for their livelihood. The arrival of powdery mildew in Madeira in the 1850s destroyed the economy, leading to widespread starvation and mass emigration (Large 1940; Ordish 1987; and Pinney 1989).

As with many other new diseases, the causes and workings of powdery mildew remained unknown for several years while researchers and growers directed their efforts to learning the disease's pathology and to combating it. There were many false leads. In Italy, the appearance of the disease coincided with that of the first railroads. Peasants, putting these things together, blocked new construction and tore up miles of rails already laid to fight the disease (Pinney 1989). But others were both more scientific and successful in their approach. A.M. Grison and Pierre Ducharte in Versailles, J.H. Léveillé in Paris, the Reverend M.J. Berkeley and E. Tucker in England, and Giovanni Zanardini in Venice are all credited with making headway in combating the disease (Large 1940; Ordish 1987; and Barnhardt 1965).

By the early 1860s most French vines were regularly being sprayed with sulfur-based solutions, and by this time the knowledge of how to control powdery mildew was commonplace in California. The relatively late expansion of the grape acreage in California, the early use of sulfur, coupled with the relatively dry climate, probably account for the fact that the state's agricultural press recorded little damage from powdery mildew. This represents an example of scientific breakthroughs coming in time to ward off a potential crisis for the Golden State. Europe's experience with mildew was but a prelude to a far more devastating American invasion, and this time California's vineyards would not get off so easily.

Phylloxera

Phylloxera is a form of plant aphid that, like powdery mildew, was endemic in the eastern United States. The insect feeds on the vines' roots, weakening and eventually killing the plant. Phylloxera was first identified in Europe (where it was accidentally introduced with imported American rootstock) in 1863. It first appeared in California about a decade later.[2] By the mid-1870s the disease was ravaging the prime grape-growing areas of northern California. According to Vincent Carosso, more than 400,000 vines were dug up in Sonoma County alone between 1873 and 1879 to combat the pest. By 1880, phylloxera outbreaks had occurred in all of the state's wine grape-growing regions except Los Angeles (Carosso 1951; Pinney 1989). The future looked dire for California's vineyards.

As with the case of powdery mildew, advances in scientific knowledge eventually gave growers the upper hand in the battle against phylloxera, but the costs were staggering. Experiments conducted in both France and the United States during the 1870s and 1880s investigated literally hundreds of possible chemical, biological, and cultural cures. Most techniques, including applying ice, toad venom, and tobacco juice, proved ineffective. Four treatments appeared to offer some hope: submerging the vines under water for about two months, using insecticides (namely carbon disulfide and potassium thiocarbonate), planting in very sandy soils, and replanting with vines grafted onto resistant, native American rootstocks.[3] Only replanting on resistant rootstocks proved economically feasible, and even this course of action required an extraordinary investment. In the age before the biological revolution, often identified as beginning with the diffusion of hybrid corn in the 1930s, the vast majority of the vines of Europe and of California were systematically torn out, and the lands were replanted with European varieties grafted onto American rootstocks. This was a slow and painful process that resulted in severe hardship in the winemaking areas of the world. But the battle against phylloxera also represents an incredible biological feat; today most of the world's more than 15 million acres of vineyards are the product of the scientific advances and investments made in the 19th century. A few details of this story will offer a better sense of the achievement.

A number of early American growers had hit on the idea of grafting foreign vines on American rootstock. But grafting had no effect on black rot and the various mildews, which typically killed vinifera in the eastern and midwestern states well before the phylloxera had time to do its damage. This, along with the generally unfavorable climate in the eastern states, meant that grafting was not widely pursued. The idea of grafting onto American rootstocks to resist phylloxera reemerged in the 1860s and 1870s with the pioneering works of Charles V. Riley in Illinois and Missouri, Eugene Hilgard in California, and George Husmann in Missouri and California (Morton 1985; Ordish 1987; Carosso 1951; Pinney 1989).

Once the general principle of replanting on American rootstocks was established, much tedious work remained to be done and many detours and blind alleys had to be explored. The key problem was to discover which American varieties were in fact more resistant to phylloxera, which would graft well with European varieties, and which would flourish in a given region with its particular combinations of soil and climate.[4] In addition, grafting techniques had to be perfected. As with the initial attempts to introduce new grape varieties into myriad and largely unknown geoclimatic regions of California, the pursuit of information about the best grafting combinations required considerable trial and error as well as intensive scientific investigations.[5] In California, scientists working for the University of California, the Board of State Viticultural Commissioners, and the United States Department of Agriculture (USDA) all conducted experiments on a wide variety of

vines and conditions. Similar efforts took place across Europe. As a result of the initiatives of Riley, Husmann, and others in Missouri, that state's nurseries became the leading producers of resistant rootstock for farmers across Europe. By 1880, "millions upon millions" of cuttings had already been shipped to France. Ordish estimates that France, Spain, and Italy together would have required about 35 billion cuttings to replant their vineyards (most of these would have been grown in European farms and nurseries after the first generations were supplied from America). To better appreciate the physical magnitude of this undertaking, 35 billion cuttings would have required roughly 12 million miles of cane wood—enough to circumnavigate the earth about 500 times (Pinney 1989; Carosso 1951; Ordish 1987).

In California, the very real threat that phylloxera would wipe out the state's vineyards played a major role in generating the political support for funding the institutions that would contribute immensely to the state's agricultural productivity. Most important was the work of the College of Agriculture of the University of California. In addition, as a direct response to the epidemic, the state founded the Board of State Viticultural Commissioners in 1880. After years of denial and foot dragging by grape growers, the new Board of State Viticultural Commissioners took aggressive action. It surveyed the infested areas; it made and published translations of the standard French treatises on reconstituting vineyards after phylloxera attack; and it tested the innumerable "remedies" that had been hopefully proposed since the outbreak of the disease in France (Pinney 1989). In 1880 the State Legislature also appropriated $3,000 for the University of California to expand its efforts in the fight against phylloxera. (As Pinney and others have noted, the relationship between the board and university researchers was seldom harmonious and often outright hostile.) Under Hilgard's enlightened leadership, the university spearheaded an impressive variety of research and outreach programs, including the dissemination of knowledge already gained in France. But in the 1880s the battle against phylloxera was still in its infancy. The general principles were understood, but detailed information on the best procedures and varieties for each microregion of the state had to be laboriously compiled, and the costly process of ripping out vines and transplanting onto the recommended rootstocks was only beginning. It was not until 1904 that the USDA initiated a systematic program of testing throughout the state. By 1915 about 250,000 acres of vines had been destroyed, but relatively little land had been replanted with resistant rootstock (Pinney 1989).

Pierce's Disease

In the late 1990s Pierce's disease emerged as a serious problem in California, causing a reported $40 million loss in recent years. Up and down the state nervous grape growers were demanding that something be done. In October 1999, the University of California announced the formation of a task force to mobilize the University's scientific, technical, and information outreach expertise to help the state's grape growers combat Pierce's disease. Amid much fanfare, California Governor Gray Davis proposed in March 2000 spending an additional $7 million per year to combat the disease.[6] A brief account of earlier outbreaks of Pierce's disease sheds light on the potentially devastating nature of this threat.

The historical accounts of the attacks of powdery mildew and phylloxera tell a story of how scientists created new information, technologies, and methods that allowed farmers to coexist, albeit at an enormous cost, with the diseases. The story of Pierce's disease is altogether different. It represents a frightening case study in which the early research efforts offered little or no support to the state's farmers. The disease systematically and totally destroyed the vineyards in what at the time was the heart of the state's wine industry, dramatically altering the fortunes of thousands of farmers and reshaping the agricultural history of California. Farmers in the infected areas had no recourse but to abandon their vineyards and search for other crops.

The story begins in the German colony of Anaheim, now in the shadow of Disneyland's majestic Matterhorn in the Santa Ana Valley. This agricultural community started with the organization of the Los Angeles Vineyard Society in 1857 with a capital stock of $100,000. After overcoming early organizational prob-

lems, the settlement began to flourish. The first vintage in 1860 yielded about 2,000 gallons. Production increased rapidly, from nearly 70,000 gallons in 1861 to over 600,000 gallons in 1868. By 1883 the valley was home to 50 wineries with about 10,000 acres of vines and a production of about 1,250,000 gallons of wine (along with a sizeable quantity of brandy and raisins; Pinney 1989). Prospects for the southern California wine industry looked bright. However, lady luck dealt the valley a cruel blow with the sudden emergence of an unknown affliction originally termed the Anaheim disease.

> The vineyard workers noticed a new disease among the Mission vines. The leaves looked scalded, in a pattern that moved in waves from the outer edge inwards; the fruit withered without ripening, or sometimes, it colored prematurely, then turned soft before withering. When a year had passed and the next season had begun, the vines were observed to be late in starting their new growth; when the shoots did appear, they grew slowly and irregularly; then the scalding of the leaves reappeared, the shoots began to die back, and the fruit withered. Without the support of healthy leaves, the root system, too, declined, and in no long time the vine was dead. No one knew what the disease might be, and so no one knew what to do. It seemed to have no relation to soils, or to methods of cultivation, and it was not evidently the work of insects (Pinney 1989).

Within a few years most of the vines had died. Prosperity had turned to economic ruin. The disease soon spread with varying severity to neighboring regions, contributing to the eventual demise of grape growing in what now comprises Los Angeles, Orange, Riverside, San Bernardino, and San Diego counties.

Even identifying the disease was a slow process, and after over 100 years farmers are still waiting for a cure. At first, several growers thought the vines might be succumbing to phylloxera, but careful investigation soon dispelled this notion. As more and more vines became infected, vineyardists asked the public authorities for expert opinion. Thus the Board of State Viticultural Commissioners and the University of California had to redirect scarce resources away from the phylloxera campaign to investigate the new Anaheim disease. In August 1886, Hilgard sent F.W. Morse, a chemist who had been working on phylloxera, on an inspection trip to the Santa Ana Valley. In his report, Morse described the conditions of the affected vines, the soil, the weather, and other conditions. However, he failed to detect any insects or microscopic organisms that could be held responsible for the mysterious disease. Thus, he erroneously concluded that the disease was probably due to particular weather patterns and that conditions probably would return to normal. Hilgard shared this optimistic prediction and so informed local farmers (Smith et al. 1946; Gardner and Hewitt 1974). Further studies by Morse and other agents of the Board of State Viticultural Commissioners were no more enlightening. The failure of state officials to identify the problem stimulated vineyardists to appeal to the federal government (Carosso 1951). Consequently, in 1887 the USDA dispatched one of its scientists, F.L. Scribner, to the infected area and enlisted the aid of Dr. Pierre Viala, an eminent French researcher who accompanied Scribner. After eight days examining the vines, they too were baffled by the affliction. Scribner concluded that a fungus did not cause it, and that the disease appeared in the roots. Viala suspected that a parasite might be at fault (Gardner and Hewitt 1974). When Anaheim disease appeared in the San Gabriel area in 1888, the Board of State Viticultural Commissioners, at the urging of one of its prominent members, J. De Barth Shorb, hired a "Microscopist and Botanist," Professor Ethelbert Dowlen. Shorb provided Dowlen with laboratory equipment and an experimental greenhouse on his estate. For several years Dowlen studied the problem, but without much success. He tentatively, but mistakenly, concluded that a still-unidentified fungus caused the disease.[7] Numerous other experts came and went, but the vines kept dying. Diagnosis ranged from plant sunstroke to root rot. Every manner of spray, dust, and pruning method was recommended and tried, but to no avail. These efforts were generally less outlandish than the reasoning that led Italian peasants to tear up the train tracks to fight powdery mildew, but they were no more effective.

It remained for another USDA scientist, Newton B. Pierce, to identify the disease. Pierce arrived at Santa Ana in May 1889. He imported 200 healthy vines from Missouri and planted some on the Hughes ranch in Santa Ana, where he located his experiment station. After several years of study that included a five-month stint in France investigating known vine diseases, Pierce was able to reject most popular theories (Smith et al. 1946). In 1891 he concluded that the disease was not anything already known, that it was probably caused by a microorganism, and that there was no known cure. By this time the wine industry had disappeared from the Santa Ana Valley. More generally, the spread of Pierce's disease in southern California was an important factor contributing to the shift in the center of the state's wine production. Between 1860 and 1890, Los Angeles County's share of production fell from 66 percent to 9 percent. In contrast, the share produced in the San Francisco region rose from 11 percent to 57 percent over these three decades (Pinney 1989).

Pierce's study closed the investigations of this vine disease for almost half a century. The hiatus was partly due to the difficulty of the task, but also because the malady mysteriously ceased being a serious problem. As a postscript, the identification of the bacteria responsible for the disease as well as a precise diagnosis of how it is transmitted has only been achieved in recent years. Research has shown that the disease is caused by a bacterium (*Xylella fastidiosa*) that is transmitted by a number of leafhoppers, including the smoke tree sharpshooter, the blue-green sharpshooter, and most importantly, the newly introduced glassy-winged sharpshooter. This latter insect is a far more effective vector than the other sharpshooters because it is larger, can fly further, and is more adept at boring into the vine's wood. When the sharpshooter feeds on a vine, it transmits bacteria that multiply and inhibit the plant's ability to use water and other nutrients. The disease is inevitably fatal. The incidence of the disease varies with the geographical characteristics of the surrounding countryside, because the sharpshooter thrives in wet sites with abundant weedy and bushy growth. It is now thought to exist in every county of the state. At present, short of attacking the vector (which most scientists think is at best a delaying action), there still is no effective method to control the disease. As with the battle against phylloxera, a successful strategy will probably depend on genetically altering the plant to better resist the disease.

Threats to the State's Tree Crops

The grape industry was by no means exceptional in its susceptibility to what at the time were exotic pests and diseases. Most fruit and nut crops faced similar onslaughts as new and often mysterious invaders took a terrible toll until methods could be developed to limit the damage. As noted earlier, when California gained statehood in 1850, the area was relatively free of pests and plant disease problems. Rampant and uncontrolled importation of biological materials changed all that, and by about 1870 a succession of invaders had attacked the state's crops, threatening the commercial survival of many horticultural commodities. In addition to grape phylloxera, some of the major pests that were introduced or became economically significant between 1870 and 1890 were San Jose scale, woolly apple aphid, codling moth, cottony cushion scale, red scale, pear slug, citrus mealybug, purple scale, corn earworm, and Hessian fly. Among the diseases to emerge in the 1880s and 1890s were "pear and apple scab, apricot shot hole, peach blight, and peach and prune rust" (Smith et al. 1946). Large orchards of single varieties added to the problem by creating an exceptionally receptive environment for the pests, and the state's nurseries further contributed to the difficulties by incubating diseases and spreading infected plants. Thus, within a few decades, California's farmers went from working in an almost pristine environment to facing an appalling list of enemies in an age when few effective methods had been developed anywhere for cost-efficient, large-scale pest control.

There was a general pattern to the appearance, spread, and control of new pests and diseases. At first the afflictions were not well understood, and the losses were often catastrophic. This led to the tearing out and burning of orchards, to quarantines, to the development of chemical controls, to a worldwide search for parasites to attack the new killers, and to the eventual developments of new cultural methods and improved varieties that were resistant to the pests or diseases. The University of California and government scientists spearheaded these various efforts and together made numerous stunning breakthroughs that fundamentally altered the course of agriculture. To illustrate, let us offer some historical detail on just two of the invaders—San Jose scale and cottony cushion scale.

San Jose Scale

San Jose scale (*Aspidiotus pernicious*) was first discovered in San Jose in the orchard of James Lick in the early 1870s. Lick, who is best known for the observatory he funded, was an avid collector of exotic plants. Most historical accounts suggest the scale hitched a ride on trees Lick imported from Asia. From his property it spread slowly to nearby farms and eventually to other parts of California. By the 1890s it had reached the East Coast and was active in all the main deciduous fruit-growing regions of the Pacific Coast. The fact that San Jose was a center for commercial nurseries undoubtedly hastened the scale's spread. At first, farmers were slow to respond to the new scale, in part because the pest took time to multiply and growers tended to attribute their losses to other causes because of its innocuous appearance. By 1880, farmers and scientists recognized San Jose scale as a grievous problem.[8]

The pest attacks all deciduous fruit trees, many ornamental and shade trees, and selected small fruits, especially currants (Marlatt 1902; Quaintance 1915). The scale infests all parts of the trees that are above ground, including the leaves and the fruit. If uncontrolled, San Jose scale could mean financial ruin to orchardists. On mature trees, the scale scars and shrivels the fruit, in many cases rendering it worthless. It can also stop growth and cause a systemic decrease in vigor, reducing the yield of the tree. Eventually, the tree dies prematurely, long after it has become economically unprofitable. If left untreated, most varieties of fruit trees infested at the nursery would not survive to bearing age (Quaintance 1915). The problem in the 1870s was that little was known about the scale and the technologies for dealing with it were not yet developed. Thus, as was the case when phylloxera began destroying the world's vineyards, the very future of the deciduous fruit industry seemed in doubt. Hundreds of thousands of trees were destroyed, property values in infected areas stagnated or fell, the development of new orchards temporarily stalled, and the agricultural press lamented the deterioration in fruit quality.

From the perspective of hindsight, the response to this and the other new pests of the period was truly remarkable. The university and USDA scientists were methodical in their search for biological and chemical controls. Coupled with these efforts, a new chemical industry with its own research, manufacturing, and sales forces came into being, and with it developed the modern agricultural spraying equipment industry. The relatively little attention that San Jose scale receives today is a testimony to the success of those efforts. But writing in 1902, one of America's foremost entomologists noted that "the fears aroused by this insect have led to more legislation by the several States and by various foreign countries than has been induced by all other insect pests together." (Marlatt 1902) At a time when California producers were beginning their struggle to gain access to international markets, more than a dozen countries, including Canada and many of the leading nations of western Europe, imposed restrictions or outright bans on the importation of American fruit because of the San Jose scale (Morilla, Olmstead, and Rhode 1999; Morilla, Olmstead, and Rhode 2000; Marlatt 1902). In California, San Jose scale was one of the proximate causes underlying the creation of the State Board of Horticultural Commissioners in 1883 and the passage of the state's first horticultural pest control and quarantine law (Smith et al. 1946). These measures had an important impact on the development of the state's horticultural sector.

The fight to control the scale took two separate and at times competing tracks—biological and chemical. The discovery of biological controls was a high priority for the USDA. "The importance of discovering the origin of this scale arises from the now well-known fact that where an insect is native it is normally kept in check and prevented from assuming any very destructive features (or at least maintaining such conditions over a very long time) by natural enemies, either parasitic or predaceous insects of fungous or other diseases" (Marlatt 1902). The USDA's entomologists-turned-detectives focused their search on Asia, given the knowledge that James Lick had imported plants from Asia and that the disease was not known in Europe. By careful observation and deduction, they one by one eliminated Australia, New Zealand, the Hawaiian Islands, and Africa. Evidence appeared to point to Japan as the scale's home. But in 1901 and 1902 one of the USDA's entomologists, C.L. Marlatt, spent over a year exploring the farmlands and backcountry of Japan, China, and other Asian countries. His findings showed that the scale almost

surely originated in China. He also found what he was looking for—an Asian ladybird beetle (*Chilocorus similis*) that feasted on the scale. Marlatt sent boxes of the beetles to his experimental orchard in Washington, D.C. Only about 30 survived the journey and only 2 of those made it through the first winter. With this breeding stock and fresh imports from Asia, the beetle population was increased and studied. Subsequently, roughly 20 other insect predators were identified and studied. Other researchers investigated controlling the scale with fungal diseases (Marlatt 1902; Quaintance 1915).

Although the attempts at biological control appeared promising, in the end they were not successful. Reflecting on these efforts, A.L. Quaintance (1915) of the USDA noted that "the combined influence of these several agencies [insects] is not sufficient to make up for the enormous reproductive capacity of this insect (San Jose scale)." A number of factors accounted for this setback. The primary agent, the Asiatic ladybird, often fell victim to native insects that preyed on its larvae. In addition, the practice of spraying to combat the scale killed potential predators and their food supplies.

The inability to perfect reliable biological controls encouraged farmers to rely on spraying as their primary defense against San Jose scale. The first insecticides used were mainly lye solutions to which several substances were added, such as soap, kerosene, tobacco, sulfur, carbolic acid, and crude petroleum. At first, the common practice was to spray the trees' foliage, but eventually farmers discovered that if they applied the chemicals during the dormant season they did not need to be as careful, and they could apply stronger doses without damaging their trees. About 1886 the lime-sulfur spray began replacing other washes, becoming a leading fungicide as well. The formulas were improved, and homemade concentrates started being replaced by standard commercialized preparations (Smith et al. 1946). As previously noted, the developments in the chemical industry and the spray equipment industry in the fight against San Jose scale would prove valuable in fighting other pests. In addition, many cultural methods learned in the fields, such as short pruning and shaping of trees to facilitate pest control, proved valuable in improving quality and reducing harvest cost (Marlatt 1902).

Cottony Cushion Scale

The history of the campaign against the cottony cushion scale (*Icerya purchasi*) represents one of the truly fascinating stories in the state's agricultural development. The cottony cushion scale sticks in bunches to the branches and leaves of citrus with devastating effects if uncontrolled. This scale was first observed in California in 1868 in a San Mateo County nursery on lemon trees recently imported from Australia. The scale first appeared in southern California's citrus groves during the industry's infancy in the early 1870s, and by the 1880s the damage was so extensive that the entire industry appeared doomed. Growers burned thousands of trees and helplessly watched their property values fall. The early attempts to control the scourge only increased anxiety (Stoll 1995).

Growers tried all manner of remedies, including alkalis, oil soaps, arsenic-based chemicals, and other substances that were being tested in the fight against San Jose scale, but the pest continued to multiply. Apparently, the cottony waxy covering of the scale protected it from the killing power of these liquid poisons. In desperation, both the USDA and the University of California pursued fumigating experiments for several decades. Fumigation involved the costly process of covering the trees with giant tents and pumping in various toxic gases. Experiments with carbon disulfide began in 1881. By the end of the decade hydrocyanic acid had emerged as the most promising treatment. Potassium cyanide, sodium cyanide, liquid hydrocyanic acid, and calcium cyanide all gained favor at one time or another in the pre-1940 era. Whereas these fumigation experiments were first aimed at cottony cushion scale, with the discovery of biological controls of that insect, the primary target eventually shifted to other pests (Smith et al. 1946).

Aware that cottony cushion scale existed, but did little damage in Australia, American scientists turned their attention to discovering why. They surmised that the scale was native to Australia and that natural predators limited its spread. Incredibly, bureaucratic and financial obstacles initially prevented the USDA from sending one of its scientists to Australia. Undaunted, Charles V. Riley, the chief of the USDA Division of Entomology, and Norman Colman, the California Commissioner of Agriculture, persuaded the U.S. State Department to allocate $2,000 to send USDA entomologist Albert Koebele to Australia, ostensibly as part of the delegation to the 1888 International Exposition in Melbourne. Koebele's true mission was to search for predators of the cottony cushion scale. He hit the jackpot on October 15, 1888, with the discovery of a ladybird beetle (vedalia or *Rodolia cardinalis*) feeding on the scale in a North Adelaide garden. Koebele sent a shipment of 28 ladybird beetles to another USDA entomologist, D.W. Coquillet, stationed in Los Angeles. Many more would follow. Coquillet experimented with the insects, and by the summer of 1889 the beetles were being widely distributed to growers. Within a year after general release, the voracious beetle had reduced cottony cushion scale to an insignificant troublemaker, thereby contributing to a threefold increase in orange shipments from Los Angeles County in a single year. According to one historian of this episode, the costs were measured in thousands and the benefits of the project were undetermined millions of dollars (Smith et al. 1946; Graebner 1982; Doutt 1958; Marlatt 1940).

This success encouraged Koebele to make another journey to Australia where he discovered three more valuable parasites helpful in combating the common mealybug and black scale. Other entomologists made repeated insect safaris to Australia, New Zealand, China, and Japan, as well as across Africa and Latin America. There were many failures, but by 1940 a number of new introductions were devouring black scale, yellow scale, red scale, the Mediterranean fig scale, the brown apricot scale, the citrophilus mealybug, the long-tailed mealybug, and the alfalfa weevil. In addition, scientific investigations led to improved ways of breeding various parasites so that they could be applied in large numbers during crucial periods (Smith et al. 1946). As with Koebele's initial successes, the rate of return on these biological ventures must have been astronomical.

Collective Action

These battles against plant pests and diseases represented classic cases of a geographically dispersed and economically diverse population trying to grapple with the problems of externalities and public goods in a democratic society.[9] Externalities are present when all the costs and benefits derived from an individual action are not completely borne or captured by the agent undertaking the action; in this case an agent's actions positively or negatively affect other economic actors. As a result, there is a gap between the costs and benefits to an individual agent (the private costs and benefits) and those to society as a whole (the social costs and benefits). The public goods problem arises from the lack of rivalry and excludability in consumption.[10] A successful eradication plan for a pest such as San Jose scale required protecting all the orchards in an infected area to prevent infestation. Because pest control displays characteristics of a public good and has positive externalities, leaving it to private individual initiative would likely encourage too little pest control, as reflected in the investments in research and in the application of prevention and eradication methods. In this situation, there is a case for public authorities to intervene by coordinating and leading individual efforts into a collective action cause.

Under these conditions, finance of eradication programs by voluntary contributions would allow individuals to benefit even though they do not contribute to the cost of the program and may not even cooperate with the pest control measures. This, in turn, creates a demand for collective action to employ the state (or some form of contractual authority) to coerce compliance in both the financing and operation of the control programs. Such actions necessarily limit individual freedom. In a democratic and market-oriented society, enacting such infringements on property rights can be a difficult and costly process. The fact that farmers not only acquiesced but also actively campaigned for such controls offers strong testimony as to the severity of the threats to their livelihood.

As discussed earlier, most of the diseases had recently been introduced from other parts of the world and were therefore unknown in California when the problems arose. To eradicate the disease from their private holdings, individual growers would have had to make enormous investments to develop basic and applied research programs and eradication methods. Given costly information and the small scope for expected private benefits, such investments were probably unprofitable for individual growers. Despite the substantial monetary losses from their individual economic point of view, it would have been more efficient to let the disease destroy their crops and maybe shift to less

intensive production processes or to other crops. In fact, this was the course of action taken after the arrival of Pierce's disease, when vine growers of the Anaheim and San Gabriel Valley abandoned vines and planted citrus trees.

On the benefit side, the advantages of pest control to society as a whole are probably larger than those to individual farmers or even all farmers. Also important are the long-run or dynamic benefits derived from pest control. Practically all actions taken in this respect have had positive and significant spillovers to similar or related problems. For example, the fight against the pests and diseases of the last century led to basic and applied scientific discoveries that were crucial in improving the knowledge needed to combat other plant diseases. (In a number of cases the advances in agricultural sciences also had a direct bearing on improving human health.) The different eradication methods developed in the second half of the 1800s, such as the use of chemicals and insecticides, the breeding and grafting practices, the biological control by means of natural predators, etc., have been used extensively ever since. Similarly, much legislation concerning plant protection, such as quarantine and inspections laws, and a great part of the research and administrative institutions have their origins in the second half of the 1800s. Both the body of legislation and the state institutions detailed in Table 5.1 have effectively contributed to preventing the introduction and spread of diseases in California and elsewhere.

The efforts to combat injurious insects and diseases in California were built on earlier innovations in the understanding and control of disease. By the 1850s American agricultural leaders, including entomological and horticultural groups, were developing institutional structures that would provide the foundation for education, research, and collective action. In the 1840s, Solon Robinson and others organized the National Agricultural Society with the objective of directing the Smithsonian bequest to agricultural research. In the 1850s Marshall P. Wilder organized the U.S. Agricultural Society to lobby for the establishment of land grant colleges and the creation of a department of agriculture. The Morrill Act that granted land to the states for agricultural and industrial colleges was passed in 1862. By the early 1870s agricultural entomology courses were being offered in a number of colleges throughout the United States.

In California, important institutional structures began emerging shortly after statehood. Among the early institutions created were the State Agricultural Society and the California Academy of Sciences, organized in 1853. Both of these bodies promoted discussion and the exchange of information, but they were ill equipped to perform basic and applied research and outreach. In 1868 the University of California and the College of Agriculture were established to help fill this void. One of the college's early leaders, Eugene Hilgard, proved to be a man of enormous vision, talent, and energy. Trained in Germany as a biochemist and soil scientist, Hilgard established the policy of faculty having both research and extension responsibilities and took the lead in setting up experiment stations and a publication program aimed at communicating directly with farmers.[11] Much of the technical and research work on plant pathology that would lead to major breakthroughs in plant protection was undertaken at the university. Gradually, other state boards and institutions designed to deal with particular problems came into existence. One of the most important and active boards was the State Board of Viticultural Commissioners, created in 1880. This agency worked to provide information on phylloxera and supported research that tried to curb the ravages of Pierce's disease. But its legacy is tarnished, in part, by a long and often vitriolic squabble with Hilgard and other university scientists.

Quarantine and inspection laws provided another important tool in the arsenal to control pests and diseases. Here, California was a pioneer, enacting its first quarantine legislation in 1881. The legacy of these early efforts is with us today. Even the casual tourist entering the state by car encounters the state agricultural inspection stations designed to block pests and diseases that might hitchhike a ride into the state's fields. For most states it would be nearly impossible to stop the migration of pests and diseases from neighboring states. But California's long coast to the west and mountains and deserts to the north, east, and south offer natural barriers to migrating insects and diseases. With improvements in transportation and the increased mobility of people and commodities, the challenge of preventing new infestations has become even more daunting. But all future efforts, be they biological, chemical, or administrative in nature will be much easier to envision and implement because of the scientific and institutional foundations laid in the 19th and early 20th centuries.

Notes

[1]Our account is cursory in that it only touches on the problems of the horticultural sector and ignores the enormous problems that pests and diseases created for field and row crops and for livestock. Whereas California was a pacesetter in dealing with pests and diseases in the horticultural sector, the experiences with problems with other crops and livestock were important, but in many ways similar to what occurred in other states.

[2]Carosso (1951, p. 110) dates the arrival in Europe between 1858 and 1863. According to Pinney (1989, p. 343), "the disease had been discovered as early as 1873 in California", but this was when it was first positively identified by the Viticultural Club of Sonoma. Carosso maintained that the "disease was known to have existed in California before 1870 . . ." and vines on the Buena Vista estate probably had shown signs of infestation as early as 1860. See Carosso (1951, pp. 109-111); Butterfield (1938, p. 32).

[3]Ordish (1987, pp. 64-102) and most others use arcane 19th century terminology, labeling *carbon disulfide* (CS_2) as *carbon bisulfide* or *carbon bisulphide* and *potassium thiocarbonate* (K_2CS_3) as *sulphocarbonates of potassium*.

[4]"Resistance" is not a sure thing. When replanting onto apparently identical resistant rootstock, it is expected that about 20 percent of the plantings will be susceptible to phylloxera. In addition, over time the insects evolve to be able to overwhelm plants that previously had been resistant. Thus, the initial spread of phylloxera represented a watershed in the history of grape growing, and ever since it has been necessary to develop new resistant varieties to stay ahead of the insect.

[5]As an example, the first U.S. varieties shipped to France were labrusca and labrusca-riparia hybrids that had a low resistance to phylloxera. In California the initial recommendation that growers use *Vitis californica* for rootstock proved to be a mistake (Pinney 1989, pp. 345, 394; Carosso 1951, p. 125; Ordish, 1987, pp. 116–119).

[6]*The Washington Post*, March 27, 2000.

[7]Pinney 1989, p. 307; Gardner and Hewitt 1974, pp. 18-96. Dowlen reportedly had studied botany at the South Kensington School in London with Thomas Huxley and billed himself as a French expert on vine disease.

[8]Marlatt 1902, p. 156. It was in this year that it received its official name of *Pernicious*.

[9]For more on the economics of exotic pest and disease principles see Chapter 2.

[10]There is "rivalry" in the consumption of a good or service when the consumption by one agent pre-

vents others from enjoying it as well. This is not the case of a pest control program. Two farmers can simultaneously enjoy a plan's benefits without imposing additional costs on each other. "Excludability" exists when one can limit the access to a good. This is true of most goods sold in the marketplace. However, when a pest control program is under way, it may be hard to exclude any one farmer from benefiting from eradication efforts on nearby farms.

[11]Eugene Hilgard earned his Ph.D. in organic chemistry at the University of Heidelberg.

References

Barnhart, J.H. 1965. *The New York Botanical Garden Biographical Notes Upon Botanists*, vols. 1 and 3. Boston: G.K. Hall.

Butterfield, H.M. 1938. *History of Deciduous Fruits in California*. Sacramento: Inland Press.

Carosso, V.P. 1951. *The California Wine Industry. A Study of the Formative Years*. Berkeley and Los Angeles: University of California Press.

Doutt, R.L. 1958. "Vice, Virtue, and the Vedalia." *Bulletin of the Entomological Society of America*. 4(4):119–123.

Essig, E.O. 1940. "Fifty Years of Entomological Progress, Part IV, 1919 to 1929." *Journal of Economic Entomology*. 33(1):30–58.

Gardner, M.W., and W.B. Hewitt. 1974. *Pierce's Disease of the Grapevine: The Anaheim Disease and the California Vine Disease*. Davis and Berkeley: University of California Press.

Graebner, L.A. 1982. "An Economic History of the Fillmore Citrus Protective District." Ph.D. Dissertation, Department of Economics, University of California, Riverside.

Large, E.C. 1940. *The Advance of the Fungi*. New York: Henry Holt.

Marlatt, C.L. 1902. "The San Jose Scale: Its Native Home and Natural Enemy." *USDA Yearbook*. Washington, D.C. Yearbook of the United States Department of Agriculture: GPO.

Marlatt, C.L. 1940. "Fifty Years of Entomological Progress, Part I, 1889 to 1899." *Journal of Economic Entomology*. 33(1):8–15.

Morilla, J., A.L. Olmstead, and P.W. Rhode. 1999. "Horn of Plenty: The Globalization of Mediterranean Horticulture and the Economic Development of Southern Europe, 1880-1930." *Journal of Economic History*. 59(2):316-352.

Morilla, J., A.L. Olmstead, and P.W. Rhode. 2000. "International Competition and the Development of the Dried-Fruit Industry, 1880-1930." In S. Paumuk and J.G. Williamson, Eds., *The Mediterranean Response to Globalization Before 1950*. London and New York: Routledge. pp. 199–232.

Morton, L.T. 1985. *Winegrowing in Eastern America*. Ithaca, N.Y.: Cornell University Press.

Ordish, G. 1987. *The Great Wine Blight*. London: Sidwick & Jackson.

Pinney, T. 1989. *A History of Wine in America. From the Beginnings to Prohibition*. Berkeley and Los Angeles: University of California Press.

Quaintance, A.L. 1915. "The San Jose Scale and Its Control." *Farmers' Bulletin*, no. 650. Washington, D.C.: USDA.

Ryan, Harold J. et al. 1969. "Plant Quarantines in California." University of California: Berkeley, California.

Smith, Ralph E. et al. 1946. "Protecting Plants from Their Enemies." In Claude B. Hutchison, Ed., *California Agriculture*. Berkeley and Los Angeles: University of California Press. pp. 239–315.

Stoll, S. 1995. "Insects and Institutions: University Science and the Fruit Business in California." *Agricultural History*. 69(2):216–239.

The Washington Post. "Deadly Pest Sours Vintner's Grapes," March 27, 2000, pp. A3, A12.

Weber, G.A. 1930. *The Plant Quarantine and Control Administration. Its History, Activities and Organization, Institute for Government Research*. Service Monographs of the United States Government, no. 59. Washington, D.C.: The Brookings Institution.

PART II

Exotic Pest and Disease Cases: Examples of Economics and Biology and Policy Evaluation

6

Bovine Spongiform Encephalopathy: Lessons from the United Kingdom

José E. Bervejillo and Lovell S. Jarvis

Introduction

Bovine spongiform encephalopathy (BSE), also known as "mad cow disease," is a slowly progressive degenerative disease affecting the central nervous system of adult cattle. It is inevitably fatal. BSE was first identified in the United Kingdom in 1986. Since that time, the disease has been found in cattle in other European and Asian countries, with transmission probably linked to the feeding of rendered parts of BSE-infected animals to other cattle in the form of meat and bone meal. More than 95 percent of the approximately 185,000 cattle diagnosed with BSE to date have been found in the United Kingdom.

Although BSE would have been an important disease had it only affected cattle, the disease appears transmissible to humans. In 1996, 10 human patients in the United Kingdom were identified to have a variant of Creutzfeldt-Jakob disease (vCJD), a chronic, neurodegenerative, fatal disease that appears linked to the consumption of BSE-infected meat. Measures have been taken in the United Kingdom and other countries to identify and destroy BSE-infected animals and to close all of the pathways by which the disease is passed from animal to animal. Substantial time passed, however, between the emergence of BSE as a widespread animal disease and the identification of a threat to humans. During that period, a substantial number of humans consumed BSE-infected meat. Little is understood about the relationship between the consumption of BSE-infected meat and the probability of developing vCJD. The prediction of the number of future vCJD cases also depends on the length of the incubation period, which is still unclear. It has thus been difficult to predict how many people will die as a result of having consumed BSE-infected meat. To date, fewer than 125 cases of vCJD have been reported. It appears that all persons who have contracted vCJD have a common genetic trait. Although this information suggests that the human impact could remain small, estimates of the total expected human cases vary from only slightly more than those that have already occurred to 136,000 in the United Kingdom alone.

This chapter examines the risk to the United States from the potential introduction of BSE, based on the United Kingdom experience and preventative measures taken to date in the United States. Our analysis is intended only to provide an overview and summary conclusion. It relies heavily on several previous studies. The major reference to the British case is contained in *The BSE Inquiry Report*, published by the United Kingdom government in October 2000 (Phillips et al. 2000). For a more detailed study of risks currently facing the United States, we recommend the three-year study by the Harvard Center for Risk Analysis (Cohen et al. 2001). Similarly, the U.S. General Accounting Office (GAO 2002) Report reviews the U.S. policies implemented to reduce the risk of damage from BSE and offers recommendations for further policy changes.

To date, no case of BSE has been diagnosed in the United States. The BSE crisis in the United Kingdom has served as a learning experience for the U.S. industry and government regulatory agencies. Considering all the current regulatory policies with respect to BSE prevention and the surveillance system that has been developed in the last 12 years, and provided the U.S. livestock industry keeps a high standard of compliance with current regulations, it is possible to conclude that the risk of contracting the disease will remain at a minimum expression.

Epidemiology

BSE is considered a disease of the group of transmissible spongiform encephalopathies (TSEs) that are fatal degenerative brain diseases common in sheep and several other animal species, e.g., elk, deer, cats, and minks. One human form of TSEs is *kuru*, identified in Papua, New Guinea where, in the past, natives had a tradition of eating the brains of the deceased elders. Another human form is the Creutzfeldt-Jakob disease (CJD), which can take different forms: the familial form that is inherited; the sporadic form, which is the most common; and the form acquired from surgical instruments or corneal transplants. BSE has been linked to a new variant of CJD, or vCJD.

To date, there is no viable treatment for a TSE, and, indeed, there is no validated live-animal diagnostic test. Autopsies of animals that have died from a TSE show spongelike lesions in the brain. Considerable medical controversy was created by a search for the BSE infective agent. S. Prusiner (1995) proposed that the agent is an abnormal form of the prion protein, which does not act as a conventional agent of disease because it seems to replicate without the need of DNA. Although the abnormal prion protein theory is now widely accepted, some scientists argue that the abnormal prion protein could be a result rather than a cause of the TSE infection (Chesebro 1998). The causal relationship between BSE and vCJD is also questioned by P. Brown (2001), who argues that an environmentally determined imbalance between copper and manganese is the key factor in explaining the appearance of any form of TSE.

In general, it appears that transmission occurs via the consumption of parts of an animal that is affected by abnormal prions. Because the abnormal prions are very resistant to destruction by heat, chemicals, disinfectants, extreme pH, and radiation, the only known guarantee by which transmission can be stopped is to preclude consumption of any part of an infected animal. If the disease is transmitted, the disease passes through an incubation period before manifesting itself clinically. The incubation pe-

riod for BSE in cattle is typically from three to five years. An affected animal will show central nervous system signs, with disorientation, paralysis, and death.

Introduction and Spread of BSE in United Kingdom and Continental Europe

The origin of BSE is still disputed. It is generally accepted that transmission occurred after cattle were fed meat and bone meals (MBM) that contained contaminated material. It is most likely that the spongiform encephalopathy in sheep and goats known as *scrapie* somehow was altered to become BSE and was transmitted to cattle via MBM from rendered sheep and goats. Horn et al. (2001), however, have stated that BSE could have been caused by an unmodified scrapie agent. A different theory sustains that the BSE agent resulted from a random point mutation that occurred probably during the early 1970s (Phillips et al. 2000). It is also possible that contaminated MBM came from some imported meat of a wild ruminant. Horn et al. (2001) also stress that transmission from mother to calf, or through pasture contamination, or the use of veterinary preparations, might have played a small role in BSE dissemination. This issue has become more pointed since recent cases of BSE have come from a cattle population that has not been fed MBM. In any event, infected cattle carcasses were in turn rendered and fed to bovines as MBM, eventually causing a full-scale epizootic.

The epizootic started in 1986 and the number of new cases reached its peak in 1993. The number of BSE affected animals detected in the United Kingdom and Europe since 1986 is shown in Table 6.1.

In September 1985, the Pathology Department of the Central Veterinary Laboratory (CVL) of Great Britain investigated the death of a cow caused by what was—at the time—an unrecognizable disease. This was probably the first known case of BSE. Two further cases were diagnosed as TSE by the end of 1986. During 1987, the CVL concluded that the new disease, called BSE, was caused by the consumption of MBM made from carcasses of infected animals. Because it was not clear, however, that the meat from BSE-infected cattle

Table 6.1 Reported cases of BSE since 1986, by country

Country	1986-1995	1996-2001	Total cases
UK	161,322	20,804	182,126
Ireland	115	722	837
Portugal	32	583	615
France	13	502	515
Switzerland	186	222	408
Germany	4	134	138
Spain	0	84	84
Other countries[a]	2	165	167
TOTAL	161,674	22,897	184,571

[a]Includes other European countries, plus three cases in Japan.

Source: OIE (www.oie.int).

posed a risk to human health, no further action was taken. This policy was only very gradually changed, with ultimately dire consequences. For example, in March 1988, the United Kingdom Ministry of Agriculture, Fisheries, and Food recommended that animals showing signs of BSE should be destroyed and compensation paid to their owners, and the Department of Health was also asked to evaluate the disease for its human health implications. Nonetheless, a compulsory slaughter and compensation scheme was not introduced until August 1988.

In February 1989 an official investigation concluded that the risk of transmission of BSE to humans was remote and that BSE thus was unlikely to have human health implications. However, the report's risk assessment was later considered inadequate. Had this report been evaluated more critically, actions to prevent human health problems might have been taken earlier. In fact, no major precautions were recommended except that manufacturers should not include ruminant offal and thymus in baby food (Phillips et al. 2000). In June 1989, certain types of cattle offal thought most likely to be infectious were banned from use in human food. However, the government allowed mechanically recovered meat to be used for human food until December 1995. Mechanically recovered meat, or MRM, involves the removal of meat scraps left attached to the vertebral column of slaughtered animals using jets of water. MRM was a potential pathway for humans to be infected with BSE because MRM sometimes includes spinal cord and dorsal root ganglia. The latter was thought to be innocuous at the time,

but has since been demonstrated to be infectious in the late stages of incubation (Phillips et al. 2000). Thus, BSE-infected meat was still in the human food chain through 1995.

The United Kingdom was also slow in restricting the use of MBM as an animal feed. MBM produced from rendering carcasses of cattle and sheep was used in the United Kingdom until it was banned in July 1988. MBM production occurs when carcasses are milled and cooked (boiled at atmospheric or higher pressures), resulting in a protein solution covered by a layer of fat or tallow. The protein compound was then used in the manufacture of feedstuffs for cattle, sheep, pigs, poultry, zoo animals, and pets. TSE has been present in sheep herds for centuries, and MBM had been used as a cattle feedstuff for 50 years without any sign of BSE. In the early 1980s, however, the rendering process was changed to allow the use of lower temperatures. This change may have opened a window for the infectious agent to get into cattle. Although the rendering process was never considered an effective mechanism of inactivating TSE infectious agents (Phillips et al. 2000), the change in the rendering process may have changed the agent's infectivity.

The use of MBM probably carried greater risk in the United Kingdom as opposed to the United States for several reasons, although international comparisons are difficult to establish on this issue (Ardans et al. 1999; Brown et al. 2001; Horn et al. 2001). MBM was used more extensively in cattle feed in the United Kingdom, especially for young dairy calves, and calves are more susceptible to infection with BSE than are adult bovines. Dairy calves comprise a higher proportion of the herd in the United Kingdom than in the United States. MBM has not been as important a feedstuff for beef as for dairy calves, and consequently beef cattle have shown significantly lower levels of the disease. Also, although MBM has been used in many other countries, the ratio of sheep to cattle is higher in the United Kingdom than in the United States or mainland Europe, and the incidence of scrapie in sheep is also high in the United Kingdom.

Existing conditions for BSE dissemination in United Kingdom were aggravated by operational factors or policy implementation problems. The July 1988 ban on the use of MBM as a livestock feed was not completely enforced for several months. Moreover, the United Kingdom government gave the animal feed industry a "period of grace" of some weeks to clear existing feed stocks before the ban took effect. Some firms continued to clear stocks after the ban came into force. Farmers in their turn used up the stocks that they had purchased. This led to thousands of animals being infected after the ruminant feed ban was declared. The government also did not rigorously evaluate how much infective material was needed to transmit BSE. It concluded that there was little risk of cross-contamination in feed mills, i.e., from pig or poultry feed to cattle feed. However, cross-contamination occurred, resulting in the infection of thousands of cattle. Since it takes, on average, five years after initial infection for the clinical signs of BSE to become apparent, this cross-contamination issue was not appreciated until 1994 (Phillips et al. 2000).

The ban on MBM also did not preclude United Kingdom manufacturers from exporting MBM elsewhere. As a consequence, contaminated MBM continued to be used as a cattle feedstuff in many countries long after the United Kingdom ban in July 1988. In fact, non-European countries imported increasing amounts of British MBM from negligible volumes in 1988 to about 30,000 tons in 1992 and again in 1993. Outside the European Union, Indonesia, Israel, and Thailand were the principal importing countries between 1988 and 1996, during which time exports of MBM totaled almost 214,000 tons (Waterhouse 2001). The recent cases detected in Japan and Israel might have originated in the consumption of contaminated British MBM, although Japan imported only 333 tons of MBM from the United Kingdom between 1988 and 1996. During 2000-2001, most Japanese imports of European MBM originated in Denmark and Italy.

The linkage between BSE and the vCJD was officially recognized in March 1996. The presumption that BSE originated from scrapie and that scrapie was not a human pathogen delayed considerably the formal recognition of the linkage, even though there were experimental results that showed an altered host range after the passage of scrapie through a different species. For instance, mouse-adapted strains of scrapie that passed through hamsters changed their transmissibility upon back passage to mice

(Brown et al. 2001). The source of contamination in humans must have been contaminated beef. Brown et al. (2001) point out the following possible ways of contamination: cerebral vascular emboli from stunning instruments, contact of muscle with brain or spinal cord tissue by saws or other tools used in the slaughterhouse, inclusion of paraspinal ganglia in cuts with vertebral tissue, and fragments of spinal cord or paraspinal ganglia in MRM. MRM used to be a standard ingredient of sausages, hot dogs, bologna, and other meat preparations.

New cases of BSE have been decreasing since 1993. The MBM ban is thought to have been a major factor in controlling the disease in the United Kingdom, after having reached a peak of more than 1,000 cases per week. However, the number of detected cases in continental Europe was significantly higher in 2000 than in 1999, and again higher in 2001 than in 2000. Furthermore, in September 2001, a five-year-old dairy cow was diagnosed with BSE in Japan. This was the first detected case of BSE in a nonimported animal outside Europe. In June 2002, an animal was diagnosed to have BSE in Israel. Thus, it appears that the epizootic has not yet been controlled. It is difficult to determine the causes for new cases. There could be some BSE-infected material that remains in storage and is being fed to animals, animals could still be developing the disease after having been infected in the past, or there may be mechanisms of BSE transmission that are not understood and thus are not yet controlled. In fact, a number of the new cases of BSE diagnosed in the United Kingdom are animals that were born after the 1988 ban and even after the 1996 extension of the ban. The reason for these occurrences is still disputed, since maternal transmission has been challenged by some studies (Wilesmith and Ryan 1997).

Intervention Strategies

The major policy measures taken in the United Kingdom, the European Union, and the United States to control BSE and eliminate the animal and human health risks are summarized in this section. The United Kingdom government introduced a number of changes into the beef industry to diminish the risk of BSE transmission. Other countries have followed similar paths, gaining from the experience accumulated in the

United Kingdom aiming toward the same goal. The United States began introducing regulatory policies as early in the process as 1989. The rapid response has benefited the general public, reducing the level of risks to a minimum.

United Kingdom

BSE has been a "notifiable" disease in the United Kingdom since June 1988, at which time a surveillance program was initiated requiring a histologic examination of the brains of all fallen stock. A fallen animal is one that, though alive, has fallen and is unable to regain its feet. This condition may appear for numerous reasons, only one of which is BSE. In July 1988, the United Kingdom government banned the use of ruminant protein (MBM) in ruminant feed, although this prohibition was not completely enforced until several months later. This ban was extended to all protein materials of mammalian origin in November 1994, although blood, milk, and gelatin were excluded. In April 1996, all MBM of mammalian origin was banned from all farm animal feed and from fertilizer.

Measures taken to protect humans began in August 1988, with a decision to slaughter and destroy all BSE-affected cattle, and later extended to destroy milk from affected cattle, except for milk fed to calves. Human consumption of specified bovine offal (SBO), such as brains and spinal cords, was prohibited in November 1989. Intracommunity exports of SBO continued until April 1990, while extracommunity exports of SBO and MBM continued until 1996. Exports of live cattle older than 6 months were suspended in June 1990. The list of SBO considered risky was expanded on several occasions after 1992. In March 1996, after the recognition of the direct relationship between BSE and vCJD, the government ordered the destruction of all cattle older than 30 months, with compensation to the owner (over 30-months scheme, or OTM). Excluded from this scheme were the grass-fed cattle, from domestic or foreign origin, that entered the Beef Assurance Scheme. In this case, animals can be raised up to 42 months old and still be destined to human consumption. At the same time, a selected cull program was implemented, under which all animals belonging to the same cohort as a BSE-diagnosed animal born between July 1989 and June 1993 were also slaughtered. This program

was later extended to include all animals that had access to the same feed as the BSE-infected animal during the first 6 months of life, even if they were not born in the same herd. Calves born from confirmed BSE-infected cows were also sacrificed.

During the year 2000, the United Kingdom government went through a number of institutional changes aimed to improve the effectiveness of BSE surveillance, control, and eradication, including the creation of the Food Standards Agency (FSA), which is now responsible for food safety policies. In a report published in December 2000, the FSA identified the main BSE control policies as being the animal feed ban, the removal of specified risk materials from sheep, goats, and cattle that are not allowed to enter the food chain, and the OTM scheme.

European Union

European Union (EU) member countries did not always adopt policies regarding BSE at the same time, nor did they enforce adopted regulations with the same effectiveness. We provide the details only on the EU's common policies since those followed by single countries (other than the United Kingdom) are beyond the scope of this section.

The first measure approved by the EU was a prohibition on imports of cattle born in the United Kingdom prior to the July 1988 MBM feed ban. BSE was declared a notifiable disease in April 1990, and a BSE surveillance program was initiated in May 1990. A ban on the use of mammalian MBM as a feed for ruminants was adopted in July 1994. In January 2001, this prohibition was extended to include all farm animals and the use of blood. The use of MRM was banned in March 2001. In March 1996, the EU imposed a ban on United Kingdom exports of all live cattle, beef and beef products, and tissues, a measure that is expected to be adjusted as time passes (DEFRA 2001). The use of central nervous system tissues in the manufacture of cosmetics was prohibited in January 1997, and the ban was later extended to a broader list of risky materials. In December 2000 the EU approved the purchase for destruction scheme, a program equivalent to the British OTM: all cattle over 30 months old are tested for BSE before their meat can enter the food chain. Carcasses are destroyed if the animal tests positive.

Policies applied throughout the EU faced a number of serious logistical problems. The countries' testing capacity has been limited. Control measures were taken before standardized test protocols existed. Slaughterhouses have had limited capacity to process the volume of carcasses and the so-called specified risk materials (SRMs), particularly after the inclusion of intestines as a SRM. Consequently, some countries have had to store carcasses or export SRMs to other EU countries for rendering or incineration. The removal of the vertebral column at the slaughter plant has proved troublesome. There have been problems also with the ban on MBM use in animal feed, because not all countries were doing the same thing at the same time. Recalling processed animal proteins from mill intermediaries and farms could not be completed in a short time. Countries also did not have the capacity to store and dispose of MBM or to get authorization for using these materials for feeding non-food-producing animals. There was uncertainty about how to deal with small plants that produced pet food and fish food, and there are still concerns about how to detect the presence of products, such as gelatin or hydrolyzed proteins (European Commission 2001). Implementation problems throughout the EU led the European Commission to adopt a single legal text in July 2001 (Byrne 2001).

The United States

The federal agencies responsible for controlling and enforcing regulations regarding BSE are the Customs Service, the Animal and Plant Health Inspection Service (APHIS), the Food Safety Inspection Service, and the Food and Drug Administration. Additionally, the Centers for Disease Control and Prevention and the National Institutes of Health, both under the jurisdiction of the Department of Health and Human Services, are responsible for monitoring vCJD (GAO 2002).

The U.S. government imposed a ban on the importation of all live ruminants from the United Kingdom in July 1989 and also restricted the import of certain cattle products from the United Kingdom and other countries where BSE has been diagnosed. The United States Department

of Agriculture/APHIS (USDA/APHIS) began a BSE surveillance program in 1990, requiring brain tissue examinations of fallen stock and initiating an outreach program aimed at veterinarians, cattle producers, and veterinary laboratories. A restriction on imports of ruminant meat, edible products and by-products from countries known to have BSE was approved in 1991. In June 1997 the FDA prohibited the use of MBM derived from mammals (also called mammalian protein—which excludes milk, blood, and gelatin) to ruminants. In December 1997 the ban on imports of live ruminants and most ruminant products was extended to all European countries. Although there is no evidence that embryos, semen, or reproductive tissues can transmit BSE, the import of embryos from BSE-affected and high-risk countries has been suspended. In December 2000 imports of rendered protein and wastes from Europe were prohibited.

Since 1993 the FDA has requested on several occasions that manufacturers of regulated products, cosmetics, or dietary supplements that use bovine-origin materials restrict their purchases to BSE-free countries. This request was extended in 1997 to include gelatin in the manufacture of indictable, implantable, or ophthalmic products. Blood donors must have not lived in United Kingdom for more than 6 months during the period 1980-1996.

The Cost of BSE in the United Kingdom and Affected Parties

This section contains an overview of the costs of the BSE crisis in the United Kingdom. Data come mostly from Phillips et al. (2000) and from official statistics of the Department for Environment, Food & Rural Affairs (DEFRA).[1] It is important to keep in mind that the cost of controlling and eradicating the disease in the United Kingdom has no direct translation into a hypothetical cost for the United States in case BSE is found in this country. The existing preconditions are completely different (regulations that are now in place in the United States that were nonexistent in the United Kingdom 10 or 15 years ago), and the profile of the cattle sectors of each country are also not comparable. Also, the United States surveillance system already has accumulated experience that was not available for the British animal health system of the mid-1980s. The public response would likely be different.

The official estimated cost of the BSE crisis in United Kingdom since it started in 1986 to the end of the fiscal year 2001-2002 is £3.7 billion (roughly $5.6 billion at the current exchange rate).[2] Most of this cost occurred after March 1996. For example, while the 1995-1996 annual cost was estimated at £20 million, the 1996-1997 cost jumped to £850 million (Phillips et al. 2000). The total cost includes public expenditure on research and development, compensation paid to private agents, the loss experienced by the beef industry, and the disruption of the international beef trade. Currently, the total annual cost directly associated with BSE control at the private and government levels is estimated at £552m (FSA 2000), which is equivalent to 12 percent of the gross value of production of beef cattle and dairy sectors taken together.

Table 6.2 summarizes the United Kingdom government's expenditure on TSE-related research, including the budget of the Department of Health on the CJD surveillance program. In nominal terms, the total is £169 million in 15 years, almost two-thirds of which was spent during the last 5 years.

The BSE crisis significantly affected the beef industry. The 1989 prohibition on using SBO for human consumption forced slaughter plants to separate their processing systems, thus increasing costs. The feed ban ultimately eliminated the MBM product line that rendering plants used to produce from otherwise unusable slaughter by-products and sell to feed mills. Unable to profit from this processing, rendering firms transformed themselves into waste disposal firms that charged slaughterhouses for removing animal wastes from their premises, while retaining their market for tallow (Phillips et al. 2000).

A number of businesses that had some connection with the beef and cattle industry were affected little or not at all. For example, the pet food industry was already using very few beef products because it had already switched to alternative sources of protein. When the medical and cosmetics industries were precluded from using domestic beef industry by-products as raw materials, they switched to foreign suppliers, without a significant change in their production costs (Phillips et al. 2000).

Conversely, the poultry and pig-meat industries benefited from the BSE crisis, as consumers lost confidence in beef and switched to other meats (Burton and Young 1997). Firms that supplied quality assurance and inspection services, required by the new regulations, may also have benefited (Phillips et al. 2000).

The economic impact of the BSE crisis until March 20, 1996, was significant, according to the conclusions reached by *The BSE Inquiry Report*, but the costs "were minor in relation to the economy of the UK as a whole and each of the industry sectors" (Vol. 10 # 4.34). Indeed, the BSE crisis seemed to have accelerated some of the changes that the beef industry was already undergoing as a result of excess capacity and food safety concerns. Some of the induced effects of the crisis, namely, the loss of jobs in the beef industry, were offset within a year after March 1996 due to an increase in the output of other meat industries (i.e., pig, poultry). In some cases, layoffs were only temporary (DTZ-Pieda 1998). However, the serious impact came with the explosive increase in public expenditures (OTM scheme; law enforcement costs) and the loss of export markets.

Compensation paid for slaughtered BSE infected animals, from 1988-1989 until 1995-1996, was £135 million in nominal terms. Surveillance, the disposal of carcasses, and administrative costs added a further £87 mil-

Table 6.2 United Kingdom government cumulative expenditure on TSE-related research (thousands of current pounds per 5-year period)

Funding agency	1986/87–1990/91	1991/92–1995/96	1996/97–2000/01	Total cumulative
MAFF[a]	4,400	27,500	60,161	92,061
Other[b]	6,938	22,122	47,904	76,964
Total	11,338	49,622	108,065	169,025

[a]Ministry of Agriculture, Fisheries and Food.
[b]Dept. of Health, Research Councils.
Source: Phillips et al. (2000).

lion. Total costs increased dramatically from 1988 to 1994, even on a per animal basis. In nominal values, per animal costs increased from £736 to £1,776 from 1988–1989 to 1993–1994 and then decreased only slightly to £1,707 in 1995–1996. The cost per confirmed case seems to have evolved in a pattern similar to that of the epizootic itself. If the cost per animal after 1996 was about the same as the average for the previous period—£1,475—then total compensation costs for the period 1996–2001 might have reached £26.5 million.

In 1996, the United Kingdom government implemented three new compensation schemes: the OTM, the Selective Cull Program, and the Calf-Processing Aid Program. As of 2001, more than 4.7 million cattle have been sacrificed under the OTM scheme, 77,000 animals have entered the Selective Cull Program, and more than 12,000 calves have been sacrificed because they were the offspring of BSE-infected cows. Annual slaughter under all these programs has represented between 7 and 9 percent of the United Kingdom cattle inventory. Total cumulative costs of these programs exceeded £2 billion by the end of 2001. The annual cost of the compensation scheme is shown in Table 6.3. This table does not include the costs of carcass disposal for animals whose beef was not allowed to enter the food chain or the incineration of the resulting MBM and tallow. In the first four years of the scheme, the disposal cost amounted to £575 million (FSA 2000). Currently, the total annual cost to the United Kingdom government for compensation and disposal is estimated at £400 million (FSA 2000).

Note that subsidies to the beef cattle sector already amounted to about 12 percent of output value in the early eighties, before the BSE crisis began. Total government subsidies rose to 51 percent of output value in 1996. Compensation schemes accounted for 43 percent of total subsidies in 1996.[3] In subsequent years, the level of subsidies slightly decreased (DEFRA 2002).

Table 6.4 shows United Kingdom beef and live animal exports in selected years. During the first half of the 1990s, exports actually increased 73 percent in nominal values, thanks in part to a decreasing domestic consumption. After 1996, cattle exports virtually disappeared, although exports of sheep and goats continued in about the same level as before. During the 12 months after March 1996, it is estimated that total market value (domestic and foreign) for United Kingdom beef fell by 36 percent in real terms (DTZ-Pieda 1998). Assuming that United Kingdom exports would have followed the EU trend during the second half of the 1990s, had BSE not been linked to the vCJD, UK exports might have been about $800 million higher in 1999, or about 12 times the actual value. It is difficult to establish the value of lost exports for the period 1996–2000, but if the United Kingdom had kept the share of the EU-15 exports at the 1990-1995 average level (40 percent), then this value of forgone exports could be set close to $2 billion.

Although per capita beef consumption in the United Kingdom has been decreasing since the 1970s, the BSE crisis accentuated this long-term trend (see Figure 6.1). Per capita consumption fell by 27 percent from 1994 to 1996, when concern for human health was at its

Table 6.3 United Kingdom expenditures on compensation schemes, in pounds (nominal)

| Year | Cost of Each Program per Animal (£/head) | | | Gross total (million £) |
	OTM	Selective cull	Calf-processing aid	
1996	463.7	..	92.0	562
1997	425.4	1,368.4	90.6	494
1998	266.4	2,263.2	77.5	334
1999	275.3	..	65.1	289
2000	288.1	..	..	281
2001	253.6	..	..	157
1996–2001 Cumulative (million £)	1,829	124	164	2,117

Source: DEFRA (2002).

Table 6.4 United Kingdom exports of beef, beef preparations, and live animals (cattle, sheep, and goats) in selected years (thousand $)

Year	Beef	Live animals	Total	Share of EU-15 exports (%)
1991	421,395	154,901	576,296	30
1995	851,583	143,275	994,858	52
1999	38,309	27,292	65,601	3

Source: Based on FAO-STAT (2002). (http://apps.fao.org).

height, but then largely recovered by 2000. The BSE issue is probably not the major factor in explaining the current relatively low levels of beef consumption in the United Kingdom. There is a long-term trend of substitution of pig and poultry that is related to changes in consumer preferences, demographics, and, especially, in relative prices (Burton and Young 1997; Verbeke et al. 2000; Lloyd et al. 2001). The sharp decrease in consumption in 1996 is probably attributable to the food scare because of the media coverage of meat health-related issues and miscommunications between industry, government agencies, and the public. During the period 1996-2000, however, the real domestic price of beef fell roughly 25 percent (see Figure 6.2), and the decline in the price of beef

encouraged consumers to consume more beef, despite any increase in concern for human health. The domestic supply of beef decreased more than 25 percent between 1992 and 1996, and then partly recovered by 2000 to a level 8 percent below the 1992 figure. Conversely, the domestic supply of other meats increased consistently by 16 percent from 1992 to 2000.

Major Risk Factors and Current U.S. Regulations

The United States established a BSE surveillance plan in May 1990. Surveillance efforts focus on those animals that are thought to be at greatest risk, i.e., adult animals that manifest neurological abnormalities or are nonambulatory ("downers") at slaughter. As of December 2000, the brains of about 12,000 such animals had been examined for BSE or any other form of TSE in cattle. No evidence of either condition has been detected. Passive surveillance is also carried out through data collected by veterinary schools, the Food Safety and Inspection Service, necropsies performed at zoos, and a veterinary laboratory diagnostic system. Private practitioners provide another source of passive surveillance. Currently, the surveillance level

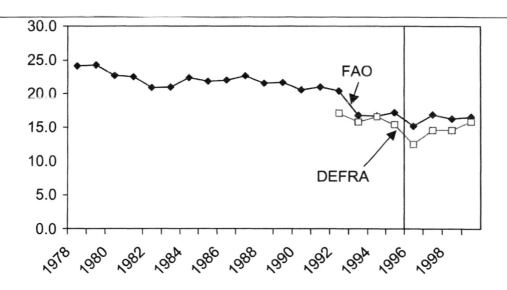

Figure 6.1 United Kingdom bovine meat domestic consumption (kg/year/person).
Sources: ➡ FAO-STAT, 2002
▱ DEFRA, 2002

Figure 6.2 United Kingdom cattle prices [constant 1995 £(p/kg)].
Source: DEFRA (2002).

carried out in the United States is considered well above the level recommended by the Office International des Épizooties.

The government carried out an initial risk assessment in 1991 and updated the assessment in 1993 and 1996. The overall risk of BSE occurring in the United States is considered "extremely low and decreasing." The principal risk, according to the recent study by the Harvard Center for Risk Analysis, is that someone will fail to comply with current regulations, e.g., by importing infected animals, supplying prohibited MBM to ruminants, mislabeling feed that is prohibited for cattle, or inadequately disposing of animals that die on the farm. However, the probability of a large-scale crisis such as the one that occurred in the United Kingdom is extremely low. The Harvard report concluded that even if BSE were to be introduced in the United States, the disease should be rapidly controlled and eradicated by application of the measures currently in place (Cohen et al. 2001).

BSE could occur in the United States in the following hypothetical situations: (1) spontaneous cases in the domestic cattle herd; (2) cases where cattle are exposed to other domestic or wild species of mammals that carry other types of spongiform encephalopathies, especially sheep, deer, elk, and mink; (3) imports of infected cattle or livestock feed; and (4) the recycling of various cattle by-products that may be infected (P. Brown 2001; Cohen et al. 2001).

If BSE in cattle behaves in a form parallel to CJD in humans, then we should expect to observe a few cases per year. Cohen et al. (2001), for example, estimated that "spontaneous" BSE could result in only one or two cases per year, which would have almost no consequences for the human population, given current regulations. A spontaneous BSE case might pass undetected into the food chain because of the length of the incubation period. This could happen by the consumption of risk tissues such as brains, the use of MRM in the manufacture of meat preparations, or the spread of the infectious agent over assumed risk-free tissues. MRM is still used up to a concentration of 30 percent in the manufacture of hot dogs, sausages, soups, and stews, among other meat preparations.

Spongiform encephalopathies are present in the United States in sheep (scrapie), minks (transmissible mink encephalopathy, TME), and in deer and elk (chronic wasting disease, CWD). Debate exists regarding the risk posed by pigs and poultry, since these species are considered free of TSEs, or at least very unlikely to become infected. Until 1997, mammalian protein could be added to cattle feeds, so the same window that allowed the infectious agent to pass from an infected sheep tissue to a bovine in the United Kingdom could have occurred in the United States. Indeed, rendering processes in the United States are not required to reach the BSE-agent sterilization levels that are required in Europe. In fact, cross-contamination could have occurred in any country where MBM was manufactured with sheep tissues infected with scrapie and subsequently fed to cattle. However, the probabilities that BSE would spread elsewhere as it did in United Kingdom are considered low because the ratio of sheep to cattle and the use MBM in cattle feed are considered the two main factors specific to the United Kingdom's cattle production. It is believed that cross-species transmission will not occur, so the fact that scrapie or CWD is found in the United States does not increase the likelihood of BSE occurrence. Even in the case that exposure to CWD could cause BSE, the species barrier between these two TSEs is considered impenetrable (Cohen et al. 2001). The species barrier between cattle and minks seems less strong, but current FDA regulations prohibit the use of mink meat in the manufacture of cattle feed (Cohen et al. 2001).

Imports of live ruminants and most ruminant products from countries known to have BSE have been prohibited since 1989. Cohen et al. (2001) estimated that in a hypothetical scenario where 10 infected animals were to be imported into the United States, only three new cases of BSE would be discovered, and the disease would be eradicated in 20 years, provided the conditions for dissemination remain unchanged. For instance, MBM's use as a cattle feed has been banned since August 1997, and the import of all European rendering wastes and rendered protein has been banned since December 1997. Thus, even if an animal with BSE is eventually imported and ends up in a rendering plant, the disease should not disseminate by means of infected MBM in cattle feeds. However, human error or negligence is always possible.

The risk of importing infected animals is also considered negligible, although imports of live animals have been allowed from countries that have been considered BSE free in the past, but that have now had cases of BSE. For example, the United States imported 242 bovines from Japan between 1993 and 1999 (GAO 2002), though Japan discovered a case of BSE in September 2001. The use of MBM in Japan was allowed until a ban was imposed in September 2001. There is a need for improvement of the tracking system, or traceability, that would improve the ability to quickly locate and dispose of any imported animal deemed to pose a risk. In fact, some of the animals imported in the last 15 years have not been found (GAO 2002).

Recycled materials used as cattle feed could open a route for infection, although they are all considered very low risk. The most important are plate waste, milk products, blood and blood products, and mammalian protein of porcine or equine origin. Ninety percent of plate waste is bakery products, and only a small fraction is meat products (Cohen et al. 2001). It is still possible to consume brains and other risky tissues in the United States, and the use of wastes from restaurants is exempted from the mammal-to-ruminant feed ban. Although the volume of risky materials from plate waste might be considered negligible, if restaurants serve what are considered risk materials, some control over the wastes should be carried out to reduce risks. There is no evidence that any TSE can be transmitted through the consumption of milk or milk products. However, in the EU, milk from BSE-affected cows is not allowed to enter the food chain.

Open Questions

The United Kingdom's experience has created considerable concern in other countries, particularly since the United Kingdom exported BSE-infected MBM and live animals to many other countries before the problem was fully recognized and exports stopped. Numerous other countries, in Europe and elsewhere, have recently encountered cases of BSE. Although the likelihood is very low, an outbreak still could occur in the United States, and it is worth considering the likely economic consequences of such an outbreak.

The principal effect of a BSE outbreak that is small in magnitude and quickly eliminated is that the United States would lose its designation as a BSE-free country, at least for some period of time. It is conceivable that importing countries would either officially restrict beef imports from the United States or, even if they did not, that foreign consumer demand for our beef would significantly decline. This could be an important loss for the U.S. beef industry. Exports of cattle, fresh beef, offal, meat preparations, and by-products such as fats and tallow amount to $4 billion annually. Plus, there are a number of food preparations that contain bovine meat (usually from MRM), and exports of these products may be estimated as near $500 million. The United States is the world's largest beef exporter, accounting for about 15 percent of all beef traded worldwide.[4] A disruption of U.S. exports would affect the world market, altering prices and trade flows. Importers would most likely suspend all their purchases from the United States until they were convinced that its beef products were safe. If the outbreak was rapidly controlled, trade in beef, particularly fresh boneless beef, might resume. Importers, however, would likely be reluctant to buy risky products, such as brains or by-products that contain nervous tissue or food preparations made of MRM, and uncertainty would remain regarding the origin of the infection and whether there were more cases that had passed undetected into the food chain.

Alternatively, if a case of BSE resulted in the removal of SBO and an inability to sell rendered products, the economic loss would be approximately 3.5 percent of cattle receipts, or approximately $1.4 billion annually.[5] Thus, although a minor outbreak of BSE almost certainly would not cause catastrophic loss for the U.S. beef industry, it could cause a significant economic loss and a restructuring of parts of the industry.

The potential loss from BSE suggests two recommendations. First, United States and foreign consumers of U.S. beef should be continually informed of the measures being taken to protect and monitor the U.S. beef herd and to prevent any BSE-infected beef from entering the human food chain. These efforts ought to extend to the governments of countries importing U.S. beef. Such information should help to prevent panic and a major loss of consumer demand in the event that a case of BSE occurs. Second, although much has been done to protect the United States from BSE, there is still much scientific uncertainty regarding the origins of BSE and its modes of transmission, both among animals and from cattle to humans. The United States should support research that seeks to unravel these secrets, and it should continue to refine its policies, particularly regarding cattle slaughter and processing, to ensure that these reflect current scientific information.

Although the probability of a BSE outbreak in the United States is thought to be very small, it is hard to be completely sanguine because many uncertainties remain regarding the origin and dissemination of BSE among cattle. Furthermore, several avenues exist by which BSE could still contaminate the food chain. For example, central nervous system tissues are considered the most risky material, but the stunning methods used in slaughterhouses could result in infected tissues that are not part of the central nervous system. Stunning is carried out using compressed-air guns that send a bolt into the brain of the cow being slaughtered. The force of the bolt can cause the spraying of infectious brain tissues into the blood stream and thus contaminate hearts, lungs, and livers. The hole in the animal's head may also become a way of contamination by dripping of infectious material onto other parts of the carcass. Similarly, although experimental data suggest a very small probability of maternal transmission of BSE (Donnelly et al. 1997; European Commission 2002), a number of the animals diagnosed with BSE in the United Kingdom were born after the extended prohibition of 1996 (FSA 2000). It is unclear how these animals became infected.

Human exposure also depends on the level of compliance with current regulations that govern slaughter and rendering plants, the disposal of animals that die on farms, and the treatment of restaurant wastes and other by-products. The effectiveness of the feed ban needs evaluation, considering the current problems with the enforcement system (GAO 2002). Furthermore, the effectiveness of the feed ban depends on how feedstuffs are labeled and how they are handled at the farm level. Mislabeling and/or supplying prohibited feed to cattle can dilute the effectiveness of the feed ban (Cohen et al. 2001). Labeling imported products poses another set of questions. A product may come from a BSE-free country, but this country could have imported the raw materials from a BSE-infected country. Fortunately, the United States is not a significant importer of MBMs.

Conclusions

The economic impact of BSE on the United Kingdom was large, partly because of the fear created among domestic consumers that stressed the long-term decline in demand for beef in the United Kingdom, but especially because beef exports ended almost entirely. The prices of beef and live animals fell sharply, and the distribution and marketing chain in the United Kingdom was severely damaged. The major cost has been the loss of exports and the increase in producer subsidies, including compensation schemes paid for animals that are banned from entering the food chain. The cost of the BSE crisis in the United Kingdom since 1986 is probably close to $7 billion,[6] more than twice the value of the annual production of the cattle sector in the United Kingdom. As a result of that experience, considerable concern has been expressed that BSE could spread to the United States, with a similar effect on its beef industry and human population.

We conclude, as did the Harvard study, that the risk of a BSE outbreak in the United States is now low. Similarly, were an outbreak to occur, we believe that it could be contained without significant spread into the human food chain. Accordingly, it would appear that the economic risk posed by a BSE outbreak in the United States is much smaller than that which was witnessed in the United Kingdom. It is smaller because it appears that the transmission of BSE from one herd to another is relatively easy to control and that, with appropriate measures regarding the treatment of slaughtered and dead animals, the transmission from infected animals to humans is unlikely.

Nevertheless, even when the risks of BSE infection and transmission are believed to be small, a single case of BSE could cause significant damage to the U.S. cattle industry if the fear resulting from such a case were to cause a reduction in either domestic or foreign demand for U.S. beef and/or animals. Foreign consumers, who would have to pay higher costs for beef from other countries, would also suffer significant losses. The United States should thus strive to ensure that domestic and foreign consumers (governments) can correctly believe that the risk of BSE dissemination throughout the U.S. beef herd and, thus, into the human food supply, is minimal. If that is believed, and if their consumption decisions are based on that belief, the damage occurring from an outbreak, should it occur, will be much reduced. At this time, there is scientific support for such a view. Additional research must be continued to ensure that the view is not taken for granted.

Notes

[1] Formerly Ministry of Agriculture, Fisheries and Food.

[2] Official statistics do not account for the whole cost of the campaign; e.g., the disposal of cattle and administrative costs are not disclosed for the period 1996-2001. Assuming a similar cost structure as during the previous period, total cost may have reached £4.6 billion already.

[3] Excluding Selective Cull expenditures. The Selective Cull program is not considered a subsidy.

[4] Dollar value, only fresh and processed beef.

[5] Value of specified offal and recovered meat, estimated as a percentage of the value of cattle, at the wholesale level.

[6] There is no official, comprehensive cost estimate. The estimate provided does not account for human mortality.

References

Ardans, A., A. Thurmond, D. Klingborg, and L.S. Jarvis. 1999. "Case study: BSE." In Raymond H. Coppock and Marcia Kreith, Eds., *Exotic Pests and Diseases, Biology, Economics and Public Policy*. UC-AIC. pp. 201–206.

Brown, David R. 2001. "BSE Did Not Cause vCJD: An Alternative Cause Related to Post-Industrial Environmental Contamination." *Medical Hypothesis.* 57(5):555–560.

Brown, Paul. 2001. "AFTERTHOUGHTS about BSE and vCJD." *CDC-Emerging Infectious Diseases.* 7(3). Commentary. U.S. Dept. of Health. Downloaded from http://www.cdc.gov on 9/21/01.

Brown, Paul, Robert G. Will, Raymond Bradley, David M. Asher, and Linda Detwiler. 2001. "BSE and vCJD: Background, Evolution, and Current Concerns." *CDC-Emerging Infectious Diseases.* 7(1). U.S. Dept. of Health. Downloaded from http://www.cdc.gov on 9/21/01.

Burton, Michael, and Trevor Young. 1997. "Measuring Meat Consumers' Response to the Perceived Risks of BSE in Great Britain." *Risk Decision and Policy.* 2(1):19–28.

Byrne, David. 2001. "The Importance of Having Comprehensive Safety Legislation on BSE in Place." European Commission, Meeting of the

Ministers of Agriculture of the CEFTA Countries, Bratislava, September 2001.

Chesebro, Bruce. 1998. "BSE and Prions: Uncertainties About the Agent." *Science.* 279:42–43.

Cohen, Joshua T., Keith Duggar, George M. Gray, and Silvia Krendel. "Evaluation of the Potential for BSE in the United States." Harvard Center for Risk Analysis/Harvard School of Public Health. USDA/APHIS, November 2001.

DEFRA. 2001. "BSE Progress Report." Department of Environment, Food and Rural Affairs, London, June 2001. Downloaded October 2001 from http://www.defra.gov.uk/animalh/bse, publications.

DEFRA. 2002. "Economics & Statistics." Department of Environment, Food and Rural Affairs. Downloaded May 2002 from http://www.defra.gov.uk/esg/default.htm.

Donnelly, C.A., S.M. Gore, R.N. Curnow, and J.W. Wilesmith. 1997. "The Bovine Spongiform Encephalopathy Maternal Cohorty Study: Its Purpose and Findings." *Applied Statistics.* 46(3):299–304.

DTZ-Pieda. 1998. "Economic Impact of BSE on the U.K. Economy." DTZ-Pieda Consulting, London, United Kingdom.

European Commission 2001. "Report on Implementation of Latest BSE Control Measures in Member States." Health and Consumer Protection Directorate-General, European Commission Press Release. Downloaded in February 2001 from http://europa.eu.int/comm/dgs/health_consumer/library/press/press104_en.html.

European Commission. 2002. "The Safety of Bovine Embryos." Scientific Steering Committee, Health and Consumer Protection Directorate-General, European Commission, May 2002.

FAO-STAT. 2002. "Statistical Database." Food & Agriculture Organization. Downloaded May 2002 from http://apps.fao.org.

FSA. 2000. "Review of BSE Controls. Final Report." Foods Standards Agency, U.K. Government, December 2000. Downloaded on June 2002 from http://www.bsereview.org.uk/data/report.htm.

GAO. 2002. "Mad Cow Disease: Improvements in the Animal Feed Ban and Other Regulatory Areas Would Strengthen the U.S. Prevention Efforts." Report to Congressional Requesters. GAO-02-183. Washington, D.C.: U.S. General Accounting Office. January 2002.

Horn, Gabriel, Martin Bobrow, Moira Bruce, Michael Goedert, Angela McLean, and John Webster. 2001. "Review of the Origin of BSE." DEFRA Report. Downloaded October 2001 from www.defra.gov.uk/animalh/bse.

Lloyd, T., S. McCorriston, C.W. Morgan, and A.J. Rayner. 2001. "The Impact of Food Scares on Beef and Inter-Related Meat Markets." AAEA Annual Meeting Selected Paper. Chicago, IL.

OIE. 2002. "Online Database." Office International des Epizooties. Downloaded June 2002 from http://www.oie.int.

Phillips, Lord, June Bridgeman, and Malcolm Ferguson-Smith. 2000. "The BSE Inquiry Report." U.K. Government, October 2000. Downloaded in February 2001 from www.bse.org.uk.

Prusiner, Stanley. 1995. "The Prion Diseases." *Scientific American.* 272(1):48–57.

Verbeke Wim, Ronald W. Ward, and Jacques Viaene. 2000. "Probit Analysis of Fresh Meat Consumption in Belgium: Exploring BSE and Television Communication Impact." *Agribusiness: An International Journal.* 16(2):215–234.

Waterhouse, Rosie. 2001. "British Firm Linked to Global BSE." *Sunday Times*, February 4.

Wilesmith, John, and J.B.M. Ryan. 1997. "Absence of BSE in the Offspring of Pedigree Suckler Cows Affected by BSE in Great Britain." *Veterinary Record.* 141:250–251.

7

Evaluating the Potential Impact of a Foot-and-Mouth Disease Outbreak

Javier Ekboir, Lovell S. Jarvis, and José E. Bervejillo

Introduction

An outbreak of an exotic animal disease such as foot-and-mouth (FMD) can cause major economic losses in a previously unexposed population. To prevent introduction of highly contagious exotic animal diseases, the United States uses trade restrictions as well as border controls on travelers coming from infected countries or regions. However, these measures do not constitute a perfect shield against possible outbreaks. During 2001, a major FMD outbreak occurred in Great Britain, and minor outbreaks occurred in France and the Netherlands. Argentina and Uruguay, who had recently eradicated FMD, were reinfected. The introduction of FMD into California's livestock population is a real threat.

FMD is probably the most contagious of all animal diseases. Since FMD does not affect humans, its consequences are only economic, yet an outbreak can still have devastating consequences. Constant monitoring and surveillance, rapid diagnosis and preparedness for control and eradication are required to minimize the *probability of occurrence* and, in case it happens, the *cost of an outbreak*. The probability of occurrence of an FMD outbreak has changed in recent years, since new potential routes of entry have developed. Historically, it was assumed that the importation of animals and animal products was the most likely source of infection. However, import regulations and border controls, as well as FMD eradication in neighboring countries, have reduced this risk to negligible levels. However, there is probably an increased risk of FMD introduction from travelers coming from countries with FMD, smuggling of infected animal products by such travelers, the disposal of garbage transported in planes and ships, and from ecoterrorism. These alternative paths should be taken into consideration in establishing an appropriate state and national FMD policy.

The cost of an outbreak would come from three different sources: (1) control and eradication efforts, including the vaccination of or the slaughtering of infected or exposed animals, compensation for destroyed animals, cleaning and disinfecting of infected premises, and quarantine enforcement; (2) losses that arise from the interruption of production processes and its effects on upward and downward linkages; and (3) changes in trade flows that result from import restrictions that would be imposed by FMD-free trading partners having zero tolerance for the disease.

Epidemiology of FMD

The FMD virus is an *Aphtovirus* within the Picornaviridae family. The most important characteristics in the epidemiology of the disease include the rapid growth of the virus, its stability under a variety of conditions and the occurrence of serotypes (Donaldson 1991). There are seven serotypes and several subtypes within each. The infections caused by different serotypes are clinically indistinguishable. The animals that survive an FMD infection become permanently immune to the particular strain that caused the infection; however, there is no cross-protection between stereotypes.

FMD rarely affects humans,[1] but rather attacks all cloven-hoofed animals. In the United States these animals include cattle, sheep, goats, pigs, camels, deer, and bison. Cattle, in particular, are important in the epidemiology of FMD because of their high susceptibility to airborne virus, because they may excrete the virus for at least four days before the first symptoms appear, and because of their economic importance. Even though sheep and goats can also be infected, their symptoms are often less severe or are subclinical. Pigs are the most important source of air dissemination of the virus; once infected, they excrete vast quantities of the virus. They also have a high susceptibility to infection by the oral route (Donaldson and Doel 1994). Thus, pigs can be described as amplifying hosts and cattle as indicators. Sheep can be described as maintenance hosts because they quite often have mild or unapparent signs that can easily be missed (Donaldson 1994). Despite its infectiousness, FMD may infect some susceptible species and spare others in the same area (Dunn and Donaldson 1997).

The primary methods of FMD transmission are aerosols, direct and indirect contacts with infected animals, and ingestion. It is generally accepted that the virus most commonly infects by way of the respiratory route, especially in ruminant species where very small doses can initiate infection (Donaldson 1994). Of all FMD transmission mechanisms, movements of infected animals are by far the most important, followed by movements of contaminated animal products (Donaldson 1994). Humans may inhale and harbor the virus in the respiratory tract for as long as 24 hours and may serve as a source of infection to animals (APHIS 1991). Trucks and feed products can transport the virus after entering an infected farm. Garbage containing uncooked meat scraps and bones from infected animals has been a source of infection in pigs.

Virus excretion occurs before infected animals manifest clinical signs. With natural routes and high-exposure doses, the incubation period can be as short as two to three days, but can take up to fourteen days (Donaldson 1994). The airborne virus is emitted over a four- to five-day period by an infected animal, and excretion of the virus may start up to four days before the onset of the first clinical signs. Airborne virus is believed to have spread 60 km over land and 200 km over sea (Moutou and Durand 1994; Donaldson 1991). Factors favoring airborne spread of FMD virus are low to moderate wind speed, high humidity, stable atmospheric conditions, particularly a temperature inversion, absence of heavy precipitation,

and high stocking density of cattle downwind (Donaldson 1994).

Apart from the respiratory route, less frequent routes of infection are breaks in an animal's integument, i.e., the skin or mucous membrane. Thus, the injection of faulty FMD vaccines, foot-rot in sheep, the feeding of rough fodder, harsh use of milking machines, surgical procedures, and damage caused by fingernails during nose restraint of cattle can all provide entry points for the virus. Veterinarians and artificial insemination technicians can be very important vectors in the early phases of an outbreak.

During the acute phase of the disease, which generally lasts three to four days, all excretions, secretions, and tissues contain virus. Animals in this condition are very potent spreaders of virus. Products made from such animals contain high quantities of virus, and many products must be decontaminated or destroyed. However, matured and deboned meat has been shown to be free from the virus, which is inactivated by the drop in pH during rigor mortis. The virus survives in the bone marrow and lymph nodes (Donaldson and Doel 1994).

FMD infection results in low mortality for adult livestock that usually does not exceed 5 percent. Mortality is much higher in young animals, especially under conditions of dense stocking, reaching up to 90 percent (Donaldson 1994). In addition, infected adults lose weight because they stop eating, cows lose their heat, and milk production drops considerably. After a relatively short period (between two to three weeks), most infected adult animals recover from the lesions and become productive again, unless secondary bacterial infection occurs. In some cases, a permanent reduction in productivity has been observed. Tongue lesions in pigs are much less dramatic and heal much more rapidly (Donaldson 1991). Losses are lower for herds in which FMD is endemic as a result of building up resistance. However, total production losses in herds with endemic FMD can amount to 10 percent of annual output.

After recovery from FMD, up to 80 percent of ruminant animals may become persistently infected. It is believed that these carriers can initiate fresh outbreaks when brought into contact with fully susceptible animals. Vaccinated or immune animals exposed to infection may also become carriers. The duration of the carrier state varies according to the species involved, the strain of the virus, and probably other unidentified factors. The maximum recorded periods that infected animals of different species have acted as carriers are over three years for cattle, nine months for sheep, four months for goats, five years for the African buffalo, and two months for water buffalo (Donaldson 1994).

The latest epidemics in Taiwan, Italy, and the UK illustrate that the FMD virus is extremely contagious. In the first case, FMD was detected in a farm on March 14, 1997. New cases were reported on March 17 and 18. On March 20 the government imposed restrictions on the movements of animals, after receiving the results from the tests that confirmed the presence of FMD. By this time the disease was present in 28 farms, and one week later it had been confirmed in 217 farms. The number of herds infected reached 1,113 in the fifth week. Mass vaccination was started on March 29. Eventually, depopulation comprised more than 4 million pigs, about 38 percent of Taiwan's inventory of pigs. The eradication campaign was unable to keep pace with the dissemination of the disease (Yang et al. 1999).

On March 11, 1993, a premise in southern Italy was identified as infected, and it was reported that a truck with cattle had left the premise on March 3. The infection was confirmed at the truck's destination in the northern province of Verona. A private veterinarian noticed FMD symptoms in a group of calves and ordered the appropriate diagnostic tests, but the veterinarian then continued with his daily routine through a number of other farms in the zone. No further action was taken, except for monitoring of that farm, until the test results confirmed the presence of FMD virus a week later. By that time, several other farms visited by the veterinarian on the day the symptoms had been discovered were reporting symptoms of FMD on their herds. Three more outbreaks were confirmed in this region between March 11 and March 27. The spread of the disease was attributed to the veterinarian, a beet delivery truck, and air dissemination. Stamping out was applied on infected animals and those considered "dangerous contacts." Nine hundred animals were sacrificed, and, officially, no vaccine was used. Depopulation of the 900 animals required three days. Immediately after confirmation of the

outbreak, a 3- to 5-km radius protection zone and a 10-km radius surveillance zone were established. A spatial simulation model was used to determine the probability and direction of airborne dissemination. When the simulations showed that the risk of airborne infection was low, daily control of the herds was restricted to the protection zone, because the available resources were not enough to monitor the surveillance zone. The area was declared FMD free on May 1 (Maragon et al. 1994; Kitching 1998).

On February 20, 2001, an outbreak of FMD was confirmed in an abattoir in Essex, England. On the same day, an outbreak was confirmed on a neighboring farm. Two days later, an outbreak was confirmed on another neighboring farm, and three days later the disease was confirmed on a sow-fattening unit. In subsequent weeks, outbreaks were confirmed all over the country, stemming largely from movements of sheep and some cattle through major markets. Within three weeks, more than 250 outbreaks had been confirmed and more than a million livestock (mainly sheep) had been slaughtered or marked for slaughter. Outbreaks subsequently occurred in France and the Netherlands, related to the export of sheep from Britain (Harvey 2001). The epidemic in the United Kingdom was controlled by October 2001, and the United Kingdom was declared again free of the disease in January 2002, 11 months after the first case was detected. The crisis affected 10,500 farms, and ultimately resulted in the slaughter of 6.1 million animals, 4.1 million under the disease control campaign, and 2 million under welfare programs, out of a total inventory of about 60 million animals (sheep, cattle, pigs, goats, and deer). Most of the animals slaughtered were sheep (almost 5 million) and cattle (761,000). The direct cost of compensation schemes,[2] during the year 2001, reached $1.8 billion (DEFRA 2001). The total economic cost of the FMD outbreak goes beyond the compensations paid to farmers, because it affects livestock-marketing firms, the rural tourism sector, and related industries. Nonetheless, Harvey suggests that the net loss to the United Kingdom economy will be partly offset because most of the losses to the rural tourism sector will be compensated as consumers spend elsewhere in the economy. England exports only a small amount of livestock products, so the loss of international markets is a relatively small factor.[3]

These cases illustrate that several factors can affect the spread of the disease:

- Weather influences airborne dispersion.
- Animal density affects dissemination. Larger infected herds shed more virus into the air, and animals living in cramped conditions are more susceptible to infections since they are more stressed.
- Husbandry methods crucially affect the rate and extent of disease spread, and, in the case of an outbreak, they are a key consideration for control and eradication strategies. Because backyard operations usually undergo little sanitary surveillance, it is difficult to identify an infection in its early stages if it occurs in such an establishment. However, large-scale livestock operations involve frequent movements of animals and people that favor rapid spread of the outbreak. Service trucks and people enter and exit several farms per day; every week, milking cows are replaced, and young animals are sent to stockers or feedlots all over the country.
- Once an outbreak occurs, its consequences depend on how fast the disease is identified and quarantine imposed, the effectiveness of the quarantine, whether animal movements can be traced, the availability of funds for depopulation and cleaning infected premises, and how rapidly depopulation can be carried out.

Control and Eradication Policies

Our analysis indicates that the cost of an FMD outbreak in California would likely be higher than the outbreaks that have occurred in Taiwan, Italy, or England, mainly because the outbreak would have its greatest effect on the California dairy sector. California's dairies are large, with herds of 1,500 not uncommon, often are closely colocated, and each dairy is visited daily by numerous contacts. Hired laborers come to work and leave, milk is picked up, feed is delivered, manure is often removed for disposal elsewhere, and veterinarians arrive to treat animals having common illnesses. Frequently, visitors to one dairy later visit another neighboring dairy. Thus, density and high-input/high-output technology create conditions conducive to the rapid spread of FMD if it is introduced.

Our analysis indicates that the time required to diagnosis FMD and initiate a stamping-out policy would be the most important factor in determining the outbreak's ultimate effect in California. Because the clinical signs of FMD are similar to those of other vesicular diseases present in the United States, and because FMD has been absent from the United States for more than seven decades, it is possible that the first cases of FMD will not be properly diagnosed. In addition, since any farm on which a vesicular disease is detected must go through a costly quarantine period, farmers may choose not to report the symptoms, hoping that it is not FMD and that it will pass promptly. Once FMD has appeared, at least four alternative control and eradication strategies are available:

1. Total stamping-out (i.e., depopulation of all symptomatic and apparently healthy animals that have been exposed, directly or indirectly, to the virus); depopulation can be complemented with ring vaccination, especially in densely populated areas;

2. Partial stamping-out (i.e., depopulation of only symptomatic animals) with early or late ring vaccination;

3. Partial stamping-out without vaccination; and

4. Eradication through vaccination only.

Total stamping out is the current U.S. strategy and thus the policy that would be implemented if an outbreak should occur in California (APHIS 1991). An outbreak could require depopulating California's entire cattle herd if not controlled effectively. Alternative policies could be a more economical way of dealing with an outbreak (Garner and Lack 1995). If it were known in advance that this result was probable, the state might find it more economical to vaccinate the entire herd and quarantine movements with the rest of the United States. Stamping out would then be applied only to animals that are clearly infected. This approach would result in depopulating many fewer animals and would thus maintain livestock production at a higher level in the years immediately following the outbreak. After two years of no visible outbreaks, vaccination would cease, and after two more years of production with no FMD outbreaks, California would be able to export beef again. However, the conditions under which alternative policies would be preferable should be evaluated in advance because once an outbreak has occurred, eradication strategies are largely irreversible.

The feasibility of stamping-out depends on the number of animals to be depopulated, because the costs and resources required for rapid depopulation escalate very fast. Vaccination could be used if stamping out becomes unfeasible, but under the present guidelines this would only be known after a substantial number of animals have been slaughtered.[4] Given the production conditions prevailing in California and the United States, the threshold above which stamping out is no longer the best policy is not known. Nonetheless, given that the United States is the world's second largest beef exporter and that the U.S. beef exports would cease so long as the presence of FMD in the United States was a threat to our trading partners, there is a strong incentive to eliminate FMD quickly. Thus, given the large economic losses due to the interruption of U.S. beef exports, it seems probable that stamping out is the preferable U.S. policy, even if it were not the most attractive policy for California, acting alone.

Estimating the Cost of a FMD Outbreak in California's South Valley

The expected cost of an outbreak is defined as the estimated cost of the outbreak multiplied by the probability of occurrence. The FMD virus could be introduced into California through a number of routes, and the risks of each of these routes are not well understood. Such risks have changed over time, and there are no up-to-date estimates of risk levels. It is important to analyze these risks since an understanding of their magnitudes is important to the appropriate design of an FMD control policy. This research, however, is concerned only with estimating the cost of an outbreak, should one occur in California.

The Model

A FMD outbreak in California's South Valley and, eventually, its dissemination throughout the entire San Joaquin and Chino valleys, were

simulated with an epidemiological model.[5] The simulation results provided inputs for an economic model that were then used to estimate the outbreak's costs.[6]

The epidemiological model is a random state-transition model developed from a Markov chain. Similar models have been used by several authors to simulate FMD outbreaks (Miller 1979; Dijkhuizen 1989; Berentsen et al. 1992; Garner and Lack 1995). Because the potential behavior of an FMD outbreak under current production conditions in California is unknown, the model's parameters were based on (1) a review of production conditions in the South Valley, (2) overseas experiences, and (3) expert opinions.

The state-transition model has two components: states and transition probabilities. The *states* are different disease-related herd categories, i.e., susceptible to the disease, latent, infected with the disease, or dead as a result of depopulation (slaughter). A *transition* probability is the probability that an individual herd will move to state j in the next period when it is presently in state i. These probabilities, and consequently the number of susceptible, latent, and depopulated herds that the model will simulate in each period, depend on production and environmental conditions and on the control strategies used.

Stamping out is currently the preferred option in dealing with an FMD outbreak in the United States and California. It requires banning all movements of susceptible animals that might have been exposed to the virus for about two weeks;[7] prompt and rigid control of the movements of animals and animal products, vehicles, equipment, and people in a surveillance area around any outbreak area; the rapid depopulation and the cleaning and disinfecting of infected or exposed premises; and intense surveilance of suspected herds. The efficiency of this policy depends on the timely availability of sufficient human, physical, and financial resources. If the policy cannot be quickly implemented with a high degree of efficiency, the final eradication cost may be higher than if alternative policies are implemented. Study of alternative policies, however, is beyond this chapter's goal.

The model estimates the number of latent infections as well as the number of infected premises. Since it is expected that the logistics associated with the depopulation of dangerous contacts could be a major constraint on the eradication of an outbreak in California, the transition state "latent to infectious" was introduced to explicitly explore the consequences of partial stamping out and of beginning depopulation at different stages of the epidemic. The results show that the extent and duration of an epidemic depend on (1) the delay between infection and diagnosis of the disease, (2) the type of control strategy applied, (3) the availability of human and financial resources, and (4) the effectiveness of animal health authorities in executing the eradication policies.

Given the high density of susceptible animals in California's South Valley and the intensive technologies used by farming establishments in the area, the transition probabilities used in this study were higher than those used in similar models developed for use in other contexts. Seven scenarios were simulated. Scenarios 1–4 use high dissemination rates that reflect the information collected in the South Valley, while scenarios 5–7 use lower dissemination rates taken from cases in other countries, as reported in the literature. Still, the lower dissemination rates used here are equal to the highest rates used in other studies. All dissemination rates were allowed to change randomly up to 30 percent in any direction during each simulation. The model was run 100 times, and the mean results for each scenario are reported.

The model incorporates three major epidemiological assumptions: (1) the outbreak is successfully contained within California's borders; (2) the disease is eradicated in a limited period of time—in other words, it does not become endemic; and (3) the outbreak is a one-time event, i.e. the disease is completely eradicated after the occurrence of one outbreak. We believe that these assumptions are optimistic. An FMD outbreak would likely spread to neighboring U.S. states and Mexico. More than one outbreak would probably occur, and FMD might well become endemic for some period of time. The results are thus a lower bound of the expected cost of an outbreak.

The economic model estimates three cost components: (1) the direct cost of depopulation, cleaning and disinfecting, and quarantine enforcement; (2) the cost of the direct, indirect, and induced losses caused by the outbreak, es-

timated with an input-output model of the California economy; and (3) the cost of the losses caused by trade restrictions. The production losses include only those relating to cattle, dairy, pigs, and directly related industries. Losses in other livestock industries, in wildlife, and in outdoor activities are not included. Similarly, economic losses of other types caused by the imposition of quarantine are also not included.

Results

The seven scenarios are presented in Table 7.1. Each scenario is characterized by a set of assumptions regarding the FMD dissemination rate, the time at which depopulation begins, and the effectiveness of depopulation of both infected and latent animals. The simulation results from the assumptions characterizing each scenario are provided in terms of the percentage of California's herds and the total number of animals that are ultimately destroyed to bring the outbreak under control.

The simulations show that the key factor determining the effect of an FMD outbreak is the rapidity with which depopulation begins after an outbreak occurs. Given the conditions prevailing in California's South Valley, a delay of one week can be decisive. Even when the dissemination rates are high, as they are expected to be in California, early intervention combined with high efficiency in identifying and depopulating herds containing animals with latent infections can substantially reduce the magnitude of an epidemic.[8]

Scenario 1 represents the worst possible set of assumptions considered in this study. Dissemination of the disease is rapid, depopulation does not begin until the third week,[9] 90 percent of infected animals are depopulated, but latent animals are not depopulated. This scenario results in the ultimate depopulation of all of California's animals. Scenarios 2–4 demonstrate that California could significantly reduce the number of animals depopulated if it is able to identify an outbreak quickly and initiate an effective control/depopulation strategy. Note, however, that depopulating latent animals has relatively little effect if depopulation begins only in week 3 and the rate of depopulation remains at only 90 percent; see scenario 1. However, if depopulation begins more quickly, and

Table 7.1 Simulation results: percentage of the South Valley herds destroyed under seven different scenarios

Scenario	Dissemination rate	Depopulation per week (%)		Depopulation starts @ week no.	Total herds destroyed (%)	Total animals destroyed (1,000's)
		Latent	Infectious			
1	High	none	90	3.0	100	808
2	High	90	90	3.0	93	750
3	High	95	95	2.5	24	200
4	High	95	95	2.0	19	151
5	Low	none	50	3.0	97	792
6	Low	none	90	3.0	87	698
7	Low	90	95	3.0	26	210

Source: Ekboir (1999a).

the depopulation rates are raised from 90 percent to 95 percent, the effect of the outbreak is reduced significantly (in terms of the total animals affected).

Scenarios 5–7 use lower dissemination rates. With lower dissemination rates, depopulation can begin in the third week and still experience success. Nonetheless, containment of an epidemic requires the rapid depopulation of dangerous contacts. For example, if infectious herds are depopulated slowly, at 50 percent per week, and latent animals are not depopulated, an outbreak still leads to the loss of 97 percent of herds even when the dissemination of the disease is slower (scenario 5). Indeed, even if the depopulation rate of infectious herds is raised to 90 percent per week, but latent animals are not depopulated, 87 percent of herds are ultimately lost (scenario 6). However, when 90 percent of latent infections are removed, as well as 95 percent of infectious herds (scenario 7), only 26 percent of herds is lost.

A key factor affecting the efficiency of eradication policies is the actual value of the dissemination rates. If dissemination rates are low, the stamping-out policy can be started later in the outbreak without disastrous consequences. If the dissemination rates are high—which is more likely in California—and if depopulation starts late, the stamping-out policy may require depopulation of all herds in the affected region. If that were known in advance, the adoption of an alternative strategy, e.g., ring vaccination combined with a slower depopulation rate, might result in a lower economic loss. In any case, it is clear from the simulations that, regardless of the dissemination rates, a high degree of preparedness, including the timely

availability of financial resources, and the ability to act decisively are necessary conditions for containment of the epidemic.

Table 7.2 shows the total, direct and indirect costs of the outbreak of each of scenarios 1–7, plus one additional scenario in which the outbreak spreads to the entire San Joaquin Valley and Chino Valley. The total cost due to the FMD outbreak in California is equal to the sum of the direct, indirect, and induced output losses, plus the cost of cleaning and disinfecting and enforcing the quarantine, plus the losses due to trade restrictions. The direct, indirect, and induced output losses were estimated as the direct output loss multiplied by the corresponding output multipliers from an economic impact assessment modeling system known as IMPLAN developed by the Minnesota IMPLAN Group, Inc. In addition to the output losses, a FMD outbreak would trigger trade losses to both California and the United States; given the difficulties in estimating the beef exports originating in California, trade losses were estimated under the assumption that export restrictions would apply to the whole United States. If the United States could be effectively zoned, exports might continue from regions other than California, and the results shown would then be too high.

The trade effects include restrictions on all meats, skins, and dairy products originating in any state in the United States. The model assumed that trade restrictions are lifted two years after depopulation of the last infected or exposed herd and that U.S. exporters regain their market share in the FMD-free market immediately. This is a very optimistic scenario because it assumes that the cleaning and disinfecting efforts would be 100 percent effective in eliminating the virus from all infected premises and that other exporters would not permanently capture a portion of the U.S. share of the FMD-free market. The trade losses arise exclusively from a lower export price. It is assumed that exporters in other states are able to maintain the volume of exports they shipped before the outbreak. This assumption is unlikely, but follows the basic assumption that the outbreak is restricted to the South Valley. It is also assumed that California does not export any pork meat and that trade restrictions on pork meat are applied only by Japan and Korea.

The calculations are based on the following assumptions: (1) the quarantines are lifted 120 days after depopulation of the last infected or exposed premise; (2) depopulated farms return to production 60 days after depopulation of the last infected or exposed premise; (3) the supply of animals outside the infected region is large enough to repopulate the quarantined premises in a short period of time; (4) the price of cattle remains at the levels prevailing before the outbreak; (5) dairies start selling milk immediately after the quarantines are lifted; (6) dairies that are not depopulated sell milk in the quarantine area without interruption at the same prices they received before the outbreak; (7) feedlots need 130 days after being repopulated to bring the animals to slaughter weight; and (8) hog facilities finish their animals in 40 days after the lifting of the quarantines. These assumptions are also considered optimistic.

Under scenario 1, the cost of an FMD outbreak is $8.5 billion, with $4.3 billion of this accruing to California and the rest to other U.S. states (the loss to other U.S. states equals column 4, less column 3). If the dissemination rates are high, a half-week delay in the start of depopulation increases the loss by $135 million (compare scenarios 3 and 4, column 5). A delay of seven days, combined with slower depopulation, increases the loss by $1.75 billion (scenarios 1 and 4, column 5). If the outbreak spreads to the entire San Joaquin Valley and Chino Valley, the loss increases by $5 billion over even the most pessimistic scenarios where the disease is confined to the South Valley (scenarios 1 and 8, column 5). The sharp increase in the estimated costs that would occur if the response to an outbreak of FMD is slowed or incomplete suggests that California might find it profitable to invest

in additional resources to monitor and respond to an FMD outbreak. However, any such judgment depends crucially on how such efforts might change the probabilities that an outbreak will occur, be monitored, and be controlled, as well as on the costs of changing these probabilities. By way of a simple example, if improved monitoring could reduce the cost of an outbreak by $1 billion, and if the probability of an outbreak occurring in any year were 1 in 10,000, California would find it profitable to spend up to about $140,000 more annually than it is currently spending to respond to an FMD outbreak.

Although additional analysis is needed to arrive at more reliable figures, the estimates obtained suggest that more attention should be paid to such preparation. For example, even in the most optimistic case (scenario 4), eradication of the outbreak would require depopulation and disposal of 149,000 cows and 2,183 pigs in the first two weeks of the eradication campaign. Past experiences indicate that it is almost impossible to develop so rapidly the capacity for depopulation and disposal.[10] If depopulation and disposal cannot be carried out as rapidly as assumed in scenario 4, it is probable that an outbreak would then spread still more rapidly throughout California and perhaps to other states with large livestock industries. Scenario 8 replicates scenario 1 (high dissemination rates, no depopulation of latent infections, and 90 percent of infectious herds eliminated each week) under the assumption that the outbreak affects the San Joaquin and Chino valleys as well as the South Valley. Total costs nearly double. Note that the costs associated with the imposition of international trade restrictions on U.S. beef are essentially constant from one scenario to another, since it is assumed that the presence of FMD anyplace in the United States will cause importers to restrict beef imports from the United States. The increase in costs brought about by the spread of the disease is almost wholly associated with the need to depopulate and dispose of a growing number of animals. Of course, if the outbreak spreads to the rest of the United States, the costs will be much larger than those shown in Table 7.2.

Policy Issues

The model simulated the economic losses caused by an FMD outbreak based on several

Table 7.2   Total costs of an FMD outbreak in California (million $)

Scenario	Direct costs[a]	Production losses[b]	CA trade losses[c]	Total U.S. trade loss[d]	Total costs[e]
1	1,428	990	1,871	6,098	8,516
2	1,345	920	1,871	6,101	8,365
3	545	251	1,871	6,107	6,903
4	476	190	1,871	6,104	6,768
5	1,462	1,239	1,984	6,282	8,983
6	1,320	1,056	1,969	6,253	8,630
7	560	259	1,871	6,113	6,934
8	4,819	2,613	1,871	6,098	13,531

[a]Direct costs include depopulation, including compensation for destroyed animals and materials, cleaning and disinfecting, and the quarantine cost.

[b]Includes direct production losses, plus indirect losses (reduced output in linked industries such as input suppliers, service providers, and milk and livestock buyers), and induced losses (a reduction in employment, sales, and consumption throughout the state's economy).

[c]State of California trade only.

[d]Trade losses from the entire United States, including California.

[e] (a) + (b) + (d) = e.

Source: Ekboir (1999a).

relatively optimistic assumptions, e.g., total quarantine effectiveness from the first day it is imposed, rapid depopulation and disposal of infected and exposed animals, comprehensive cleaning and disinfecting of infected premises to prevent future contamination, and containment of the outbreak within California. Thus, in these scenarios the disease does not become endemic, and production returns to pre-outbreak levels after a relatively short period of time. The success of a control and eradication campaign depends heavily on three factors: (1) the effectiveness of public surveillance programs that allow early identification of the disease; (2) preparedness of all personnel who would eventually be involved (veterinarians, cleaning crews, and law enforcement agents); and (3) timely availability of physical, human, and financial resources to enforce quarantines and achieve depopulation and disposal.

The effectiveness of the prevention measures depends importantly upon collective action. Producers must be willing to report the disease and cooperate fully with the disease control authority in depopulation, disposal, and decontamination. Producers are likely to be willing to cooperate only if they are adequately compensated for animals that must be destroyed and if they have faith that other producers will do the same. If other producers do not take measures to control the disease, it makes no sense for an individual farmer to depopulate his or her farm and repopulate later with unexposed animals.

The decisions taken by a single producer depend crucially on the measures taken by his neighbors. Policy, including producer education prior to and during an outbreak, is a crucial factor in determining the outcome. We turn now to the analysis of these policy issues.

Early Diagnosis

Early diagnosis constitutes the most important factor in reducing the total cost of an FMD outbreak. A major obstacle to early diagnosis was the low level of awareness among farmers and veterinarians (considering that FMD has been absent from the United States for more than 70 years) that existed prior to the recent FMD outbreak in Great Britain. That outbreak brought FMD to the forefront of world news and galvanized substantial private and public action in the United States, including California. California is much better prepared to deal with FMD today than it was a few years ago. Nonetheless, other vesicular diseases exist whose clinical symptoms are similar to those of FMD. Farmers might not recognize FMD in an initial outbreak or might delay reporting it. Indeed, the disease went unrecognized in Great Britain for several days (the disease is difficult to recognize in sheep). The law requires that any farm on which a vesicular disease is detected must be quarantined for a relatively long period. Because of the costs associated with the quarantine, farmers may decide not to report immediately a disease, assuming that it is not FMD and that the symptoms will disappear soon. Farmers need to be continually reminded of the danger from FMD and instructed regarding how to recognize its symptoms. Farmers also need to understand the immediate need to report the disease and establish total control over movements onto and off their farms in the event of an outbreak.

California also contains numerous backyard livestock operations that are less intensely managed than commercial operations, and the backyard operations often have weak biosecurity. Owners of backyard operations could try to sell a sick animal instead of calling a veterinarian, and, if they invited others to their farms or moved the animal, they could expand the outbreak before it was identified. An FMD outbreak in such an establishment is likely to be more difficult to identify and costly for public animal health programs to monitor.

Carcass Disposal

The feasibility of stamping-out policies depends on the number of affected animals that must be depopulated. Stamping out can only be implemented when the expected number of outbreaks and/or animals infected is relatively small (Donaldson 1994). The real bottleneck is the logistical and environmental problem of carcass disposal. Since carcasses cannot be left to rot in the open, the speed of depopulation is constrained by disposal capacity—and the longer depopulation is delayed, the greater is the probability that the disease will continue to spread. Another possible obstacle to depopulation of exposed herds could be lack of political support for killing a large number of apparently healthy animals, although there is probably greater understanding of the need for depopulation since the outbreak in Great Britain.

According to the Animal and Plant Health Inspection Service guidelines, three disposal methods can be used. In order of preference they are burial, burning, and rendering. Burying hundreds of thousands of carcasses in the South Valley would require excavation of miles of trenches, which could not be disturbed for several years. Burying the carcasses would put a major cost on producers because the land would be lost for most productive uses. Burning the carcasses would require massive amounts of wood or other fuel, which would probably be difficult to acquire in a short time. Burning might also create air pollution problems and could require exemptions from air quality standards. The use of an air curtain, assuming that adequate equipment is available, would reduce the quantity of fuel needed and the environmental impact of massive burning, but would increase the burning time. Disposal in landfills might be limited because carcasses would have to be mixed with waste in a fixed proportion. There is also a cost of faster filling of a landfill.

Because disposal is such a key issue, a cost–benefit analysis of alternative carcass disposal methods should be conducted to identify the best measures for California. Burial was by far the cheapest way of carcass disposal applied in Taiwan's 1997 outbreak, where rendering and burning were also used (Yang et al. 1999). Burial is the method of disposal that has been used most frequently in Britain, although burning has also been used.

Exposed animals showing no signs of infection should also be slaughtered, but under the United States Department of Agriculture plan they can be diverted to human consumption or protein utilization. However, the slaughter capacity in the South Valley probably would not be enough to process the required number of animals in a timely manner. An obstacle to this approach could be a lack of willingness by consumers to purchase beef resulting from the slaughter of "exposed" animals. Although FMD does not constitute a threat to humans, many consumers still do not understand this fact.

Compensation for Losses

In economic terms, FMD eradication is a measure taken to eliminate the externality caused by the disease's highly contagious nature and thus reduce societal costs in the long run. The application of stamping out requires total cooperation by farmers, and a basic condition of their cooperation is an adequate compensation policy for animals that are to be depopulated. If compensation is set too low, producers have less incentive to report sick animals and could try to dispose of them, thus disseminating the disease. However, compensation cannot be set too high or producers would have an incentive to introduce the disease to healthy animals in order to claim compensation for their destruction.

Under current regulations, indemnity payments cover only the direct cost of animals and materials destroyed. It has been documented, though, that because of trade restrictions, total economic losses can exceed by several times the costs covered by indemnity payments (Berentsen et al. 1990; Yang et al. 1999). Moreover, these "consequential losses" may be incurred not only by the livestock producers but also by all related industries. It is impossible, however, to estimate the consequential losses with any degree of accuracy. Therefore, these losses should be addressed by other measures such as those used to provide relief after natural disasters (e.g., low cost loans, tax relief, and special unemployment payments).

Given the expected magnitude of the consequential losses, it may be difficult for the livestock and dairy industries to return to business after the lifting of the quarantines. The industry should study the creation of a self-insurance scheme to help cover the indemnification of consequential losses. The basis could be a fund that would be invested in the financial markets until needed. Because of the low probability of an outbreak, the initial investment could be relatively small and constituted over a number of years. The main limitation in following this policy would be to convince producers of its long-term social benefits (Ekboir 1999b).

Trade Issues

Under any conceivable scenario, an FMD outbreak in California would impact significantly the state's livestock trade, on a local and regional basis, for a period that may span from a few months to several years. An outbreak would also greatly affect international trade by eliminating the FMD-free status that the United States currently enjoys and that permits the United States access to other high-priced FMD-free markets. A major FMD outbreak that resulted in trade restrictions on beef imports would cause large trade losses. The United States would be forced to sell its animal products in FMD-endemic markets, where prices are lower than in the FMD-free market. Since most U.S. beef exports are currently shipped to Japan and Korea, which do not recognize the regionalization principle, the outbreak would probably affect all U.S. exports.[11]

Local Impacts Once an outbreak is confirmed, trade routes from and into the infected area would be closed, and a quarantine area would be set up. The quarantine area could include a few miles around the infected premises, or, in case of a rapid spread of the disease, it could include the entire state or several states. Because of the quarantine and depopulation, the livestock and dairy industries in the quarantine area will be affected. In the early days of the outbreak, there would be a localized excess supply of livestock because products originating in the infected region could only be consumed in the quarantine area. This excess supply may last a negligible period and may not have any effect on prices. As depopulation advances, the supply of livestock and milk to processing plants would fall because farms cannot be repopulated until the quarantine is lifted. In order to minimize the lost revenue and to maintain their market share, processing plants might be forced to import raw materials from outside the quarantine area.

This study has not considered the costs that would be created by the disruption of social and economic activities within the region where quarantine would be imposed and where disposal of carcasses is occurring. Transportation of goods and people would be halted or caused to detour. Individuals might not be able to get to work or students to school. Burning could cause significant air pollution, and burial could threaten groundwater.

Regional Impacts California is a net importer of meat and dairy products. For this reason, the impact of an FMD outbreak would be less devastating to California than a similar outbreak would be in other regions that are net exporters. California would be able to substitute the lost production with imports, at least to a considerable extent. However, if the depopulation of herds is significant, output would decrease to a level in which large imports would be needed. These imports would have an effect on trade flows with other states and on meat and dairy prices.

The dairy and livestock industries are linked forward and backward to a number of industries, i.e., input suppliers, service providers, and milk and livestock buyers. A serious disruption of the dairy and livestock industries would also affect the linked industries. Live cattle would not be allowed to leave the quarantine area, so calves that are not depopulated and would normally be sent out of the state for raising, would have to be retained in the state. This alternative could pose serious logistic and economic problems if vaccination is applied. Policy needs to plan for these alternatives to minimize the impact of such events.

International Impacts   The Uruguay Round Agreement established a new regulatory framework for international trade. In respect to beef, trade was affected in various aspects. Over time, subsidies and tariffs are to be reduced, and protected markets approved minimum access commitments. The SPS protocol led to growing acceptance of regionalization, dependence on science-based risk assessment, and prohibition of the use of sanitary barriers as barriers to trade. The Office International des Epizooties (OIE) has become the international authority that sets sanitary standards, though countries are not forced to accept them.

Almost all countries accept the principle of regionalization. Japan and South Korea are noticeable exceptions that would play a significant role in case of an FMD outbreak in California because they purchase 62 percent of the U.S. beef and pork exports, by value. If Japan and South Korea were to ban imports from the United States, it would impact the international meat markets substantially, with significant movements in prices and trade flows.

If Japan and South Korea banned beef imports from the United States, the United States would be forced to sell in the FMD-endemic market. Considering the volumes that would have to be rerouted, beef prices in the FMD-endemic market would fall abruptly, while prices in the FMD-free market would increase. The reduced profitability of foreign markets should increase the domestic supply (as less beef would be exported), reducing the demand for Australian and New Zealand beef. The mid- and long-term impacts on world beef markets are difficult to predict, because they will depend both on the response of livestock producers in the United States and in other major exporting countries (Australia and New Zealand),

and also on changes in policy and in demand in the largest importing countries.

Would Japan and South Korea maintain a "zero tolerance" policy if the United States, their major beef provider, suffered from a major FMD outbreak? Policymakers in Japan and South Korea would surely want to protect their livestock industries, just as the United States did when the outbreak occurred in Great Britain. However, domestic producers in these countries are not capable of increasing the beef supply. Other exporters, especially Australia, would only be able to partially replace United States beef in the Asian countries in the short run. Thus, beef prices should be expected to rise sharply in the Asian markets if FMD outbreak occurs in the United States and U.S. exports are interrupted. Though domestic consumers in Japan and South Korea could substitute chicken and pork for beef, this would not wholly solve the problem. As a result, it might be that Japanese and Korean authorities could be persuaded to revise their zero tolerance policy if they could be assured that the risk of contracting FMD from U.S. imports from states outside California was minimal. It could be of benefit to all countries to develop such contingency plans.

Final Comments and Recommendations

The presence of a large number of commercial dairies and, consequently, a high-density population of animals represents a major logistical problem for California in the event of an FMD outbreak. A large number of movements of trucks transporting feed, milk, calves, replacement heifers, cull cows, plus veterinarians, artificial insemination technicians, and workers, occur in and out of these premises every day. Additionally, the high density of animals favors airborne diffusion of the infection. Because infected animals shed the virus before the onset of clinical signs, it is probable that by the time FMD is identified, several thousand animals would already be infected. The simulations presented here show that the effectiveness of the eradication campaign depends crucially on the date it is started. A one-week delay in starting depopulation could increase the proportion of infected premises from 18 percent to more than 90 percent. The costs of an FMD outbreak in

California will necessarily be high and will raise rapidly the greater the spread of the disease.

To reduce the costs of an FMD outbreak, proper surveillance mechanisms operating through government agencies, the livestock industry, and farmer's organizations are essential. The depopulation of a large number of animals would require substantial physical, human, and financial resources. Criteria regarding the compensation to be provided farmers for depopulated animals must be established and maintained at a level commensurate with current livestock market conditions. Providing additional training for animal health officials, veterinarians, producers and their employees would increase the effectiveness of the first efforts (which are crucial in reducing the damage caused by the outbreak). Ensuring the availability of the requisite resources, including individuals and equipment to undertake the slaughter of animals and to dispose of the carcasses, is a requisite. Establishing a tight quarantine is essential to halting the spread of disease, though this will be difficult, because it will affect other economic activities. Individuals throughout society should be educated as to the importance of cooperation. Law enforcement agents and technicians need to be trained how to properly clean and disinfect vehicles.

The California livestock industry contains a number of small backyard operations. The backyard operations could be a potential entry route for the FMD virus because they have lax biosecurity. From a policy point of view, these operations constitute a problem because they are difficult to identify and costly to monitor. Current programs that target small and backyard operations should be reviewed to increase their efficiency.

Considering the major obstacles that a total stamping-out policy would face, other policies, such as ring or even total vaccination, could be more efficient. Further research is needed to determine the conditions under which each policy would be preferable.

An outbreak of FMD in California would have a major impact on the livestock and related industries throughout the United States through the disruption of domestic and international livestock trade flows. Contingency plans should be developed to minimize these disruptions in the event of an outbreak.

Notes

[1] When humans are infected, the symptoms are very mild and without consequences.

[2] Compensation to farmers was paid under two programs or schemes: the disease control and the livestock welfare disposal scheme. The former was used to compensate farmers for the loss of (infected) livestock, the latter to compensate for impeded production that resulted from the movement restrictions adopted during the campaign.

[3] Beef exports by the United Kingdom were reduced to a minimum after the "mad cow" crisis in 1996.

[4] Vaccination can be used if (a) the disease has not been contained within six months of the outbreak; (b) the outbreak reaches epidemic proportions (25 percent of the susceptible population in areas of high density livestock); (c) the cost/benefit ratio of the slaughter program approaches 1:2; (d) FMD becomes endemic in wildlife of three or more states; (e) legal restrictions prevent carrying out the slaughter program (APHIS 1991).

[5] The South Valley includes Fresno, Kern, Kings, and Tulare counties.

[6] For a more detailed description of the model and its assumptions, refer to Ekboir (1999a, Chapter 7).

[7] Under current production practices, this would involve all major livestock states.

[8] FMD is so infectious that exposed animals have a very high probability of contracting the disease. Infected animals will produce virus for several days before clinical signs of FMD become apparent. Thus, other animals will have been exposed by the time an infected animal is diagnosed. Accordingly, animals exposed through direct contact and indirect contact (animal and human movements) and contiguous location must also be depopulated, even though they do not (yet) show clinical signs of the disease.

[9] The outbreak is assumed to begin when an animal is infected on day 1. It begins to secrete virus immediately and thus to infect other animals, but the animal is only diagnosed to have FMD subsequently. After diagnosis, resources are marshaled to control the outbreak. If depopulation is begun between day 8 and 14, it has begun in the second week. If depopulation begins between day 15 and 21, it has begun in the third week.

[10] Eradication of 900 cows in Italy took three days. Depopulation of more than 4 million pigs in Taiwan required about a month after the army was called in; 130,000 pigs were killed per day at the peak of the depopulation campaign, after the soldiers had gained considerable experience in killing the pigs and disposing of the carcasses.

[11] The regionalization principle allows for continued beef exports from clean areas within a country where a FMD outbreak has occurred, if the outbreak can be contained within a quarantined area. Quarantine effectiveness is evaluated by importing countries.

References

APHIS. 1991. "Foot and Mouth Disease Emergency Disease Guidelines." Animal Plant Health Inspection Service, U.S. Dept. of Agriculture, Washington D.C.

Berentsen, P.B.M., A.A. Dijkhuizen, and A.J. Oskam. 1990. "Foot and Mouth Disease and Export: An Economic Evaluation of Preventive and Control Strategies for The Netherlands." Agriculture University, Wageningen, The Netherlands.

Berentsen, P.B.M., A.A. Dijkhuizen, and A.J. Oskam. 1992. "A Dynamic Model for Cost-Benefit Analyses of Foot and Mouth Disease." *Preventive Veterinary Medicine.* 12:229–243.

DEFRA. 2001. "Agriculture in the United Kingdom, 2001." Department for Environment, Food and Rural Affairs, London, United Kingdom. Downloaded from http://www.defra.gov.uk/esg/econfrm.htm on June 2002.

Dijkhuizen, A.A. 1989. "Epidemiologic and Economic Evaluation of Foot and Mouth Disease Control Strategies in The Netherlands." *Netherlands Journal of Agricultural Science.* 37:1-12.

Donaldson, A.I. 1991. "Foot and Mouth Disease." *Surveillance.* 17(4):6–8.

Donaldson, A.I. 1994. "Epidemiology of Foot and Mouth Disease: The Current Situation and New Perspectives." ACIAR Proceedings (50), Canberra, pp. 9-15.

Donaldson, A.I. and T.R. Doel. 1994. "La Fièvre Aphteuse: le Risque pour la Grande-Bretagne après 1992." *Ann. Méd. Vét.* 138:283-293.

Dunn, C.S., and A.I. Donaldson. 1997. "Natural Adaptation of Pigs of a Taiwanese Isolate of Foot and Mouth Disease Virus." *The Veterinary Record.* 141(7):174–175.

Ekboir, J. 1999a. *Potential Impact of Foot and Mouth Disease in California.* Agricultural Issues Center, Div. of Agriculture and Natural Resources, University of California, Davis.

Ekboir, J. 1999b. "The Role of the Public Sector in the Development and Implementation of Animal Health Policies." *Preventive Veterinary Medicine.* 40:101–115.

Garner, M.G., and M.B. Lack. 1995. "An Evaluation of Alternate Control Strategies for Foot and Mouth Disease in Australia: A Regional Approach." *Preventive Veterinary Medicine.* 23:9–32.

Harvey, D.R. 2001. "What Lessons from Foot and Mouth? A Preliminary Economic Assessment of the 2001 Epidemic." Working Paper 63. The Center for Rural Economy, The University of Newcastle upon Tyne, Newcastle, England, March.

Kitching, R.P. 1998. "A Recent History of Foot and Mouth Disease." *J. Comp. Pathol.* 118(2):89–108.

Maragon, S., E. Fachin, F. Moutou, I. Massirio, G. Vincenzi, and G. Davies. 1994. "The 1993 Italian Foot and Mouth Disease Epidemic: Epidemiological Features of the Four Outbreaks Identified in Verona Province." *The Veterinary Record.* 135:53–57.

Miller, W. 1979. "A State-Transition Model of Epidemic Foot and Mouth Disease." In McCauley et al., Eds., *A Study of the Potential Economic Impact of Foot and Mouth Disease in the United States.* University of Minnesota, APHIS-USDA.

Moutou, F., and B. Durand. 199). "Modelling the Spread of Foot and Mouth Disease Virus." *Vet. Res.* 25:279–285.

Yang, P.C., R.M. Chu, W.B. Chung, and H.T. Sung. 1999. "Epidemiological Characteristics and Financial Costs of the 1997 Foot and Mouth Disease Epidemic in Taiwan." *The Veterinary Record.* 145(25):731–734.

8

Risk Assessment of Plant-Parasitic Nematodes

Howard Ferris, Karen M. Jetter, Inga A. Zasada, John J. Chitambar, Robert C. Venette, Karen M. Klonsky, and J. Ole Becker

Introduction

In this chapter, we consider plant-parasitic nematodes as exotic pests and as representatives of other exotic soilborne pests and diseases. The species in this study are selected for current policy and trade barrier implications and for biological significance.

The process of invasion of an exotic nematode species has four phases: arrival, establishment, integration, and spread. Arrival is the introduction of a new population into a community of established species, typically by some factor other than individual locomotion. Associations of plant-parasitic nematodes with imported agricultural commodities suggest that the probability of arrival is high. The fact that nematodes are small, spatially aggregated, and difficult to identify increases the probability of their arrival. Establishment occurs when the newly arrived species maintains a population through local reproduction, not continuous immigration. Nematode species with a broader geographic range are more likely to become established once they have been introduced. However, the introduced species may not reach economically damaging levels; that depends on conditions at the invasion site. Integration may require ecological and evolutionary changes in both the invading population and the resident community. Integration is not a rapid process; it may proceed over a long period. Spread is the movement and redistribution of a species by active or passive means. Active dispersal of plant-parasitic nematodes through locomotion is most important at the scale of individual fields. Passive means of movement, including wind, water, contaminated equipment, or transportation of infested commodi-

ties, are important at the field, regional, and national scales.

Understanding of the phases of invasion allows assessment of the threat that exotic plant-parasitic nematodes might pose to California agriculture. The fact that a nematode species is a serious pest elsewhere does not mean that it would be equally damaging in California. The challenge is to determine the probability that the nematodes could arrive, become established, and cause significant economic, ecological, or societal damage. These risks must be weighed against the cost, and probability of success, of intervention.

All the nematode representatives selected are A-rated pests and either do not occur in California or have established limited infestations. They represent different life history strategies, host ranges, and modes of dispersal. The question is not whether these A-rated nematodes will be introduced into the state, but whether they can be eradicated, contained, or will become established after introduction.

Despite quarantine and containment programs, plant-feeding nematodes spread to new countries and new locations. Many important exotic nematode pests play major roles in California agriculture and urban landscapes. Introduced nematodes of economic importance include the dagger nematode (*Xiphinema index*), the sugarbeet-cyst nematode (*Heterodera schachtii*), and the citrus nematode (*Tylenchulus semipenetrans*). They were probably introduced and spread with planting material or equipment. Most seem distributed with their primary agricultural hosts and are not prevalent or even present in natural habitats (although we have not surveyed exhaustively).

The representative nematode exotic pests considered in this chapter, with rationale for their selection, follow.

Burrowing nematode (*Radopholus similis*), Thorne, 1949: The burrowing nematode is a migratory endoparasite of roots of over 200 woody and herbaceous perennials, including important commercial crops such as citrus. All life stages are readily transported within plant tissues and associated soil. Due to its wide host range, it is one of the most economically important plant-parasitic nematodes in tropical and subtropical regions of the world. Burrowing nematodes have been found in soil and plant materials destined for California during border inspections, and an eradication program was completed for an isolated urban infestation in 1996. The establishment of burrowing nematodes would result in quarantines by importing countries.

Reniform nematode (*Rotylenchulus reniformis*), Linford and Oliviera, 1940: The reniform nematode is a sedentary semiendoparasite in the adult female stage of over 200 tropical plants, including commercial crops such as cotton, grapes, and citrus. It is readily transported in roots and associated soil. In 1960, reniform nematodes were eradicated from an isolated San Diego infestation after a quarantine shipment of ornamental date palms tested positive. Due to the widespread establishment of reniform nematodes, importing countries would not impose quarantines.

Rice foliar nematode (*Aphelenchoides besseyi*), Christie, 1942: The rice foliar nematode is a migratory endoparasite of leaf tissue and also feeds ectoparasitically in buds and seed coats. It tolerates desiccation and is readily transported with unhulled grain. Feeding by the nematode causes white tip disease of rice. The rice foliar nematode has been detected in rice destined for export from California in recent surveys. Unmilled rice would be subject to quarantines by importing countries.

Sting nematode (*Belonolaimus longicaudatus*) Rau, 1958: The sting nematode is a migratory ectoparasite of roots of a large number of plants. It has rather specific environmental requirements. It feeds on turf grasses in sandy soil and high-value commercial crops including cotton. The sting nematode is established in several golf courses in southern California, and an internal quarantine has been imposed to minimize the probability of its spread.

Golden nematode (*Globodera rostochiensis*), Behrens, 1975: The golden nematode is a sedentary semiendoparasite with restricted host range. Its eggs are contained and protected in a hardened cyst and may survive for up to 30 years (Spears 1968; Winslow and Willis 1972). The golden nematode, *G. rostochiensis*, and the related species, *Globodera pallida* (collectively known as potato cyst nematodes), are among the most important pests of potatoes due to the severity of damage and their survival in the absence of a host (Golden and Ellington 1972). The golden nematode is also a pest of other important California commercial crops, including tomatoes. The golden nematode has never been detected in California. Should it enter, it is unlikely that it can be eradicated. If it becomes established, the U.S. Department of Agriculture (USDA) quarantine against infested regions would be extended to California.

Biology and Ecology

Four out of every five multicellular animals on the planet are nematodes (Platt 1994). They exist in almost every conceivable habitat and have a wide range of feeding habits and food sources. They include bacterivores, fungivores, carnivores, omnivores, and plant feeders. Herein we focus primarily on plant feeders of importance as pests in agriculture.

Most, perhaps all, higher plants support the feeding of a range of nematode species. Usually several species can be found feeding on the roots of a single plant. They feed on the outside of plants as ectoparasites of roots or of bud tissues, or within tissues as endoparasites of roots, stems, leaves, or seeds. In some cases, they remain migratory throughout the life cycle; in other cases they become sedentary in some life stages. The diversity of their life history, host ranges, and survival strategies contributes to the difficulty of eliminating them once established.

Burrowing Nematode

The burrowing nematode is found worldwide in tropical and subtropical regions. It occurs wherever bananas are grown, including Africa, parts of Asia, South America, and southern Europe. It also occurs in the southeastern United States and Hawaii (Ferris and Caswell-Chen 1997).

There are more than 350 known hosts of the burrowing nematode. Most banana and plantain cultivars are attacked. Other hosts include citrus, coconut, ginger, palm, avocado, coffee, black pepper, sugarcane, tea, vegetables, ornamentals, trees, grasses, and weeds.

Burrowing nematodes cause spreading decline in citrus. Symptoms usually appear about a year after infection. Infected trees have sparse foliage, retarded terminal growth, poor color, twig dieback, and a general unthriftiness (Christie 1957; DuCharme 1954). Leaves may wilt at midday, but show temporary rejuvenation with rain or irrigation. There may be little new growth during the spring flush. Trees may bloom profusely, but bear only a few small fruit. Trees will appear undernourished without exhibiting specific symptoms of malnutrition. Below ground, dark lesions appear at the site of nematode penetration; the lesions coalesce to form a canker.

In Florida burrowing nematode infestations result in citrus yield losses of 50 to 80 percent for grapefruit and 40 to 70 percent for oranges (DuCharme 1968). Grapefruit trees appear to be more adversely affected than orange trees.

Avocado trees show similar spreading decline symptoms when infested with the burrowing nematode. The nematode can also decimate production of several indoor decorative plant species. It is a severe pest of the parlor palm and may preclude commercial production.

The burrowing nematode feeds in all life stages after hatching from the egg and is able to complete its life cycle within the root cortex. The nematode is also present in rhizosphere soil. Reproduction is sexual but parthenogenesis, the production of viable eggs without fertilization, must be possible because a population can be initiated from a single egg (Orton Williams and Siddiqi 1973).

Reniform Nematode

The reniform nematode is widely distributed in many tropical and subtropical regions of the world. It has been reported in most of Africa, the Caribbean, Japan, the Middle East, South Pacific, Central America, Italy, Spain, Mexico, China, and the Far East. Within the United States, the reniform nematode is established in Alabama, Arkansas, Florida, Georgia, Hawaii, Louisiana, Mississippi, North Carolina, South Carolina, and Texas.

Over 140 plant species in 115 genera representing 46 families are attacked by this nematode (Jatala 1991). Some of the economically important host plants are banana, cabbage, cantaloupe, cassava, citrus, kale, lettuce, mango, okra, pigeon pea, pineapple, sugarcane, pumpkin, coconut, cotton, radish, cowpea, soybean, sweet potato, crimson clover, tobacco, eggplant, tomato, and guava.

Above-ground effects on host plants include dwarfing, shedding of leaves, malformations of fruit and seeds, and general symptoms of an impaired root system. Below ground, roots are discolored and necrotic with areas of decay. Plant death is possible in heavy infestations.

Reproduction and development of the reniform nematode are favored by fine-textured soils with a relatively high content of silt or clay (Koenning et al. 1996; Robinson et al. 1987). The reniform nematode reproduces sexually; however, it may also reproduce by parthenogenesis. Juveniles develop through three molts to the preadult stage without feeding. All juvenile stages and males are found in the soil. Soon after the final molt, the young adult infective stage penetrates host roots and the anterior part of the body becomes embedded within root tissue.

Rice Foliar Nematode

The rice foliar nematode occurs in most rice-growing areas of the world, including Australia, Sri Lanka, Comoro Islands, Cuba, El Salvador, Hungary, Indonesia, Italy, Japan, Madagascar, Mexico, Pakistan, the Philippines, Taiwan, Thailand, the former Soviet Union, and Central and West Africa (Bridge et al. 1990; Ou 1985). It has been reported in the southern U.S. states that produce rice.

Rice is the most important host worldwide for this nematode. The white tip disease caused by the nematode is characterized by whitening of the leaf tips, which later become brownish and tattered. The upper leaves are the most affected; the flag leaf is often twisted, hindering the emergence of the panicle. In the seedbed, emergence of severely infected seedlings is delayed, and germination is low. The most conspicuous symptoms occur early in development. Diseased plants are stunted, lack vigor, and produce small panicles. Affected panicles frequently are sterile; kernels and surrounding bracts are distorted (Bridge et al. 1990; Ou 1985; Taylor 1969).

On strawberry, the rice foliar nematode is the causal agent of "summer dwarf" or "crimp" in the United States and Australia. Other host plants include onion, garlic, sweet corn, sweet potato, soybean, Chinese cabbage, sugar cane, horseradish, lettuce, millet, many grasses, orchids, and many ornamental plants (Franklin and Siddiqi 1972; Ferris and Caswell-Chen 1997).

Nematodes become dormant under seed hulls at the end of the growing season (Taylor 1969). They become active and are attracted to the actively growing parts of the plant after infested seed is planted. During early growth, rice foliar nematodes are found in low numbers within folded leaf sheaths, feeding ectoparasitically around the growing point (Todd and Atkins 1958). Although reproduction of this nematode is predominantly sexual, parthenogenesis has been reported (Sudakova and Stoyakov 1967).

Sting Nematode

The sting nematode has been reported from the Bahamas, Bermuda, Brazil, Costa Rica, Mexico, Australia and Puerto Rico. In the United States, the nematode occurs in Florida, South Carolina, North Carolina, Virginia, Alabama, California, Mississippi, Louisiana, Texas, Arkansas, Kansas, Oklahoma, New Jersey, and Nebraska.

Sting nematodes are the most destructive nematodes in turf grass ecosystems in Florida (Busey et al. 1991). In addition to turf and other grasses, these nematodes have a wide host range that includes grapes, citrus, cantaloupes, lettuce, tomatoes, beans, onions, corn, wheat, barley, oats, forage crops, cotton, ornamentals and weeds. Based on differences in host reactions and fitness, there may be several physiological races of sting nematodes (Abu-Gharbieh and Perry 1970; Robbins and Barker 1973).

Affected plants appear stunted, yellow, and exhibit drought and malnutrition symptoms. They fail to respond to water and nutrients. Badly affected plants collapse and die. Small patches (up to several feet in diameter) of dis-

eased turf can be noticed at a distance. Below-ground symptoms include a reduced root system with stubby, coarse roots. Above ground, shoots may show stunting, premature wilting, yellowing, and, in some cases, infested plants may die. In fields, the boundary between infested and healthy plants is well-defined.

Soil texture and composition have been identified as major limiting factors for sting nematode reproduction (Perry and Rhoades 1982). The distribution of the nematode is restricted to sandy soils; in Virginia, Miller (1972) found it only in soils with 84 to 94 percent sand. Reproduction and movement are inhibited in heavier, fine-textured soils. Males are required for reproduction, but only one mating is sufficient for sustained egg fertilization (Huang and Becker 1999).

Golden Nematode

The golden nematode was originally discovered in Germany in 1913. By that time, it had spread throughout Europe (Wallace 1964). It probably originates with the potato in South America. During the 1960s and 1970s, Canada was found to have several areas of golden nematode infestation (Mai 1977). Vancouver Island is the area closest to California where the golden nematode is known to be established. In the early 1970s, scientists in Mexico discovered an infestation of golden nematodes in the state of Guanajuato, one of the major potato-producing regions in Mexico (Alvarez 1972). In North America, it was first discovered on Long Island, New York (Nassau County), in 1941 after a potato grower noticed isolated areas of poor plant growth (Mai and Lear 1953).

Approximately 90 species in the family *Solanaceae* are known to be hosts, including potato, tomato, and eggplant (Mai and Lear 1953). In addition, there are numerous weed hosts (Goodey and Franklin 1958, 1959). Of the weed hosts, bitter nightshade, silverleaf nightshade, hairy nightshade, black nightshade, and jimsonweed are all present in California. Hosts are not equally susceptible, and cultivars may differ in their susceptibility to various races of the nematode (Kort et al. 1977).

No distinct host symptoms are associated with low populations, but as populations increase, symptoms appear. A potato crop will show poor growth in small areas that enlarge with continuous cropping. Plants in infested patches exhibit symptoms of water and mineral deficiency, including chlorotic leaves and wilting. The bodies of immature females that have erupted through the root epidermis appear as minute, white specks on the roots. At extremely high nematode densities, tubers may become infected.

Survival, reproduction, and population dynamics of the golden nematode can be greatly influenced by temperature, moisture, day length, and soil factors. In general, golden nematodes will survive in any environment where potatoes can be grown. Eggs remain dormant within the dead female's body (the cyst) until stimulated to hatch by chemical stimuli from host plant roots. The nematode eggs can remain dormant and viable within the cyst for up to 30 years (Winslow and Willis 1972). While dormant in the egg stage, the golden nematode is more resistant to nematicides (Spears 1968).

When soil temperatures are above 10°C and the proper hatching signals are received, second-stage juveniles hatch from the eggs, escape from the cyst, and migrate toward host plant roots (Clark and Hennessy 1984; Ferris 1957). Juveniles penetrate the roots, establish a feeding site, and begin to feed. Those that develop into females become rounded, break through the epidermis, and are exposed on the root surface. Male nematodes develop similarly, but in the final juvenile stage they emerge as a motile worm that leaves the root and is attracted by chemical signals from females (Green et al. 1970). After mating, each female produces approximately 500 eggs, which are retained in the body (Stone 1973). After the female dies, the body cuticle forms a protective cyst.

Introduction and Spread

Factors Influencing the Introduction and Spread of Nematodes

Nematodes are generally excellent invaders of new habitats. They have evolved numerous strategies for exploiting favorable environments and withstanding harsh conditions. Their small size and the difficulty of detecting them in plant and soil material increase the probability that they will be successfully introduced. The feeding relationships of plant-feeding nematodes with host tissues and their survival capabilities contribute to the ease with which they are disseminated with plants. Once introduced, the generally nonspecific nature of symptoms increases the probability that their presence may go undetected or unrecognized for considerable periods. Their dispersal in and around plant tissues and throughout the soil contributes to the difficulty of targeting them in management or eradication programs.

The most important determinant of rate of spread in agriculture is the movement of infested plants and propagative material. Especially important is material that will be propagated and distributed as nursery stock. Sale and movement of infested nursery stock, seed, or turf immediately spreads the nematode pest to uninfested areas and distributes it throughout the planting site. Depending on the area serviced by a nursery, spread from an infested source may be local, statewide, or even across state boundaries.

Significant movement of nematodes is also generated by natural and human forces. Nematodes with stages that are resistant to desiccation, such as the reniform and rice foliar nematodes, may be spread widely and for long distances in blowing dust. Wind spread of cysts of the oat cyst nematode (*Heterodera avenae*) across desert regions between cereal production areas has been detected in Australia (Meagher 1977; Viglierchio 1991).

Many nematodes, particularly endoparasites, are consumed in plant material by birds and other animals (Martin 1969; Thomason and Caswell 1987). They successfully survive passage through the digestive tract and become point-source infestations along migration patterns or within territorial boundaries. Their introduction into the new Polder region of the Netherlands after reclamation of the land from the sea has been associated with migratory birds. Movement from field to field also occurs with contaminated soil adhering to vehicles and farm equipment. Movement of the soybean cyst nematode (*Heterodera glycines*) in the Midwest has been associated with the purchase of used equipment from established soybean areas for use in new areas of production.

Such spread results in single or multiple point-source infestations in a new field, which, left undisturbed, might take several years to become evident. However, tillage and water movement are the norm. Nematodes are readily and rapidly spread throughout a field and among fields by irrigation water, surface runoff, engineered drainage systems, and land leveling (Thomason and Caswell 1987; Waliullah 1984). Such forces generate rapid broadcast distribution of the pests. In the irrigated agriculture of California, up to 11 separate tillage operations may be conducted in a field after harvest in late summer to prepare it for the next crop in the spring. Consequently, enormous movement of soil and its resident organisms occurs within a field in a single year. Spread throughout a field from a point-source infestation will probably occur in one or two years under conventional production practices in annual crops in California.

For some nematode species, a primary constraint in establishment is soil texture; for others, host availability and soil temperature may be more important. Burrowing and sting nematodes prefer coarse, sandy soils. In California, sandy soils are present in the Coachella Valley, the Bard Valley near Blythe, the Edison-Arvin citrus district of Kern County, and in streaks throughout the state. Citrus and date palms in the Coachella Valley are planted in soils subject to temperatures that would favor the development of burrowing nematode populations. Host crops found along the California coast, even when planted in sandy soil, experience temperatures favorable to the development of the burrowing nematode for only a few months of the year. On the other hand, reproduction and development of the reniform nematode is favored by fine-textured soils (Robinson et al. 1987).

The main dissemination risk for the rice foliar nematode is seed (Bridge et al. 1990). On a local scale, this nematode can be dispersed in floodwater, but survival in water decreases as temperature increases from 20° to 30°C (Tamura and Kegasawa 1958). Once introduced into a field, the rice foliar nematode may survive in plant debris (Sivakumar 1987).

In general those environments that favor potatoes and tomatoes also favor the golden nematode. The survival of the golden nematode is completely dependent on the presence of host crops; the nematode and host have coevolved over many thousands of years, resulting in specific recognition signals between host and parasite (Endo 1971).

Introductions of Exotic Nematode Species into California

All the nematodes considered in this study, except the golden nematode, have been introduced into California. Some species have been eradicated, some are of disputed presence, and one is established in limited areas.

The California Department of Food and Agriculture (CDFA) Nematology Laboratory, in collaboration with county agricultural commissioners, has made 70 detections of the burrowing nematode since 1995 in shipments destined for California. It has been discovered and eradicated in commercial nurseries. In 1996 it was discovered in a residential area in Huntington Beach and, due to the early detection and isolated nature of the infestation, eradicated. The source of the infestation was an illegal shipment of banana corms from Louisiana (Chitambar 1997a).

Since 1989 the CDFA Nematology Laboratory has made 64 detections of reniform nematodes in quarantine shipments. A reniform nematode infestation of ornamental date palm plants was detected in San Diego in 1960. The plants were established in a residential property before a confirmed diagnosis of the pest was completed. Subsequently, the plants were removed from the infested site, and all plants and soil were fumigated with a nematicide. As in the case of the 1996 burrowing nematode infestation, an eradication program was biologically feasible due to early detection and the isolation of the infestation.

Infestations of reniform nematodes on established yucca plants were first detected in 13 residential properties in Highland, San Bernardino County, during a residential grid survey in 1967. The infestation was traced to yuccas brought into California from Harlingen, Texas, and planted in the subdivision. The infested areas were treated with dibromochloropropane (DBCP Nemagon). In 1971 the nematode was detected again in the same locality. Despite a second treatment of Nemagon, it was still present in 1973 and 1974. After subsequent treatment, the reniform nematode was declared eradicated from the infested areas on December 31, 1978. In 1980 the nematode was detected again from the same region. The current status of the San Bernardino infestation is not known (Chitambar 1997b).

From 1959 to 1996 the rice foliar nematode was detected only twice in California by the CDFA Nematology Laboratory. The first was in 1959 in strawberries that originated in Oregon; the second was in 1963 in a fungal culture collected from a Butte County field. Attempts to find the nematode from the same field were unsuccessful. In response to Turkish requirements for phytosanitary certification of rice shipments from California, a survey for rice foliar nematodes was initiated in 1997. One confirmed and three suspected detections of rice foliar nematode were made in samples collected from two counties. These locations tested negative when examined a second time. Consequently, the government of Turkey now requires certification of California rice on a per shipment basis; each shipment must be sampled and found free of rice foliar nematode. Three detections have been made since 1998.

The CDFA Nematology Laboratory detected the sting nematode in 1962 on Bermuda grass from Georgia, in 1967 on roses from Texas, and on coconut palm from Mexico, and in 1983 and in 1987 in soil from Florida. The sting nematode was detected 84 times and *Belonolaimus* spp. twice between June 1992 and December 1993. During the last week of May 1992, a sod sample from a Coachella Valley (Riverside County) golf course tested positive for sting nematodes.

Intervention Strategies

The first step in preventing the establishment of exotic nematodes is by excluding their entry into the pest-free regions. Exclusion may be done through cultural methods and quarantines. Should an exotic nematode enter, it may be prevented from establishing through the completion of a successful eradication program. If eradication is not feasible, then containment efforts may be undertaken to prevent further spread. If an exotic nematode becomes established, growers would have a variety of control methods available, including chemical treatments, developing resistant varieties, crop rota-

tions, soil solarization, changing cultural controls, and developing biological control programs.

Exclusion

Cultural Methods Avoiding infestations by exotic nematodes is the highest priority. The use of certified nematode-free planting stock is critical. The movement of soil from infested fields must be avoided.

The most effective means of controlling the rice foliar nematode is through seed treatments. Both chemical and hot water treatment of seed can be used to kill nematodes (Atkins and Todd 1959; Pinherio et al. 1997). Although there is some risk of reduced germination using hot water treatments, careful management of treatment temperatures and immediate planting of treated seed minimize deleterious effects (Taylor 1969).

Quarantines In countries that are free of exotic nematode pests, quarantines can lower the probability of their introduction. In countries where nematodes are localized, quarantines can reduce further spread. Although all the nematodes in these case studies are A-rated pests in California, they are subject to different quarantine regulations on the basis of their biology, sources, and historical factors.

The CDFA has external quarantine programs for the burrowing and reniform nematodes to reduce the probability of their introduction through infested plant and associated materials in shipments to California. It also has an internal quarantine against the sting nematode. Entry is restricted from all areas under quarantine of soil and potting media, plants and plant parts with roots, parts of plants produced below ground or at soil level, and plant cuttings for propagation.

In addition to the CDFA's quarantine programs, the burrowing nematode nursery certification program serves as another means of protection. Certification of nursery stock is mandatory if the stock is being marketed for farm planting. The nursery has the option (voluntary) to sell noncertified stock if it will not be used for farm planting.

Formal quarantine regulations have not been implemented against the rice foliar nematode at the state or federal level. However, federal regulations have prohibited the importation of paddy rice into the United States since November 23, 1933. Shipments of rice from the southern United States to California are not restricted.

The Animal and Plant Health Inspection Service (APHIS) of the USDA enforces a federal quarantine on the golden nematode. Interstate movement of the following materials from New York state is restricted: soil, plants, grass sod, plant crowns, roots for propagation, bulbs, corms, rhizomes, root crops, small grain and soybeans (unless in approved containers), hay and straw (unless in approved containers), plant litter, corn (except shucked corn), used farm materials and equipment (unless free of soil), and seed potatoes. Potatoes for consumption grown in fields certified free of golden nematode (or receiving applications of required soil fumigants) may be transported if free of soil and moved in approved containers.

Eradication

Unless infestations are quickly identified, eradication of nematodes is extremely difficult, if not impossible. For a very small, isolated infestation, excavation of all plant material and soil and their removal to a protected area for treatment have been feasible for nematode eradication in California. Using this method, burrowing and reniform nematode introductions have been declared eradicated.

For larger infestations, soil removal and fumigation for nematode eradication is difficult. An alternative is to remove or destroy all roots and other plant material, treat the soil with nematicides, and maintain it plant free for two to three years. Due to the wide host ranges of most of the plant-feeding nematodes considered in this study, growing a nonhost crop would require elimination of host weeds. This approach was attempted in Florida for eradication of the burrowing nematode; it failed both as an eradication strategy and as a containment strategy (Noling 2001). Eradication of the golden nematode through nonhost rotation is unlikely due to the extreme longevity of eggs protected in cysts.

Containment

The primary focus in containment is to minimize the potential spread of the nematode. Considerations include restricting movement of plant material, soil, and drainage water from the infested area. In established perennials, preventing disruption of root-to-root contact is important. In the Florida program to restrict spreading decline of citrus (caused by burrowing nematode), trees and roots are removed in buffer zones two trees wide around infested sites. The buffer zones are treated with nematicides to reduce the probability of nematode spread. Decontamination of equipment and footwear is essential. Fencing of the area may be necessary to minimize animal and human traffic.

Management of Established Infestations

The use of pesticides, resistant varieties, crop rotation, soil solarization, and other cultural controls is effective to various degrees in controlling infestations and, in some cases, preventing further spread (Evans and Brodie 1980). For greatest effect, they are applied in strategic combinations targeted at the life cycle and biology of the introduced species.

Chemical Control   Chemicals used to control nematodes (nematicides) can be classified according to their volatility as either fumigants or nonfumigants. Depending on the concentration used, many fumigant nematicides are general biocides that kill many soil organisms, including nematodes, fungi, bacteria, plants, and insects. In contrast, some nonfumigant nematicides more specifically target nematodes. Some nonfumigant nematicides are nonphytotoxic and can be used to manage nematodes in perennial crops.

Pesticides are subject to review of their registration status. Environmental quality and health concerns have resulted in limits being imposed on some nematicides. Methyl bromide, for example, will not be available after 2005. For 1,3-dichloropropene (Telone II), the amount that may be applied annually per township in California is restricted. In addition, all organophosphate and carbamate pesticides are subject to evaluation under the 1996 Food Quality Protection Act.

Resistance   Plant-breeding programs seek to develop crop varieties that are resistant to nematodes. Preferably, the developed cultivar should be resistant to the target nematode and also to other major disease problems; the resistance should be uniformly inherited; and the cultivar should have desirable horticultural characteristics. Resistant planting materials can substantially reduce losses due to nematodes. However, there are several examples of plants selected for resistance to one nematode species that have elevated susceptibility to another species. Broad and durable resistance is a desirable but difficult goal in plant-breeding programs. After a suitable source of resistance is identified, it may take five to seven years to develop a resistant crop variety that is compatible with California production practices and market requirements.

Crop Rotation Crop rotation using poor hosts or nonhosts is useful for nematode control in annual cropping systems. In the absence of their food, nematodes starve and populations decline. Effectiveness of this type of control depends on availability of appropriate nonhosts. Weed hosts must be eliminated during the non-host rotation.

Soil Solarization In soil solarization, clear plastic film is laid over moist soil during periods of high solar radiation and air temperature. The resulting soil temperature elevation may be sufficient to kill pest species in upper layers of soil (Katan 1984; Stapleton and DeVay 1986). In Egypt, for example, soil solarization reduced population levels of the reniform nematode for 60 days after planting. Soil solarization has been used to reduce population levels of the golden nematode under New York field conditions (LaMondia and Brodie 1984). Since many regions of California have higher air temperatures and more solar radiation during the summer months than New York, control of the golden nematode by soil solarization may be more effective under California conditions (Pullman et al. 1984). However, constraints of solarization include nematode survival below the affected layer and the opportunity cost of removing land from production during the several weeks of the treatment period.

Cultural Control To offset sting nematode damage in turf systems, certain cultural practices, including enhancing soil aeration and moisture and close mowing, are useful (Nutter and Christie 1958; Giblin-Davis et al. 1991). Numbers of sting nematode were reduced in

soils amended with alfalfa meal, cottonseed meal, or rice straw (Tomerlin 1969). Soil amendments have also been effective in control of the reniform nematode (Badra et al. 1979; Amin and Youssef 1998).

Biological Control Nematodes have natural enemies that can reduce their ability to survive and reproduce. Several have been studied as potential biological control agents of the sting nematode in turf grass (Grewal et al. 1997; Giblin-Davis 1990; Bekal et al. 1999). Fungi and other organisms have been investigated for their potential to control the golden nematode (Jatala et al. 1979).

Parties Potentially Affected by Nematodes

Agricultural industries, including growers and the marketing sector, consumers and taxpayers may all be affected by exotic nematode infestations. Potential effects include crop loss, increased control costs, change in cultivars grown, change in crop rotations, delay in replanting of perennials, reduced interstate commerce, and trade barriers for exported crops and plant materials, increased consumer prices, and increases in regulatory costs. Such problems already exist in California. For example, in some areas sugar beet growers can only grow sugar beets once every five to seven years due to the sugar beet cyst nematode.

Agricultural Industries

Many of the commodities at risk from at least one of the nematodes under consideration in this chapter are among the highest-grossing agricultural industries in California (Table 8.1) (California Agricultural Statistics Service 2000). Among the top 10 agricultural industries in California, grapes, nursery products, lettuce, citrus, cotton, strawberries, and alfalfa would be affected by at least one of the nematodes in this study. Overall, the annual value of the commodities potentially at risk was $18.3 billion in 2000. This represents 61 percent of the total value of agricultural production in California (Table 8.1).

As the leading agricultural producing region, the absolute value of affected commodities is greatest for the San Joaquin Valley. The affect-

Table 8.1   California commodities most likely to be affected by exotic nematodes by region

Region[a]	Total value of production ($ million)	Number of top 10 commodities affected	Value of affected commodities (in millions)	Percent of total value	Major crops affected
San Joaquin Valley	14,412	6	7,249	50	Grapes, cotton, citrus, alfalfa, tomatoes (proc.), nursery[b]
Central Coast	5,610	8	4,706	84	Lettuce, grapes, nursery[b], broccoli, strawberries, unspecified vegetables, flowers, cauliflower
South Coast	3,675	9	3,137	85	Nursery[b], flowers (foliage and cut)[b], strawberries, citrus, avocados, vegetables—unspecified, broccoli, lettuce, grapes
Desert	2,588	7	1,321	51	Alfalfa, citrus, lettuce, nursery,[b] grapes, carrots, unspecified vegetables
Sacramento Valley	2,294	5	1,409	61	Rice, tomatoes (proc.), grapes, nursery,[b] peaches
Mountains	508	6	231	45	Alfalfa, nursery,[b] pasture, grapes, rice, potatoes
North Coast	256	3	145	57	Grapes (wine), nursery,[b] pasture
Total[c]	30,017	7	18,348	61	Grapes, nursery,[b] lettuce, citrus, cotton, strawberries, alfalfa

[a] Counties in each region are: San Joaquin Valley—Fresno, Kern, Kings, Madera, Merced, San Joaquin, Stanislaus, Tulare; Central Coast—Alameda, Contra Costa, Lake, Marin, Monterey, Napa, San Benito, San Francisco, San Luis Obispo, San Mateo, Santa Clara, Santa Cruz, Sonoma; South Coast—Los Angeles, Orange, San Diego, Santa Barbara, Ventura; Desert—Imperial, Riverside, San Bernardino; Sacramento Valley—Butte, Colusa, Glenn, Sacramento, Solano, Sutter, Tehama, Yolo, Yuba; Mountains—Amador, Calaveras, El Dorado, Inyo, Lassen, Mariposa, Modoc, Mono, Nevada, Placer, Plumas, Shasta, Sierra, Siskiyou, Trinity, Tuolumne; North Coast—Del Norte, Humboldt, Mendocino.

[b] Includes both host and nonhost commodities.

[c] The total is greater than the sum of counties because the value of production for some commodities was not specified by region.

ed crops in the San Joaquin Valley have a value of $7.25 billion. Even though the value of the affected commodities is less for the central and south coast regions, these regions have the largest percentage of total value of production that is potentially affected by at least one of the nematodes in this case study. The affected commodities account for over 80 percent of the total value of agricultural production in these regions.

Growers   Growers are affected through reduction in crop yields and increases in pest management costs needed to prevent crop damage or other degradation in quality. Growers are also indirectly affected by any changes in market prices for widespread establishment of exotic nematodes. With widespread infestations,

yield losses or increased costs of production may cause market prices to rise, thereby offsetting the lost revenues or increased costs.

Marketing Sector   Restrictions on plant imports from California may be imposed by other states or countries. These may be mitigated by demonstration that the containment is effective and the infestation localized, or by undertaking control measures to ensure that nematodes are not transported into uninfested regions.

The burrowing nematode is a pest of concern and will become an issue in the export commodities to at least Japan, Taiwan, and the European Union. Japan has published a list of the plants reported to be hosts of the burrowing nematode and requires that these plants be accompanied by phytosanitary certification that they are free of the nematode (M. Guidicipietro 2000). Citrus and strawberry nursery stock, carrots, and other root crops could be prohibited for export to these countries, or regulatory treatments could be required. Arizona has already indicated intent to regulate California nursery stock as a response to any modification of California's Burrowing Nematode Exterior Quarantine.

Because the reniform nematode is widespread, no impact to foreign exports has been identified. It is expected, however, that quarantine action would be taken by other states to regulate reniform nematode host material. Should the golden nematode become established, the federal quarantine on this pest would be expanded to include California.

Other Related Agricultural Industries The economic impacts go beyond crop loss and control costs at the farm level. For widespread infestations, market prices may rise and quantities may be reduced. As growers shift out of production of crops susceptible to exotic plant-feeding nematodes, industries supplying inputs (such as labor, seed, etc.) for the production of those crops may also be affected. How they would be affected depends on the inputs required by the replacement crops.

Consumers

Consumers would be affected by higher food costs. Both increased production costs and yield losses put upward pressure on consumer prices. In addition, people purchasing nursery plants for landscaping would face higher prices for those commodities.

Taxpayers

Intervention strategies for exotic nematodes (exclusion, eradication, containment, and suppression) each can be funded and administered by either the private or public sectors or some combination. When public regulatory agencies are involved, issues related to taxes and budget allocations come into play. Taxpayers may fund border control measures to prevent the entry of exotic nematodes. Public programs may be necessary to eradicate small infestations or undertake plant-breeding programs to develop resistant varieties suitable to California. Other,

important research areas include developing new nematicides, identifying natural enemies, and decontaminating plants in nurseries.

Policy Scenarios

Once a nematode has entered California, an eradication program through soil removal and nematicide treatment may be attempted if the infestation is small enough. For larger infestations, a chemical eradication program may be attempted. Usually, public agencies will mandate and conduct the eradication program. However, in some instances, such as the chrysanthemum white rust eradication program in California during the late 1990s, growers are required to complete an eradication program themselves. In this section, an analysis of a grower eradication program will be completed for the rice foliar, sting, reniform, and burrowing nematodes. Due to the long survival period of the golden nematode, eradication is not biologically feasible; therefore, only the costs of establishment are calculated for this pest. The pests and crops considered in these analyses are the rice foliar nematode on rice in the Sacramento Valley; the golden nematode on fresh and processed tomatoes in the San Joaquin Valley; the sting and reniform nematodes on cotton in the San Joaquin Valley; the reniform nematode on table, raisin, and wine grapes in the San Joaquin Valley, and wine grapes from Sonoma County; and the burrowing and reniform nematodes on oranges in the San Joaquin Valley, lemons from San Diego County, and grapefruit from Riverside County. The diversity of nematodes, crops, and regions allows us to compare how differences in input costs affect the decision to eradicate or manage an infestation and the optimal management alternatives to use.

If the eradication program fails or if it is not feasible, then the pest is considered to be established. Eradication costs will vary depending upon where the nematode is found. Eradication costs are also influenced by the cropping sequence, soil, and climate of the infested site. Due to the wide disparity in costs of eradication, sample costs per acre are presented for agricultural infestations in rice, tomatoes, cotton, citrus, and grapes.

The costs of eradication will be compared to the expected losses due to establishment. Those losses can vary significantly, depending upon the pest and the area in which it becomes established. For these analyses, the lowest treatment cost alternative for selected commodities will be determined and the costs per acre estimated. The per acre infestation costs will be aggregated over different infestation sizes to reflect the potential increase in grower costs for a specific agricultural industry. Given the number of pests considered in these analyses, market effects are not estimated. Due to the large number of options, depending on farm-specific characteristics, a policy of containment should eradication not be feasible is not pursued in this analysis. If the grower costs of establishment are less than the costs of eradication, then the implications for an eradication program by public regulatory agencies will be discussed.

Economic Analysis

The economic analysis first examines the grower costs of an eradication program, then the grower costs of establishment using the least-cost control method, and then compares the costs of eradication to the costs of establishment. Data on preinfestation levels of grower costs for both the eradication and establishment scenarios are available from University of California Cooperative Extension budgets (1997-2001). Where costs of nematicide treatments are considered, specific chemicals that have been demonstrated to be effective are selected as examples. Their selection for these analyses does not imply endorsement of specific products. The chemicals considered in the analyses include 1,3-dichloropropene (Telone II), metam-sodium (Vapam), aldicarb (Temik), and fenamifos (Nemacur 3).

Eradication

Methodology The same eradication strategy is used for each crop. Eradication takes place over a two-year period. At the start of the period one soil fumigation treatment of Telone II, at 35 gallons per acre, is completed, followed by one treatment of metam-sodium (Vapam) at 25 gallons per acre, not exceeding a concentration of 250 ppm. A second and third treatment of Telone II and Vapam are applied at annual intervals. Both Telone II and Vapam are custom applied, and Vapam is applied with 6 acre-inches of water. For reniform nematodes,

an additional pretreatment irrigation of 6 acre-inches is required to bring the nematode out of dormancy prior to soil fumigation. Application and materials for the three Telone II/Vapam treatments cost $3,615. Due to the wide host ranges of the nematodes in this study on both commercial crops and weeds, the land must remain fallow during the two-year eradication program and maintained plant free with herbicides. Herbicide treatments are $39 per acre over the two-year period. Eradication costs to growers also include lost revenues and interest on idle capital. For this study an average value of $500 per acre per year is used for a total loss of $1,000. This cost is invariant to the commodity under consideration because no crops are grown on the land. If an alternative crop is possible, then the cost would be the difference between profits earned with the original crop and profits earned with the replacement crop.

Results The eradication costs per acre range from $4,729 for the sting nematode on cotton produced in the San Joaquin Valley to $6,454 for the reniform nematode on lemons grown in San Diego County (Table 8.2). The differences in costs are due to varying regional water prices and the type of nematode eradicated.

The cost of water in California varies dramatically depending on where crops are grown. In the San Joaquin Valley, different water districts charge different prices, and the cost of water for the crops in this study ranges from $3.14 an acre-inch for raisins to $5.63 an acre-inch for table and wine grapes. In contrast, water costs $13.33 an acre-inch for grapefruit in Riverside County and $50.00 an acre-inch for lemons in San Diego County.

The type of nematode that is being eradicated also influences total eradication costs. Because reniform nematodes require an irrigation treatment before the soil is fumigated with Telone II and Vapam, reniform eradication costs are higher than the costs for the rice foliar, sting, and burrowing nematodes, all other factors being held constant. Application rates of nematicides may differ with soil texture. Lower rates than those used in these analyses may be effective in coarse-textured soils. Always, attention must be paid to soil moisture and temperature conditions, which can significantly influence efficacy of nematicides.

Table 8.2 Total eradication costs per acre[a]

Nematode	Crop	Telone II/Vapam	Herbicides	Water	Income and interest	Total cost
					($)	
	Annual Crops					
Rice foliar	Rice	3,615	39	81	1,000	4,735
Sting	Cotton	3,615	39	75	1,000	4,729
Reniform	Cotton	3,615	39	150	1,000	4,804
	Perennial Crops					
Reniform	Wine grapes, Sonoma	3,615	39	217	1,000	4,871
	Wine grapes, San Joaquin Valley	3,615	39	203	1,000	4,857
	Table grapes	3,615	39	203	1,000	4,857
	Raisin grapes	3,615	39	113	1,000	4,767
	Oranges	3,615	39	191	1,000	4,845
	Lemons	3,615	39	1,800	1,000	6,454
	Grapefruit	3,615	39	480	1,000	5,134
Burrowing	Oranges	3,615	39	96	1,000	4,750
	Lemons	3,615	39	900	1,000	5,554
	Grapefruit	3,615	39	240	1,000	4,894

[a]All costs are for the two-year program.

Management of an Established Infestation

The alternative to eradicating a newly introduced nematode is to allow it to become established and then to manage the population. The establishment scenario estimates the cost of a nematode infestation without any pest control measures adopted and compares that cost with the costs of control using a chemical treatment. When no control measures are used, yields decrease, and the cost per unit of production increases. When chemical controls are used, production costs increase, but yields are maintained, and the costs per unit of production again increase. For each crop, the point at which yield decline would be enough to warrant treatment is estimated to determine if growers should undertake control measures.

Methodology The potential decline in yields and the appropriate nematicide treatment alternative were determined for each crop in this study (Table 8.3). Yield reductions were provided for resistant and nonresistant varieties, when available. While yield reduction figures are given for resistant varieties, many resistant varieties would not be suitable for California. Due to variations in population densities from year to year and agro-climatic differences between regions, a minimum and maximum range of yield decreases is given. For perennial crops, the analysis is completed for the planting of a new orchard or vineyard.

In addition to yield reductions when no treatment is undertaken, we estimate that there will be a delay of one year before perennial crops start producing and that the productive life span of the plants will be reduced to half that without the exotic nematode infestation. The delay in bearing fruit postpones the revenues that a grower receives. We calculate the losses associated with postponed revenues as the discounted difference between what the grower would have received if no nematodes were present and what the grower receives with nematodes.

The costs of a shorter production life span are estimated by amortizing the costs to establish a grove or vineyard over half the original expected life of the vineyard or grove using the formula

Annual amortization costs =
$$\frac{\text{Total Establishment Costs} * r}{(1 - (1 + r)^t)}$$

where r is the interest rate and t is the expected life.

If a chemical treatment is used to manage an exotic nematode infestation, the appropriate nematicide treatment depends on whether the crop

Table 8.3 Exotic nematode treatment scenarios

| Nematode | Crop affected | No treatment reduction in yield (%) | | Chemical treatment | | | Other practices |
		Resistant	Nonresistant	Pesticide	Other considerations	Years effective	
Rice foliar	Rice	24	17–54	Telone II	None	1	Use clean seed
Golden	Fresh and processed tomatoes	N/A	10–30	Telone II	If treat, rotation period is 2–3 years	Pretreat when in rotation	If no treatment, use a long rotation of 5–6 years
Sting	Potatoes	0–100	10–30	Telone II	None	1	
	Turf grass/sod	Some grasses more resistant	0–100	Telone II	None	1	Temik
Reniform	Cotton	N/A	60–80	Telone II	None	2	Temik
	Cotton	N/A	40–60	Telone II			Temik
	Citrus and grapes, before planting	N/A	40–80	Telone II and Vapam	A pretreatment irrigation of 6 acre-inches to bring the nematode out of dormancy	4-5 years, then after plant treatments	
Burrowing	Grapes, after planting	N/A	40–80	Nemacur 3	None		N/A
	Citrus, before planting	N/A	40–80	Telone II and Vapam	None	4–5 years	N/A
Burrowing and reniform	Citrus, after planting	N/A	40–80	Nemacur 3	None	2	N/A

is an annual or perennial (Table 8.3). For annual crops, a preplanting soil fumigation with Telone II is recommended. To reduce surviving nematodes in debris from a previous rice crop, and for reniform nematodes in land used for cotton, the treatments must be done before each planting of the host crop. If rice was not grown the previous year, no soil treatment would be needed before planting with clean rice seed. For sting nematodes on land in a cotton rotation, treatment is every other year. The price for materials and application are given in Table 8.4. Nematode control costs for annual crops are reflected as an annual increase in the costs of production.

Treatments are more aggressive in fields that will be planted with a perennial crop to allow the roots of nursery stock to become well established before nematode populations build up again in the soil. Preplanting treatment of such fields consists of one treatment of Telone II, followed by one treatment of Vapam (Table 8.3). Telone II and Vapam are both custom applied (Table 8.4). After the pest population recovers, biennial treatments of Nemacur 3 are used to manage nematodes.

For perennial crops, the preplant nematode control costs are reflected as increases in the establishment costs of a grove or vineyard and then amortized over its expected productive lifetime using Equation 8.1. Even though Nemacur 3 is applied after the vineyard or grove becomes established, the net present value of the costs to apply it over the productive lifetime is also amortized into an annual expense.

Once the annual increase in costs was determined, the break-even yield loss value was calculated as

$$\text{Break-even yield loss value} = \frac{C_N}{RY}$$

where C_N is the cost per acre of nematode treatments, R is the average annual returns per unit for the crop as given in University of California Cooperative Extension crop budgets, and Y is the yield per acre when nematodes are treated.

The establishment of an exotic nematode may cause additional costs to agricultural industries if quarantines are imposed. The extent to which any industry is affected depends on the type of crop, what percentage of total production originates from the quarantined area, and the availability of markets in regions that will not impose quarantines.

Commodities that are affected by quarantines include those that are sold as root crops or with roots attached or other commodities in which the nematode lives. Examples include potatoes, carrots, sod, nursery plants, bulbs, rhizomes, and unhulled rice. Commodities that would not be affected would be those without direct feeding by nematodes or those that receive additional processing or treatment that eliminates the nematode. Examples include fresh citrus fruit, cotton, milled rice, fresh grapes, raisins, wine grapes, and treated root crops.

The establishment of quarantine does not necessarily result in losses for an industry. If a relatively small percentage is exported to regions protected by the quarantine, or if alterna-

Table 8.4 Nematicide application costs

Nematicide	Unit	Quantity	Price per unit	Application method	Application costs	Other costs
Telone II	Gallon	10	$18.29	Custom applied	Included in per gallon costs	None
Temik	Pound	5	$4.43	Grower applied when seeding	N/A	Only suppresses, grower would need to plant in nonhost crop after 4-5 years
Telone II followed by Vapam 3 weeks later	N/A	Telone II: 35 gpa; Vapam: 250 ppm	Telone: $640; Vapam: $325	Custom applied	Telone: included in materials costs Vapam: $40 per acre-inch of water	Applied with 6 acre-inches of water
Nemacur 3	Gallon	1		Grower applied when irrigating	N/A	None

tive markets are readily available, then no industry-wide effects may occur. When relatively small quantities are affected or alternative markets are available, the costs to move commodities from quarantined markets to other markets are small. In the short run marketing costs will be incurred as product is redirected; however, once new markets are established, these extra costs will disappear. As the percentage affected increases, or if access to alternative markets is limited, marketing costs increase, and quarantines may impose additional costs on an agricultural industry. The effects of any permanent additional marketing costs or changes in demand due to quarantines would be captured at the market level and are beyond the grower level analysis being completed for this study.

Results Treatment costs for annual crops range from a low of $22 per acre to a high of $201 per acre (Table 8.5). Costs are lowest when the nematicide Temik is used; however, nematode populations will build up over time. The costs of population buildup and the need to periodically rotate to a nonhost crop are not included in the cost of using Temik. The costs to treat sting nematodes with Telone II are also lower than the annual cost to treat rice foliar, golden, and reniform nematodes with Telone II because sting nematodes in cotton need to be treated only every other year.

As was the case with the eradication scenarios, differences in the amortized treatment costs for perennial crops are due in part to variations in the cost of water between regions (Table 8.5). In addition, vineyard and grove productive life spans vary. The longer trees or vines are in production, the greater the time over which costs are spread out, and the lower the annual amortized investment cost.

There is no scenario among the nematode pests and crop combinations under analysis in this chapter where it is unequivocally the case that a nematode population should not be subjected to nematicide treatment when the yield reductions are at higher levels (Table 8.5). However, when yield losses caused by the rice foliar nematode on rice and the golden nematode on processed tomatoes are at the lower levels, treatment may cost a grower more than lost revenues. For the rice foliar nematode, if yield losses are expected to be less than 24.9 percent, the grower should not treat. Similarly, if processed tomato yield losses are expected

to be lower than 13.7 percent, the grower minimizes losses by not treating. In all other scenarios treatment minimizes losses as the gains in yields, and, consequently, revenues, are greater than the extra cost of nematode control. In general, the yield loss threshold level is lower for perennial crops than for annual crops.

What is not examined in this analysis is whether a grower would stop growing a crop due to yield losses or increases in production costs. Based on currently available crop budgets, costs may increase anywhere from 1.7 percent to 21.9 percent (Table 8.5). In general, the percentage increase in costs is greater for the annual crops treated with Telone II than for the perennial crops (Table 8.5). The greater the percentage increase in costs, the more likely it is that growers would switch to growing other crops. Interviews with county Cooperative Extension personnel indicate that many growers would stop growing cotton if exotic nematodes, such as the sting or reniform, became established in cotton fields.

Cost/Benefit Analysis

The cost/benefit analysis compares the costs and benefits of a grower eradication program for small infestations of exotic nematodes. The costs are for eradication, and the benefits are from avoiding the costs of control.

Methodology Eradication of an exotic nematode will be undertaken if the cost of investing in nematode eradication is less than the benefits of preventing the cost of control. For annual crops, the benefits of the two-year eradication program extend beyond the two-year period; thus, the total benefits are equal to the present value of all future annual benefits. The eradication costs are compared to the present value of the benefits to determine if growers of annual crops will invest in an eradication program.

For perennial crops both the costs and benefits of the eradication program are amortized into an annual value over the productive life of the vineyard or orchard. To determine if growers will invest in an eradication program for perennial crops the annual costs and benefits are compared.

Results For all crops and for all nematodes, the costs of an eradication program are greater

Table 8.5 Annual costs and yield loss threshold levels for treating an exotic nematode infestation

Nematode	Crop	Annual increase in costs for treatment	Minimum yield loss	Maximum yield loss	Yield loss threshold level	Increase in production costs	Treat?
		- - - - - - - - - - - ($) - - - - - - - - - - -			- - - - - - - - - (%) - - - - - - - - -		
	Annual Crops						
Rice foliar	Rice	201	17	54	−24.9	21.3	Maybe
Golden	Tomatoes (fresh)	201	10	30	−3.5	3.9	Yes
	Tomatoes – (processed)	201	10	30	−10.8	13.7	Maybe
Sting	Cotton (Telone II)	103	60	80	−10.3	11.2	Yes
	Cotton (Temik)	22	60	80	−2.2	2.4	Yes
Reniform	Cotton (Telone II)	201	40	60	−20.2	21.9	Yes
	Cotton (Temik)	22	40	60	−2.2	2.4	Yes
	Perennial Crops						
Reniform	Wine grapes, Sonoma	170	40	60	−1.4	1.9	Yes
	Wine grapes, San Joaquin Valley	167	40	60	−5.1	6.9	Yes
	Table grapes	167	40	60	−2.4	2.5	Yes
	Raisin grapes	167	40	60	−8.1	8.2	Yes
	Oranges	152	40	60	−3.5	2.7	Yes
	Lemons	220	40	60	−1.8	2.0	Yes
	Grapefruit	161	40	60	−4.4	3.2	Yes
Burrowing	Oranges	150	40	80	−3.4	2.6	Yes
	Lemons	192	40	80	−1.6	1.7	Yes
	Grapefruit	155	40	80	−4.2	3.1	Yes

than the benefits (Table 8.6). Costs exceed the benefits by about 50 percent for the rice foliar nematode and the reniform nematode in cotton. Costs are over 1,000 percent higher than the benefits when Temik is used to control the sting and reniform nematodes. In general, though, the costs to a grower to eradicate an exotic nematode infestation are 160 to 180 percent greater than the benefits for both annual and perennial crops. Therefore, growers would not voluntarily invest in eradicating nematodes on their own land under any scenario.

Failure to eradicate a newly introduced nematode in one field, however, will allow the nematode to spread and infest other fields. Costs then increase for other growers.

These negative spillover effects may make it cost effective for a public agency or a grower association to incur or subsidize any eradication efforts on individual farms. By preventing the spread of exotic nematodes, the whole industry benefits. When determining the public costs and benefits of a policy to eradicate, the benefits of protecting the industry need to be weighed against the costs of eradicating discrete, small infestations.

Should an infestation of only one species of nematode become established in California, the costs of production for many agricultural industries increase (Table 8.7). For example, should sting nematodes infest 10,000 acres of land used for cotton production, industry control costs are $1 million. Industry control costs for reniform nematodes would be even greater, $2 million for a 10,000-acre infestation. Costs then rise in proportion to the acreage infested (Table 8.7). However, sting nematodes prefer coarse soils and reniform nematodes prefer fine textured soils. Because these two nematodes infest different soils, if both sting and reniform nematodes become established, costs would be cumulative. For a 10,000-acre infestation each (20,000 acres total) costs are $3 million for the industry.

Costs are not cumulative, however, when nematodes infest the same soil in the same region because control measures are effective against both species. For example, if both golden and burrowing nematodes infest tomato fields, then the cost to the industry of a simultaneous 10,000-acre infestation is only $2 million (Table 8.7). This cost is the same as the cost if either nematode becomes established.

Table 8.6   Grower costs and benefits of eradicating nematodes

Nematode	Crop	Costs	Benefits	Grower eradicates?	Percent costs greater than benefits
			------- ($) -------		
	Annual Crops[a]				
Rice foliar	Rice	4,735	3,203	No	48
Sting	Cotton (Telone II)	4,729	1,634	No	189
	Cotton (Temik)	4,729	353[c]	No	1,240
Reniform	Cotton (Telone II)	4,804	3,203	No	50
	Cotton (Temik)	4,804	353[c]	No	1,261
	Perennial Crops[b]				
Reniform	Wine grapes, Sonoma	465	170	No	174
	Wine grapes, San Joaquin Valley	462	167	No	177
	Table grapes	462	167	No	177
	Raisin grapes	462	167	No	177
	Oranges	401	152	No	164
	Lemons	613	220	No	179
	Grapefruit	427	161	No	165
Burrowing	Oranges	393	150	No	162
	Lemons	528	192	No	175
	Grapefruit	407	155	No	163

[a]Total costs are compared to the present value of benefits over time.
[b]Annual amortization of eradication costs compared to the annual amortization of management costs.
[c]Does not include costs of rotation to nonhost crop every 4–5 years.

As infestations increase, a larger proportion of production is affected. Eventually, the proportion will be large enough to affect market prices and quantities. Increases in costs will raise prices and decrease quantity demanded and quantity supplied. With higher prices and lower quantities, consumers are worse off. The higher prices will mitigate grower losses to some extent. When determining whether a public eradication program should be undertaken, these additional costs and benefits must also be considered.

Table 8.7   Aggregate grower costs of widespread nematode establishment

Nematode		10,000 Acres Infested	
		Each	Simultaneously
		($ million)	
Sting	Cotton	1	N/A
Reniform	Cotton	2	N/A
Total		3	N/A
Golden	Tomatoes	2	2
Burrowing	Tomatoes	2	2
Total		4	2

Conclusions

Nematodes are excellent invaders; they have evolved numerous strategies for exploiting favorable habitats and withstanding harsh conditions. The degree of risk differs with the life history strategy of the nematode species. Often, a lack of basic information impairs our ability to quantitatively assess that risk. In those instances, we rely on qualitative assessment, based on experiences with the pest in other areas of the world to determine whether there is cause for concern. Such assessments include a degree of uncertainty, but provide direction for immediate action and future research.

Although the direct and opportunity costs of establishment may be substantial, eradication is difficult, if not impossible, unless the nematode is detected very early. Significant costs may also accrue from the imposition of quarantine restrictions on exports of California seed, propagative material, and agricultural products. Indeed, Turkey has already imposed restrictions on the importation of unhulled rice from California because several times the rice foliar nematode has been detected.

We began this chapter with the observation that many of the important pests of California agriculture are not native to the region. They have been introduced with plant material, adhering soil, and other means. They have been spread throughout the state by our tillage, propagation, labor, irrigation, and harvesting practices. The same can happen and, indeed, has happened, with the A-rated exotic pest nematodes reviewed in this chapter and with other species that they exemplify. The challenge will continue to be to intercept their introduction, accurately identify them using the best technology available, and to eradicate, contain, or manage them as appropriate.

References

Abu-Gharbieh, W.I., and V.G. Perry. 1970. "Host Differences among Florida Isolates of *Belonolaimus longicaudatus rau.*" *Journal of Nematology*. 2:209–216.

Alvarez, M.G. 1972. "Planning for the Campaign against the Golden Nematode." XX Mexican-American Plant Protection Work Conference, Guanajuato, Guanajuato.

Amin, A.W., and M.M.A. Youssef. 1998. "Effect of Organic Amendments on the Parasitism of *Meloidogyne javanica* and *Rotylenchulus reniformis* and Growth of Sunflower." *Pakistan Journal of Nematology*. 16:63–70.

Atkins, J.G., and E.H. Todd. 1959. White Tip Disease of Rice. III. Yield Tests and Varietal Resistance. *Phytopathology*. 49:189–191.

Badra, T., M.A. Salem, and B.A. Oteifa. 1979. Nematicidal Activity of Some Organic Fertilizers and Soil Amendments. *Revue de Nématologie*. 2:29–36.

Bekal, S., R.M. Giblin-Davis, and J.O. Becker. 1999. "Gnotobiotic Culture of *Pasteuria* sp. on *Belonolaimus longicaudatus.*" *Journal of Nematology*. 31:522.

Bridge, J., M. Luc, and R.A. Plowright. 1990. "Nematode Parasites of Rice." In M. Luc et al., Eds., *Plant Parasitic Nematodes in Subtropical and Tropical Agriculture*. Wallingford, UK: CAB International. pp. 69-108.

Busey, P., R.M. Giblin-Davis, C.W. Riger, and E.I. Zaenker. 1991. "Susceptibility of Diploid St. Augustine Grass to *Belonolaimus longicaudatus.*" *Journal of Nematology*. 23(Suppl.):604–610.

California Agricultural Statistics Service. 2000. "County Agricultural Commissioners Report Data." http://www.nass.usda.gov/ca/bul/agcom/indexcac.htm.

Chitambar, J.J. 1997a. "A Brief Review of the Burrowing Nematode, *Radopholus similis.*" California Plant Pest and Disease Report. CDFA. 16:66–70.

Chitambar, J.J. 1997b. "A Brief Review of the Reniform Nematode, *Rotylenchulus reniformis.*" California Plant Pest and Disease Report. CDFA. 16:71–73.

Christie, J.R. 1957. "The Yellows Disease of Pepper and Spreading Decline of Citrus." *Plant Disease Reporter*. 41:267–268.

Clarke, A.J., and J. Hennessy. 1984. "Movement of *Globodera rostochiensis* (Wollenweber) Juveniles Stimulated by Potato-Root Exudate." *Nematologica*. 30:206–212.

DuCharme, E.P. 1954. "Cause and Nature of Spreading Decline of Citrus." *The Citrus Industry*. 35(11):6–7, 18; and 35(12):5–7.

DuCharme, E.P. 1968. "Burrowing Nematode Decline of Citrus, a Review." In G.C. Smart and V.G. Perry, Eds. *Tropical Nematology*. Univ. Gainesville, FL: Florida Press. pp. 20-37.

Endo, B.Y. 1971. "Nematode-Induced Syncytia (Giant Cells). Host-Parasite Relationships of *Heteroderidae.*" In B.M. Zuckerman, W.F. Mai, and R.A. Rohde, Eds. *Plant Parasitic Nematodes*, Vol. II. New York: Academic Press. pp. 91-117.

Evans, K., and B.B. Brodie. 1980. "The Origin and Distribution of the Golden Nematode and Its Potential in the U.S.A." *American Potato Journal*. 57:79–89.

Ferris, H., and E.P. Caswell-Chen. 1997. "Nemabase." http://www.ipm.ucdavis.edu/NEMABASE/.

Ferris, J.M. 1957. "Effect of Soil Temperature on the Life Cycle of the Golden Nematode in Host and Nonhost Species." *Phytopathology*. 47:221–230.

Franklin, M.T., and M.R. Siddiqi. 1972. "*Aphelenchoides besseyi.*" *Commonwealth Institute of Helminthology Descriptions of Plant-Parasitic Nematodes*. 1:4.

Giblin-Davis, R.M. 1990. "Potential for Biological Control of Phytoparasitic Nematodes in Bermudagrass Turf with Isolates of the *Pasteuria penetrans* Group." *Proceedings of the Florida State Horticultural Society*. 103:349–351.

Giblin-Davis, R.M., J.L. Cisar, F.G. Bilz, and K.E. Williams. 1991. "Management Practices Affecting Phytoparasitic Nematodes in 'Tifgreen' Bermudagrass." *Nematropica*. 21:59–69.

Golden, A.M., and D.M.S. Ellington. 1972. "Redescription of *Heterodera rostochiensis* (Nematoda: *Heteroderidae*) with a Key and Notes on Closely Related Species." *Proceedings of the Helminthological Society of Washington*. 39:64–78.

Goodey, J.B., and M.T. Franklin. 1958. *The Nematode Parasites of Plants Catalogued under Their Hosts*. Farnham Royal, England: Commonwealth Agricultural Bureaux. 140 p.

Goodey, J.B., and M.T. Franklin. 1959. *Supplement to the Nematode Parasites of Plants Catalogued under Their Hosts, 1955-1958*. Farnham Royal, England: Commonwealth Agricultural Bureaux. 66 p.

Green, C.D., D.N. Greet, and F.G.W. Jones. 1970. "The Influence of Multiple Mating on the Reproduction and Genetics of *Heterodera rostochiensis* and *H. schachtii.*" *Nematologica* 16:309–326.

Grewal, P.S., W.R. Martin, R.W. Miller, and E.E. Lewis. 1997. "Suppression of Plant-Parasitic Nematode Populations in Turf Grass by Application of Entomopathogenic Nematodes." *Biocontrol Science and Technology.* 7:393–399.

Guidicipietro, M. 2000. United States Department of Agriculture. Personal communication to J. Chitambar, February.

Huang, X., and J.O. Becker. 1999. Life Cycle and Mating Behavior of *Belonolaimus longicaudatus* from Sandy Soil. *Journal of Nematology.* 31:70–74.

Jatala, P. 1991. "Reniform and False Root-Knot Nematodes, *Rotylenchulus* and *Nacobbus* spp." In W.R. Nickle, Ed. *Manual of Agricultural Nematology.* New York: Marcel Dekker, Inc. 1,035 p.

Jatala, P., R. Kaltenbach, and M. Bocangel. 1979. "Biological Control of *Meloidogyne incognita acrita* and *Globodera pallida* on Potatoes." *Journal of Nematology.* 11:303.

Katan, J. 1984. "Soil Solarization for Disinfestation, Plant Disease Control." *Acta Hortic.* 152:227–236.

Koenning, S.R., S.A. Walters, and K.R. Barker. 1996. "Impact of Soil Texture on the Reproductive and Damage Potentials of *Rotylenchulus reniformis* and *Meloidogyne incognita* on Cotton." *Journal of Nematology.* 28:527–536.

Kort, J.H., H.J. Ross, Rumpenhorst, and A.R. Stone. 1977. "An International Scheme for Identifying and Classifying Pathotypes of Potato Cyst-Nematodes *Globodera rostochiensis* and *G. pallida.*" *Nematologica.* 23:333–339.

LaMondia, J.A., and B.B. Brodie. 1984. "Control of *Globodera rostochiensis* by Solar Heat." *Plant Disease.* 68:474–476.

Mai, W.F. 1977. "Worldwide Distribution of Potato-Cyst Nematodes and Their Importance in Crop Production." *Journal of Nematology.* 9:30–34.

Mai, W.F., and B. Lear. 1953. "The Golden Nematode." *Cornell Extension Bulletin.* 870:32.

Martin, G.C. 1969. "Survival and Infectivity of Eggs and Larvae of *Meloidogyne javanica* After Ingestion by a Rodent." *Nematologica.* 15:620.

Meagher, J.W. 1977. "World Dissemination of the Cereal Cyst Nematode (*Heterodera avenae*) and Its Potential as a Pathogen of Wheat." *Journal of Nematology.* 9:9–15.

Miller, L.I. 1972. "The Influence of Soil Texture on the Survival of *Belonolaimus longicaudatus.*" *Phytopathology.* 62:670–671.

Nemabase. Database of the host status of plants to nematodes: http://ucdnema.ucdavis.edu/imagemap/nemmap/accessdatabases.

Noling, Jerry W. 2001. University of Florida, Cooperative Extension. Personal communication to K.M. Jetter.

Nutter, G.C., and J.R. Christie. 1958. "Nematode Investigations of Putting Green Turf." *Proc. Fla. Sta. Hort. Soc.* 71:445–449.

Orton Williams, K.J., and M.R. Siddiqi. 1973. "*Radopholus similis.* Commonwealth Institute of Helminthology Descriptions of Plant-Parasitic Nematodes." *Set* 2:27.

Ou, S.H. 1985. *Rice Diseases*, 2nd ed. Kew, Surrey, England, Commonwealth Mycological Institute. 380 p.

Perry, V.G., and H.L. Rhoades. 1982. "The Genus *Belonolaimus.*" In R.D. Riggs, Ed. *Nematology in the Southern Region of the United States.* Southern Cooperative Series Bulletin 276. Arkansas Agricultural Experiment Station, Fayetteville. pp. 144–149.

Pinheiro, F.P., R.P. Vianello, F.S. Ebeidalla, and R.C.V. Tenente. 1997. "Thermal Seed Treatments to Eradicate *Aphelenchoides* from *Brachiaria dictyoneura.*" *Nematologia Brasileira.* 21(1):92–97.

Platt, H.M. 1994. "Foreword." In S. Lorenzen, Ed., *The Phylogenetic Systematics of Free-living Nematodes.* London: The Ray Society. pp. i-ii, 383 p.

Pullman, G.S., J.E. DeVay, C.L. Elmore, and W.H. Hart. 1984. "Soil Solarization, a Nonchemical Method for Controlling Diseases and Pests." Cooperative Extension, University of California, Division of Agriculture and Natural Resources, Leaflet 21377. 8 p.

Robbins, R.T., and K.R. Barker. 1973. "Comparison of Host Range and Reproduction among Populations of *Belonolaimus longicaudatus* from North Carolina and Georgia." *Plant Disease Reporter.* 57:750–754.

Robinson, A.F., C.M. Heald, and S.L. Flanagan. 1987. "Relationships between Soil Texture and the Distributions of *Rotylenchulus reniformis*, *Meloidogyne incognita* and *Tylenchulus semipenetrans* in the Lower Rio Grande Valley, North America." *Journal of Nematology.* 19:553.

Sivakumar, C.V. 1987. "The Rice White Tip Nematode in Kanyakumari District, Tamil Nadu, India." *Indian Journal of Nematology.* 17:72–75.

Spears, J.F. 1968. "The Golden Nematode Handbook: Survey, Laboratory, Control, and Quarantine Procedures." USDA Agriculture Handbook No. 353. 81 p.

Stapleton, J.J., and J.E. DeVay. 1986. "Soil Solarization: A Non-Chemical Approach for Management of Plant Pathogens and Pests." *Crop Protection.* 5:190–198.

Stone, A.R. 1973. "*Heterodera pallida.* Commonwealth Institute of Helminthology Descriptions of Plant-Parasitic Nematodes." *Set* 2:16–17.

Sudakova, I.M., and A.V. Stoyakov. 1967. "On Reproduction and Life History Duration of *Aphelenchoides besseyi* Christie, 1942." *Zool. Zh.* 46:1097–1099.

Tamura, I., and K. Kegasawa. 1958. "Studies on the

Ecology of the Rice Nematode, *Aphelenchoides besseyi* Christie. II. On the Parasitic Ability of Rice Nematodes and Their Movement into Hulls." *Japanese Journal of Ecology.* 8:37–42.

Taylor, A.L. 1969. Nematode Parasites of Rice. In J.E. Peachey, Ed. *Nematodes of Tropical Crops.* St. Albans, England: Commonwealth Bureau of Helminthology. pp. 264–268.

Thomason, I.J., and E.P. Caswell. 1987. "Principles of Nematode Control." In R. H. Brown and B. R. Kerry, Eds., *Principles and Practice of Nematode Control in Crops.* New York: Academic Press. pp. 87–130.

Todd, E.H., and J.G. Atkins. 1958. "White Tip Disease of Rice. I. Symptoms, Laboratory Culture of Nematodes and Pathogenicity Tests." *Phytopathology.* 48:632–637.

Tomerlin, Jr., A.H. 1969. "The Influence of Organic Amendments on Numbers of Nematodes and Other Microorganisms in the Soil." Ph.D. Dissertation, University of Florida, Gainesville. 56 p.

University of California Cooperative Extension. 1997–2001. Crop Budgets. http://coststudies.ucdavis.edu/.

Viglierchio, D.R. 1991. *The World of Nematodes.* Davis, Ca.: published by the author.

Waliullah, M.I.S. 1984. "Nematodes in Irrigation Water." *Nematologia Mediterranea.* 12:243–245.

Wallace, H.R. 1964. *The Biology of Plant Parasitic Nematodes.* New York: St. Martin's Press. pp. 152–162.

Winslow, R.D., and R.J. Willis. 1972. "Nematode Diseases of Potatoes. II. Potato Cyst Nematode, *Heterodera rostochiensis*." In J. Webster, Ed., *Economic Nematology.* New York: Academic Press. pp. 18-34.

9

Ex-Ante Economics of Exotic Disease Policy: Citrus Canker in California

Karen M. Jetter, Edwin L. Civerolo, and Daniel A. Sumner

Introduction

This chapter analyzes the effects of an introduction and establishment of the citrus canker pathogen into California. Citrus canker is a bacterial disease of most commercial *Citrus* species and cultivars grown around the world, as well as some citrus relatives (Civerolo 1984; Goto 1992a; Goto 1992b; Schubert and Miller 1999). Citrus canker is established primarily in tropical and subtropical areas where high temperatures and rainfall occur at the same time of the year (Civerolo 1984, Civerolo 1994, Stall and Civerolo 1993). However, citrus canker is also established in southwest Asia—Iran, Iraq, Oman, Saudi Arabia, the United Arab Emirates, and Yemen (Commonwealth Mycological Institute 1996). The disease is caused by *Xanthomonas campestris* (*=axonopodis*) pv. citri (*Xcc*) (Goto 1992a, 1992b; Stall and Civerolo 1991; Stall and Civerolo 1993; Vauterin et al. 1995; Young et al. 1996). However, distinct pathotypes of *Xcc* are associated with different forms of the disease (Civerolo 1984; Stall et al. 1982; Verniere et al. 1998). In addition, there are three distinct genotypes of the Asiatic strain of *Xcc* in Florida. Citrus canker probably originated in Southeast Asia or India and now occurs in more than 30 countries.

Xcc causes erumpent lesions on leaves, stems, twigs and fruit (Civerolo 1984; Goto, 1992a; Goto 1992b; Schubert and Miller 1999). Severe infections can result in defoliation, unsightly blemished fruit, premature fruit drop, twig dieback, and general tree decline (Goto 1992a; Goto 1992b).

Asiatic citrus canker (citrus canker-A) is the most widespread form of the disease globally (Goto 1992a, 1992b). Most host species and cultivars are affected and it is the most damaging.

Considerable national and international regulatory efforts are designed to prevent spread of the pathogen to, and disease establishment in, citrus-growing regions around the world where the disease is not endemic but where environmental conditions are conducive to disease development (Goto 1992b). Other forms of the disease are rarely found in nature, if at all anymore (Civerolo 1984; Verniere et al. 1998); however, all strains of the *Xcc* associated with different forms of the disease are subject to the same international phytosanitary regulations.

Despite these regulations, citrus canker-A infestations in the United States occurred in the Gulf States around 1910 and in Florida in 1986, 1995, and from 1997 to the present (Gottwald et al. 1997; Schubert and Miller 1999). The 1910 infestation was eradicated over several years with significant economic losses to producers due to lost plant and crop values, and to regulatory agencies due to eradication costs. The last detection of the 1986 outbreak was in 1992. However, the 1986 and 1997 infestations were associated with closely related strains of the pathogen. This suggests that holdover infections from the 1986–1992 infestation went undetected. The 1986 and 1995 infestations were caused by genetically distinct strains of *Xcc*. By December 1999 the disease had spread to over 400 square miles of urban areas and into commercial lime groves. An eradication program is currently underway to eliminate citrus canker infestations in Florida (Gottwald et al. 1997; Schubert and Miller 1999). In January 2000 approximately 600 acres of lime groves were burned as part of the Florida citrus canker eradication program. Through July 2002, a total of 2,238,024 residential and commercial citrus trees have been removed in the state of Florida.

California has never experienced an infestation of citrus canker. Citrus grown in California is protected under a U.S. Department of Agriculture (USDA) external quarantine against the importation of citrus fruit, stock, and other fresh or dried products from countries known to have citrus canker. It is also protected by a USDA internal quarantine around regions within the United States known to have citrus canker. In addition, California maintains its own restrictions against the importation of citrus and citrus nursery stock. Should citrus canker be found in California, public regulatory agencies would have the option to eradicate the disease or to allow it to become established. If citrus canker became established, fresh citrus from California would be subject to trade restrictions, and production costs would increase for growers.

Biology of the Disease

Symptoms

All above-ground tissues of citrus are susceptible to infection by *Xcc* (Civerolo 1984; Goto 1992a, 1992b; Schubert and Miller 1999; Stall and Civerolo 1993). Infection generally occurs through natural openings (stomates, lenticels) and wounds. On leaves, minute, blisterlike lesions appear on the lower surface initially about 7 to 10 days after infection occurs under optimum conditions. Over time, these become tan or brown with a water-soaked margin and surrounded by a chlorotic halo. The lesions become distinctly raised and have a corky appearance. At this stage, lesions are usually visible on both leaf surfaces. The lesions become erumpent, and the centers become craterlike. The centers of the lesions may fall out, creating a shot-hole effect. Severe leaf infection may result in defoliation. Citrus canker lesions on twigs and fruit are generally similar to those on leaves. Blemished fruit and premature fruit drop are major impacts of the disease if trees are left untreated.

Ecology

Several pathotypes of *Xcc* are characterized by their natural host range (Civerolo 1984; Stall and Civerolo 1993). The most virulent pathotype, *Xcc*-A, is associated with the Asiatic form of citrus canker (Civerolo 1984; Goto 1992a, 1992b; Stall and Civerolo 1993). While the host range of *Xcc*-A is broader than that of the other pathotypes, phenotypically distinct strains of *Xcc*-A (designated *Xcc*-A*) with pathogenicity limited to Mexican lime (*Citrus aurantifolia*) in India and Southwest Asia have been described recently (Verniere et al. 1998). Pathotype *Xcc*-B is associated with cancrosis B in a few countries in South America (Argentina, Uruguay, Paraguay) and has a restricted natural host range. Lemon (*Citrus limon*) is the most susceptible species, while sweet orange (*Citrus sinensis*) is little affected under natural conditions. Pathotype *Xcc*-C is associated with Mexican lime cancrosis in Brazil. Other pathogenic variants of *Xcc* may exist.

Citrus canker occurs most frequently and severely in citrus-growing areas characterized by warm, humid weather (Goto, 1992a, 1992b; Schubert and Miller 1999; Stall and Civerolo 1993). However, environmental conditions in many, if not all, citrus-growing areas are likely to be conducive to infection and disease development (Stall and Civerolo, 1993). The *Xcc* pathogen overwinters in lesions following infection in autumn on diseased leaves, twigs, and stems. In the spring, bacteria ooze out of old lesions when free water is available. These bacteria cause new infections on young leaves in the spring. Lesions on leaves are the primary sources of inoculum for fruit infection in the summer (Goto 1992a). A generalized life cycle of citrus canker is presented in Figure 9.1.

The extent of infection and severity of disease development depend on the specific *Citrus* species and cultivar, environmental conditions, and *Xcc* pathotype (Civerolo 1984; Goto 1992b; Gottwald et al. 1997; Pruvost et al. 1997). All young, developing, above-ground parts of susceptible citrus hosts can be infected (young leaves, twigs, thorns, branches, and fruit). Infection occurs primarily through stomates and other natural openings or wounds. Resistance of leaves and fruit to infection increases with maturity (Goto 1992a, 1992b; Stall and Civerolo 1993).

Xcc survives in diseased host plant tissues parasitically, on host and nonhost plants epiphytically (nonparasitically), without causing symptoms, and to a limited extent in association with plant tissue debris in the soil (Goto 1992a, 1992b). *Xcc* can survive for long periods in in-

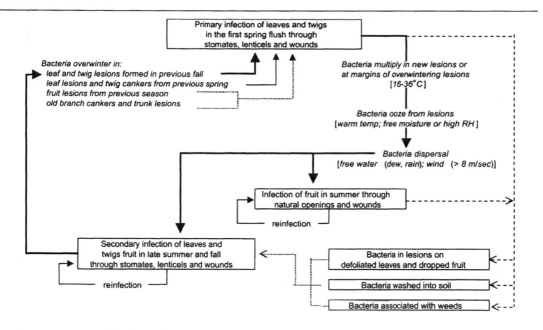

Figure 1. Disease cycle of citrus bacterial canker caused by *Xanthomonas campestris* (=*axonopdis*) pv. *citri*).

──────▶ = main infection;   ─────▶ = minor infection;   ┄┄┄┄┄▶ = neglible infection

Figure 9.1   Generalized life cycle of citrus canker.

fected bark tissue of trunks, low scaffold limbs, and lateral branches. Epiphytic survival of *Xcc* on surfaces of mature leaves and fruit is limited to only a few months during the rainy season in tropical and subtropical areas. Epiphytic populations of viable *Xcc* in semi-arid areas may be undetectable. Accordingly, this form of survival of the pathogen is not epidemiologically significant.

Introduction and Spread

Plant Introduction and Spread

Long-distance dissemination of *Xcc* occurs primarily via the movement of infected/planting stock (e.g., rootstock seedlings, budded nursery trees) and propagating material (e.g., budwood) (Civerolo 1984; Goto 1992b; Gottwald et al. 1997; Pruvost et al. 1997; Schubert and Miller 1999; Stall and Civerolo 1993). Infected fresh fruit with lesions is a potential means of long-distance spread of *Xcc*; however, there is no authenticated record that this is epidemiologically significant with respect to initiation of new infections. There is no record of transmission of *Xcc* via seed. Infested personnel, clothing, tools, equipment, boxes, and other items associated with harvesting and postharvest handling of fruits are potential means of *Xcc* dissemination over short to long distances, at least for a limited time. Long-distance dispersal of *Xcc* by animals, birds, and insects has not been conclusively demonstrated.

Short-distance spread of *Xcc* within trees, and from tree to tree, occurs primarily via wind-driven rain, especially during storms, typhoons, and hurricanes (Civerolo 1984; Goto 1992b; Gottwald et al. 1997; Pruvost et al. 1997). Strong winds that cause injuries on leaves, twigs, and fruit, and rainstorms (as well as thunderstorms, tornadoes, tropical storms, and hurricanes) that disperse the pathogen, facilitate infection. *Xcc* infection can be facilitated by feeding activities of the citrus leaf miner.

Human Introduction and Spread

Violations of current quarantine regulations excluding citrus fruit, citrus stock, and other citrus products from infested regions are a potential source for the introduction of citrus canker into California. The movement of goods often accompanies the movement of people, despite quarantine regulations against some of those goods. Either people are unaware of the restrictions, are unaware of the reasons for the restrictions, or are indifferent to them. Under those circumstances, citrus canker may be accidentally introduced as people bring in fruit or budwood from infested areas.

Citrus canker may also be introduced by people deliberately violating quarantines through commercial smuggling activities. Citrus canker has been identified on dried kaffir lime leaves smuggled into California. Dried kaffir lime leaves are a basic seasoning ingredient used in Thai and other Southeast Asian cuisine. The dried leaves do not contain any active bacteria that can lead to the introduction of citrus canker. However, concerns exist that in response to the demand for ethnic food products, budwood could be smuggled into the country. In recognition of this potential problem, Lincove, an organization that carries out research on improving citrus stock, grew and distributed about 300 kaffir lime trees to commercial citrus producers from clean citrus stock (Stutsman 1999).

Intervention Strategies

Citrus canker management is based on integrated systems of regulatory measures, disease forecasting, planting resistant or tolerant types of citrus, cultural practices, chemical sprays, and biological control (Civerolo 1984; Goto 1992b; Stall and Civerolo 1993). Intervention strategies include exclusion, eradication, and treatment should it become established.

Exclusion

Exclusion regulatory measures include state, national, and international quarantines. National and state quarantines ban the importation of citrus stock from other regions. The importation of fresh fruit, peel, and leaves from eastern and southeastern Asia (including India, Myanmar, Sri Lanka, Thailand, Vietnam, and China), the Malayan Archipelago, the Philippines, Oceania (except Australia and Tasmania), Japan, Taiwan, Mauritius Seychelles, Paraguay, Argentina, and Brazil is banned due to the presence of citrus canker in those countries. Should the exclusion regulatory measures fail, occurrence of citrus canker in California is likely to increase

regulatory activities by state and federal regulatory agencies, including (but not necessarily limited to) implementation of an action plan to delimit, contain, suppress, and eradicate any infestation.

Eradication

Should citrus canker enter California, public regulatory agencies would be responsible for its eradication, if feasible, because homeowners and individual producers would not have strong incentives to voluntarily remove trees. In the Florida citrus canker outbreak, homeowners and producers have expressed more concern about the effects to their own home and business regarding removing trees to eradicate citrus canker than about the potential effects on others. Residential tree removal is expensive, landscape values decline when trees are removed, and people would no longer be able to consume fruit from their backyards. If the private expense of removing trees is greater than the private decrease in profits, should citrus canker become established, producers typically would not choose to remove trees voluntarily.

Failure to eradicate the disease would result in negative spillover effects on other groups because citrus canker would spread to other residences and commercial operations. Eventually, the disease would spread sufficiently so that commercial production and market prices would be affected. The costs and benefits to other groups do not usually factor fully into private decisions, but they are of concern to government regulatory agencies and other affected groups. This is why state and federal regulatory agencies are responsible for eradication programs.

Establishment

Should the eradication measures fail and citrus canker become established, the cultural and marketing strategies available to tree owners would include the use of pathogen- or disease-free nursery planting stock and propagating material and pre- and postharvest chemical treatments (e.g., copper-containing sprays, sodium hypochlorite). In that case, regulatory agencies would no longer be responsible for mitigating costs associated with the disease.

Potentially Affected Parties

Parties affected by citrus canker include homeowners; citrus nurseries, producers, exporters, and fruit processors; wholesale and retail nursery outlets that include citrus, citrus relatives, and other rutaceous plants; users of citrus, citrus relatives, and other rutaceous plants that might be *Xcc* hosts for landscape and other horticultural purposes; domestic and foreign consumers of fresh citrus fruit and citrus products; and taxpayers.

Homeowners

Homeowners are affected by the possibility of tree removal during an eradication program or decreased home production if citrus canker becomes established. Many homeowners grow citrus in their backyards. In a plant survey of 3,000 backyards in northern and southern California, nearly 3,000 citrus trees were counted (California Department of Food and Agriculture 1994). The predominate type of citrus planted in backyards is lemon, followed by fresh oranges.

Producers

Citrus is one of California's largest agricultural industries. The combined crop value of oranges, lemons, and grapefruit in the 1999-2000 season was approximately $1.3 billion. The southern San Joaquin Valley counties of Tulare, Kern, and Fresno account for 66 percent of all citrus production in the state. Other counties with significant citrus production include Ventura (15 percent), Riverside (8 percent), and San Diego (5 percent) (California Department of Food and Agriculture 2000). Oranges, lemons, and grapefruit, the three major crops that constitute the California citrus industry, are augmented by limes, tangerines, and numerous hybrid citrus fruits. In addition to citrus producers, the citrus industry is interrelated with many other agribusiness industries, including producers and suppliers of inputs to citrus production, packinghouses, and processors.

Consumers

Consumption of California citrus has grown steadily during the past decade and growth trends are projected to continue. During the last nine years, U.S. per capita citrus consumption (fresh fruit and juice) has increased by 20 percent, while the population has increased by 7 percent. Similarly, foreign consumption of California citrus products is increasing. California's citrus exports to foreign countries were approximately $411 million in 1998. In 2000, protocols to export California citrus to China were approved and shipments to China began March 24, 2000 (California Department of Food and Agriculture 2000).

Taxpayers

Taxpayers and state and federal regulatory agencies would also be affected should citrus canker enter California. State and federal agencies are entrusted with maintaining biosecurity from exotic pests and diseases. Most costs associated with inspecting, surveying, monitoring, and eradicating exotic pests and diseases are incurred by taxpayers who fund these agencies.

Economic Effects of Citrus Canker Under Alternative Interventions

Should citrus canker enter California, the effect on producers and consumers depends on U.S. and California government regulatory intervention strategies, policy responses of importing regions, the time period considered in the assessment, and consumer and producer response to price changes. State and federal regulatory policies regarding citrus canker are already developed. Government intervention strategies include eradication or allowing the disease to become established. Importing regions may continue to accept California citrus with no restrictions, provided pre- and postharvest treatment conditions are met, or impose embargoes. Consumer and producer responses depend in part on the time frame considered for the analysis. In the short run, supply adjustments are limited; in the long run, acreage adjusts and the industry fully reallocates resources.

The Effects on Producers and Consumers of Eradicating Citrus Canker

The effects on producers and consumers of an outbreak of citrus canker that is subsequently

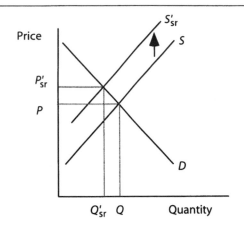

Figure 9.2 Short-run equilibrium adjustments–eradication scenario.

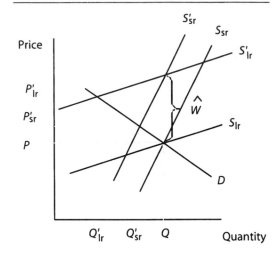

Figure 9.3 Short-run long-run equilibrium adjustments–establishment scenario.

eradicated can be shown graphically. In the short run the removal of acres due to eradication would cause the supply curve, S, to shift up to S'_{sr} (Figure 9.2).

Market supply would decrease from Q to Q'_{sr} and price would rise to P'_{sr}. Producers as a group and consumers are worse off. However, producers who do not have groves removed are better off as a result of the higher prices.

Over time, producers would respond to the higher prices by planting new groves. Production would gradually increase as those trees start bearing fruit, market supply would increase, and prices would start to fall. In the long run the supply curve would shift back to its original position and the initial equilibrium conditions between market quantity and price would be restored.

The Effects on Producers and Consumers of Allowing Citrus Canker to Become Established

If citrus canker enters and is not eradicated, it would gradually spread and become established in the state. Production costs would increase, and the short-run industry supply curve would shift up from S_{sr} to S'_{sr} (Figure 9.3). Market supply decreases to Q'_{sr} and price increases to P'_{sr}. Both producers and consumers are worse off. In the long run producers move land out of production of the infested crop and into production of other commodities. This leads to a shift up in the long-run supply curve from S_{lr} to S'_{lr}. Quantity supplied to the market further decreases to

Q'_{lr} and price increases further to P'_{lr}. Even though producers and consumers are still worse off, losses to producers are less in the long run than in the short run; however, consumers are worse off.

The short-run supply curve is steeper than the long-run curve because in the short run producers cannot easily move acreage out of production and into the cultivation of other crops. In the long run all crop production inputs can be reallocated to their most profitable use. In both the short and long run the shifting up of the supply curve is the same distance, $\hat{w}$.

If an embargo is imposed by some importing regions, the effects on markets depend on regional trade patterns. Three distinct regions make up the market (Figures 9.4 and 9.5). The first region is the infested region where the exotic pest becomes established. The second region is the market open to imports from all regions because it does not produce any host commodities. The final region is the clean region where host crops are produced, but the exotic pest is absent. The clean region imposes the embargo against the infested region to protect its agricultural industries from the entry of the exotic pest.

If the clean region is a net exporter, then quantity supplied by that region, Q_{sc}, is greater than quantity demanded, Q_{dc} (Figure 9.4). The excess supply from the clean region can be reallocated between the clean and open region until all three regions face the same price. Trade is not affected by the embargo.

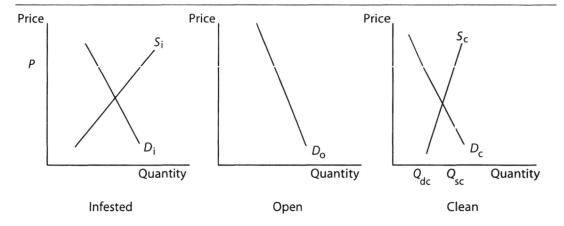

Figure 9.4   Region imposing embargo is a net exporter.

If the clean region is a net importer, then quantity supplied is less than quantity demanded (Figure 9.5). There is no longer an excess supply that can be moved between regions to equilibrate prices. In this case, trade is affected and the embargo serves to divide the one market into two separate markets. The first market is composed of producers from the infested region and consumers from the infested and open regions. The remaining market is composed of producers and consumers in the clean market (Figure 9.6).

Before the exotic pest became established, the market price was P in all regions. Production in the region that will become infested is Q_{si}. Exports are equal to $Q_{si} - Q_{di}$. This quantity is equal to the amount of imports demanded in the clean market, $Q_{dc} - Q_{sc}$. After the exotic disease becomes established, the equilibrium

quantity and price in each market is now where the supply and demand curves intersect. In the infested market (which includes the quantity demanded from the open market), the supply curve shifts up from S_i to S'_i, due to higher costs of production, production by infested growers falls to Q_i and the market price falls to P_i. However, the total quantity supplied to this market increases from Q_{di} to Q_i as embargoed goods are reallocated to the infested market. Consumers in the infested and open market are better off, while producers are worse off (Figure 9.6).

The opposite effect occurs in the clean market. Because the fresh commodity is no longer available from the infested market, total quantity supplied to the clean market falls from Q_{dc} to Q'_c (Figure 9.6). With lower market supplies, prices rise to P'_c and production by growers in

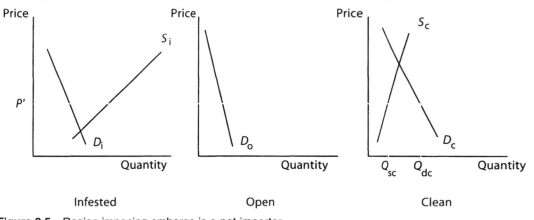

Figure 9.5   Region imposing embargo is a net importer.

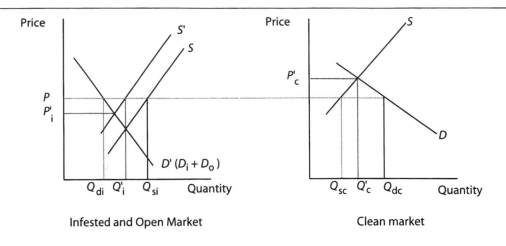

Figure 9.6 The market effects of a trade embargo.

the clean market increases from Q_{sc} to Q'_c. Producers in the clean market are better off, while consumers are worse off. Note that if the shift up of the supply curve in the infested market is enough to raise market prices so that the clean region becomes a net exporter, the embargo no longer affects trade, and prices will be the same across all regions.

Alternative Policy Scenarios

Citrus canker has never been found in California. Should it be detected before it becomes established in California, the main policy response would be to eradicate it from urban and commercial areas or to do nothing and let it become established. For the eradication scenario, additional policy options that would affect cost include whether compensation is paid and the size of the eradication boundary. If established, production costs would increase, and the regulatory response by importing regions may impose additional costs on producers and consumers. The cost of the eradication strategy is compared to the cost of establishment to determine which imposes the lowest costs for society.

Eradication Boundary Policy Alternatives

One issue that has arisen in the Florida urban eradication program is how far the tree removal boundary around an infested tree should be extended. The wider the boundary is extended, the higher the probability that the disease would be eradicated. However, eradication costs would

also be higher. Originally, the boundaries in the Florida urban eradication program were 125 feet around any infested tree found. Research later showed that the eradication boundary needed to be 1,900 feet around an infested tree to have a 95 percent chance of successfully eradicating citrus canker. For a 99 percent chance, the zone needs to be extended to a 3,000-foot perimeter. This difference in area of eradicated trees has potentially large cost consequences, depending on the size of the eradication program. It is important to remember that the conditions for pathogen spread and disease development in California are likely to be different from those in Florida. Accordingly, the size of the buffer zones in California may be different from those established in Florida.

Compensation Policy Alternatives

Federal or state governments may choose to eradicate without any compensation to homeowners or producers or offer some type of partial to full compensation. The original policy in the Florida eradication program was that no compensation would be paid to homeowners. Homeowners are now compensated with a voucher of $100 per tree removed. An alternative policy would be to compensate homeowners at the appraised value of a citrus landscape tree.

Originally, producers also did not receive any compensation from state or federal regulatory agencies for a commercial eradication program. During 1999, however, a pilot program was started offering producers the opportunity to purchase subsidized federal crop insurance for citrus canker. Producers who purchase crop insurance for citrus canker may submit claims if trees are destroyed during an eradication campaign. Producers would be compensated based on the actuarial value of a citrus tree. Currently, producers are compensated by federal regulatory agencies for trees destroyed due to the presence of citrus canker.

Trade Embargo Policy Considerations

International regulations allow foreign governments to impose pre- and postharvest restrictions and embargoes against California citrus and citrus stock should citrus canker be identified. Importing regions within the United States or foreign nations may impose restrictions on California citrus and citrus products if an outbreak of an exotic pest occurred that the importing location did not have. Internal quarantine regulations may also result in an embargo on the movement of California citrus products to other citrus-producing states and U.S. territories. Even though California currently exports citrus to regions that will embargo California fruit, whether it will affect trade needs to be determined before estimating losses due to trade restrictions.

The regions that would impose embargoes against California fresh citrus are those that currently have quarantine regulations against Florida and the U.S. states that are protected by USDA quarantine regulations from the movement of Florida fresh citrus. The foreign countries and regions most likely to impose embargoes would be New Zealand, Mexico, and the European Union. However, total exports to these countries are less than 1 percent of total California production and between 1 and 2 percent of total California exports (Agricultural Issues Center 1999). USDA internal quarantine restrictions for Florida prohibit the movement of fruit from embargoed areas into citrus-producing states.

Once the regions that would impose possible embargoes against California citrus were determined, whether they would affect trade was determined by comparing the consumption of fresh citrus to the current quantity supplied from clean regions. Consumption is estimated as 21 percent of the total U.S. supply (Table 9.1).

The 21 percent is equal to the share of the U.S. population that lives in citrus-producing states. Supply from clean regions is equal to the production of fresh citrus from Florida, Arizona, and Texas, plus imports less exports (Table 9.1). Based on the current level of production, imports and exports of fresh citrus, producers in clean regions are net exporters for all the citrus crops included in this analysis. Even if an embargo is imposed, it will not affect trade. Therefore, we do not include losses from embargoes in the estimate of the costs of establishment.

Policy Scenarios

The above policy issues were used to develop the following scenarios.

1. Urban eradication program
    (a) Eradication with a 125-foot buffer zone

Table 9.1 Determination of whether an embargo will affect trade

	Oranges	Lemons	Grapefruit
	------ *(1,000 short tons)* ------		
Total U.S. consumption	1,712	371	728
Consumption in market imposing embargo (21% of U.S. consumption)	360	78	153
Production by rest of U.S.	504	79	958
Imports	102	22	14
Total supply from clean regions	606	101	972
Embargo affects trade?	No	No	No

   (b)   Eradication with a 1,900-foot buffer zone
   (c)   Eradication with a 3,000-foot buffer zone
   (d)   No compensation to homeowners
   (e)   Homeowners compensated by issuing a voucher that can be used to purchase a replacement plant
   (f)   Homeowners compensated through payments based on the appraised value of the citrus tree removed
2. Commercial grove eradication program
   (a)   No compensation to producers
   (b)   Federal crop insurance available
   (c)   Producer compensation based on the investment value of a grove
3. Establishment
   (a)   No trade embargoes imposed

Many potential scenarios are created by these options. With three buffer zones and three homeowner compensation policies, there are nine urban eradication scenarios. The commercial eradication program adds three more scenarios, and establishment policy adds one. In addition, the effects on consumers and producers are examined under three different consumer and producer response scenarios to changes in prices.

Methodology for Estimating Economic Costs

Government Outlays of an Urban Eradication Program

The costs of eradicating citrus canker from an urban area include government outlays and private costs to homeowners. First, the cost to estimate the smallest possible infestation (one tree) in an urban area is calculated. The minimum costs are calculated for the three different eradication zone choices (125 feet, 1,900 feet, 3,000 feet) and the three different compensation policies (none, voucher, appraised value).

Urban Tree Removal Costs Government outlays are calculated as the cost per tree for removal, disposal, and compensation multiplied by the number of trees within each eradication zone, plus inspection and monitoring costs. The price to cut down and dispose of a citrus tree in an urban area was obtained from the Florida

Department of Agriculture and Consumer Services (FDACS) (Schubert 1999, personal communication; Leon Haab personal communication; and Doug Hadlock personal communication). As of December 1999 tree removal and disposal costs were $58 a tree. This price reflects a steady increase since the start of the eradication program. For our calculations, the dollar amount is rounded up to $60 per tree.

The number of trees per square mile was calculated as

$$\frac{\text{trees}}{\text{square mile}} = \frac{\text{trees}}{\text{backyard}} * \frac{\text{backyards}}{\text{square mile}}$$

where the number of trees per backyard is from backyard plant surveys competed by CDFA during a Mediterranean fruit fly urban eradication program (California Department of Food and Agriculture 1994), and the backyards per square mile is calculated by dividing the number of single homes and duplexes in urban areas by the number of urban square miles (U.S. Department of Commerce 1991).

Urban Compensation Costs Additional costs would be incurred by the government if compensation is paid to homeowners for their losses. If regulatory agencies opt to compensate a homeowner for the loss of a tree through issuing a voucher that could be redeemed at a nursery, the cost is $100, the current level of compensation in Florida. If compensation is paid according to the appraised value of a live mature tree in a home setting, the cost is estimated to be $400 per tree. This cost was estimated using the tree appraisal techniques developed by the International Society of Arboriculture (ISA 1992).

Surveillance and Monitoring Costs In the Florida citrus canker eradication program, surveillance and monitoring activities are completed six times within the eradication zone, two times within five miles of the zone, and once within 10 miles of the zone (Schubert 1999). The cost for regulatory agencies to inspect, survey, and monitor an urban eradication program is set at $90,000 per square mile within the eradication zone. This figure is calculated as the level of USDA and FDACS funding specifically for citrus canker removal, less tree removal costs, then divided by the estimated size of the infestation in Florida as of December 1999.

Once the minimum government outlays are estimated, the increase in government outlays as the size of the infestation increases is calculated. The analysis of the urban eradication cost is based on a constant rate of spread from the original infestation point. Costs are calculated as the incidence of disease spreads from 2,500 feet to 5,000 feet to 7,500 feet, etc., from the origin of the infestation. For analytical purposes, we assume that the infestation spreads outward in a circle around the original infestation point. The area of the eradication zone is then calculated as the area of a circle, πr^2 where r is how far the disease has spread.

Costs Excluded from the Urban Eradication Program Not included in this analysis are the effects on consumers who are buying citrus in stores instead of picking citrus off backyard trees. We have no data on home consumption of citrus fruit, and we assume that this is a small share of total market consumption. We also exclude the costs of homeowner resistance to the eradication of backyard trees. Homeowner resistance could result in extra public expenditures for education and outreach to ensure public support and compliance, thereby raising the total amount of government outlays to eradicate citrus canker.

The costs to remove publicly owned citrus trees are also not included in these calculations. Based on street tree inventories from seven cities in California, most cities do not plant more than a handful of citrus trees. On average there were only 1.5 citrus trees every square mile, and the average expected cost of eradication is not significantly influenced by the presence of publicly owned citrus trees. However, if citrus canker is introduced into a city having many citrus street trees, local costs would increase.

Government Outlays for a Commercial Eradication Program

The cost of eradicating citrus canker from commercial groves includes government outlays associated with the commercial eradication program and private costs to citrus consumers and producers from potential changes in market supplies and prices. First, government outlays for eradicating citrus canker as the infestation increases are calculated. Because citrus groves

are not always planted side by side, calculating the size of the infestation based on a constant spread from the initial point is not practical. Therefore, the size of the infestation was fixed at different acres, and costs were estimated for that acreage.

Commercial Tree Removal Costs Commercial grove eradication costs include tree removal, disposal (either through burning or removal off site), stump removal or treatment, and all inspection and monitoring costs. In addition, the total government eradication costs associated with the three policy choices involving compensation are estimated. As noted above, the policy choices are to pay no compensation, provide subsidized crop insurance, or pay an indemnification in an amount equal to the investment value of a citrus grove.

Costs per acre to bulldoze and burn a citrus grove were obtained from the FDACS. In 1999 these costs were between $250 and $350 an acre. A cost of $300 an acre is used in this study.

Commercial Compensation Costs The first compensation policy is to pay nothing. Crop losses then accrue to producers. The next policy is to offer producers the option to purchase subsidized federal crop insurance. The Florida Fruit Tree Pilot Crop Insurance Provisions provided the data used to estimate this cost (U.S. Department of Agriculture 1999). Based on the Florida data, the actuarial value of one citrus tree is $26. Full coverage per acre is then calculated as the number of trees per acre times $26. Based on crop budgets developed by University of California Cooperative Extension, there are 110 trees per acre for oranges and grapefruit and 136 trees per acre for lemons (O'Connell et al. 1999; Takele et al. 1997a, 1997b). Full coverage for one acre of an orange or grapefruit grove was calculated at $2,860, and for lemons it was $3,536. Because it cannot be determined in advance which groves will be infested, a weighted average was calculated for an average value of $3,000. The premium is 2.8 percent of the full coverage amount, or $84. Compensation to producers is then $3,000 less $84, or $2,916.

The final compensation policy is to compensate producers in an amount equal to the investment value of a grove. Investment values are reflected in the price to purchase land on which the grove is planted. The value of the grove itself was calculated as the average price paid for the land on which the citrus grove was planted less the average price paid for open land (California Chapter of American Society of Farm Managers and Rural Appraisers 1999). These numbers were then adjusted based on interviews with farmland appraisers in citrus-growing counties. The average investment value is estimated at $5,300 an acre. Note that this number is far higher than even the full coverage on the crop insurance scenario.

Inspection and Monitoring Costs Inspection and monitoring costs are the same as in the urban eradication scenario, $90,000 a square mile. Dividing by the number of acres per square mile results in inspection and monitoring costs of $140 per acre.

Market Effects on Fresh Citrus Fruit Consumers and Producers from an Eradication Program

Analytical Model Once government outlays are estimated, costs to consumers and producers are estimated. Consumer and producer costs are estimated as the change in welfare from changes in market quantities and prices. To estimate the changes in welfare, an equilibrium displacement model was developed for the fresh market (Alston et al. 1995) (see Appendix 9.1). Whereas citrus products go into both the fresh and the processing sectors, the processing sector is a small share, by value, of the entire citrus industry in California. California citrus products are grown primarily for the fresh market, so we focus solely on the fresh market effects.

The model shows how the equilibrium quantities, prices, and other variables respond to shocks to the system, such as the removal of acreage during an eradication program. The model is parameterized with market and biological data. Details of the equilibrium displacement model are given in the chapter appendix along with information on general data requirements and collection.

Elasticities Key parameters needed to estimate the model include the elasticities of demand and the elasticities of supply for citrus canker. Elasticities of demand are available from published reports (Huang 1993; Kinney et al. 1987).

Consumers typically adjust to price changes relatively quickly. Therefore, the short-run and long-run elasticities of demand are the same. The elasticity of demand for oranges is −0.85; for lemons it is −0.5; and for grapefruit it is −0.45.

The elasticity of supply is extrapolated from previous work (Kinney et al. 1987). Producers do not adjust to changes in supply as quickly as consumers. In the short run, few acres are replanted following the destruction of groves during an eradication campaign. Most supply adjustments in response to changes in prices are from changes in produce entering the fresh market in the short run. The short-run elasticity of supply is 0.5. We examine the annual costs of eradication for the short run only as growers would immediately begin to replant once the host-free period is complete.

Size of Infestation The market effects of the eradication program are estimated for the removal of 5,000, 10,000, and 25,000 acres for oranges and lemons, and 5,000 and 10,000 acres for grapefruit, because California does not have 25,000 acres planted in grapefruit. Because it also cannot be determined in advance which citrus crops would need to be eradicated if citrus canker invaded commercial groves, the effects on consumers and producers for each crop are presented separately and are not aggregated across crops.

The welfare effects to producers and consumers are calibrated on the recent three-year average of production levels and prices of fresh oranges, lemons, and grapefruit grown in California and the rest of the United States (USDA 2002) (Table 9.2). After the percentage change in market quantities and prices is estimated, the change in producer and consumer welfare is estimated using the methodology described in Appendix 9.1.

Methodology for Estimating the Economic Effects of Establishment

The benefits of an eradication policy are from the prevention of the costs of establishment. If citrus canker were to become established, production costs would increase, and embargoes may be imposed on fresh California citrus by importing regions. An equilibrium displacement model is used to estimate the effects of increased producer costs and embargoes on final market supply and prices (Alston et al. 1995).

Table 9.2 Average production and prices of California's top three citrus crops[a]

	Oranges	Lemons	Grapefruit
U.S. production	2,267,500	496,667	1,157,333
California production	1,756,875	417,177	199,191
Rest of U.S. production	503,601	79,331	958,102
Net exports	555,561	125,302	429,046
U.S. supply	1,711,939	371,365	728,287
U.S. acres	813,850	63,067	151,633
California acres	197,000	48,833	16,200
California yield per acre	8.92	8.54	12.30
Retail price	$1,480	$2,370	$1,172
California grower price	$175	$418	$264
Rest of U.S. grower price	$106	$356	$132

[a]All quantity data are in short tons (2,000 pounds). Prices are in $/short ton.

Once the new equilibrium quantity and price are determined, the changes in producer and consumer welfare are calculated.

Production Costs Should citrus canker become established, producers would need to treat groves with an approved pesticide (e.g., copper-containing material). In this study we assume that the copper-based fungicide, Kocide, would be applied at least four times a year. Kocide is registered for use in California on citrus and is registered as a treatment for citrus canker in other states. Kocide would be applied according to the label instructions and custom applied by a pest control company. The application costs would be $175 per acre. Total costs are divided by the tons produced per acre for each crop to determine the increase in the costs per ton.

In the short run only annual costs of production would be affected. In the long run, as new groves are planted, citrus canker control would affect the costs of establishing a grove since the disease must be treated when the grove is first planted. All costs incurred during the first four years of establishment are investment costs and amortized over the remaining life of the grove. The increase in investment costs due to fungicide treatments would result in an increase in the annual amortization costs of establishment.

Postharvest Treatments Approved postharvest treatments to control citrus canker include washing the fruit with SOPP (sodium o-phenyl phenate) or chlorine-based material (e.g., sodium hypochlorite). Citrus packing plants in California already treat fruit with these materials,

so there would be no additional postharvest treatment costs.

Elasticities The elasticity of demand and the short-run elasticity of supply are the same as those used for the eradication market effects. However, the long run effects also need to be calculated. How responsive producers are to changes in prices by making changes in production, as measured by the elasticity of supply, may affect the total costs of citrus canker establishment and who incurs those costs. Not all land is suitable for growing citrus, so in some areas producers may not be able to plant additional groves. Also, land planted in citrus may not be easily converted to other crops. To determine the effects of different supply elasticities on producer and consumer welfare, long-run supply elasticities of 1 percent and 4 percent are used in the analysis.

Related Industries

The effects on related industries depend in part on alternative uses of the land if citrus production declined as a result of citrus canker. Downstream industries include the suppliers of inputs into citrus production. Such inputs would be hired labor and suppliers of pest control services, custom crop production services, and nursery stock. It is difficult to determine what the net effect would be on the demand for hired labor. If the next best crop grown on the land that was used to produce citrus was more labor intensive, then the demand for labor would increase. If that crop used less labor, the demand for labor would decrease.

In general, the effects on citrus nursery stock inputs can be determined qualitatively, but again, the net effects are ambiguous and depend on the degree of specialization among nursery stock producers and the alternative use of land. In the short run we would expect tree fruit planting activity to rise following an eradication program. However, if citrus canker were to become established, the effects would be more difficult to determine. Possible decrease in demand from California growers and by trade embargoes against budwood lowers the demand for nursery stock. Producers then need to find alternative uses for the land. If producers were to move acreage out of citrus budwood production and into another tree crop, nursery producers may be no worse off. However, if acreage is converted into row crops, nursery producers may be worse off.

Finally, we have neglected the increased probability of adjacent states becoming infested with citrus canker should it establish in California.

Methodology for the Cost/Benefit Analysis

The cost/benefit analysis compares the costs of eradication to the benefits of preventing the costs of establishment. The analysis compares the costs and benefits to California and U.S. producers, consumers, and taxpayers.

Eradication Costs We consider eradication costs for both a relatively small and a relatively large infestation. The small infestation case is 20 square miles in an urban area and is equal to the size of the Florida infestation when it was first identified. The commercial acreage is set at 100 acres. The large infestation case is 400 square miles in an urban area and is equal to the size of the Florida infestation in December 1999. In the large infestation, the commercial acreage affected is set at 5,000 acres.

Eradication costs are equal to government outlays necessary to achieve a 95 percent or a 99 percent probability of successful elimination of citrus canker in an urban area plus the government outlays of the commercial eradication programs. Given the large number of scenarios, we report only those outlays based on compensation policies as of December 1999. For the urban eradication program, homeowners are compensated at $100 a tree. For the commercial eradication program, the outlays include paying the investment value of an orchard. In the simulations the allocation of government outlays between state and federal regulatory agencies is based on current practices for other plant diseases. Inspection, monitoring, and tree removal costs are divided equally between the federal and state regulatory agencies. Federal agencies pay all compensation.

We have assumed the outbreak occurs in lemons because it is unknown which crops will be infested with citrus canker. Lemons were chosen because that crop is more susceptible to citrus canker than oranges or grapefruit. For the small eradication program, destroying 100 acres of citrus trees would have no market effects. In the large infestation case, the loss of 5,000 acres of lemon production would affect U.S. market supplies and prices. The net present value of changes in both California and U.S. producer and consumer welfare is then added to the costs of government programs to estimate total eradication costs. The net present value of all costs is calculated for an eight-year adjustment period. Each year the elasticity of supply is raised from a starting value of 0.5 in the first year to a long-run equilibrium value of 20 in year eight. A high elasticity is used because most groves will have been replanted and in full production by year eight. At year nine the original equilibrium is restored, and costs and benefits from then on are zero. A discount rate of seven percent is used to convert future costs and benefits into current values.

Establishment Costs The net present value for costs of citrus canker becoming established is estimated in order to compare it with the net present value of all eradication costs. The net present value of the costs of establishment is also calculated for an eight-year adjustment period. Each year the elasticity of supply is raised from 0.5 in the first year to the long-run equilibrium values of 1.0 or 4.0 in year eight. The costs and benefits from year nine into perpetuity are set at the long-run equilibrium values estimated in year eight. The costs and benefits are determined for all U.S. producers and consumers of U.S. citrus products and for California producers and consumers for comparison with the eradication costs.

It is never known with certainty whether an eradication program will be successful. Therefore, the costs of the alternative eradication programs, small infestation or large, at a 95 or 99 percent probability of success, need to be compared to the expected benefits. The expected benefits of eradication are equal to the costs imposed from establishment multiplied by the probability of successfully eradicating citrus canker.

Estimation of the Economic Effects of a Citrus Canker Infestation and Discussion of Simulation Results

As stated previously, the costs of the eradication program for homeowners, tax payers, growers and consumers will be discussed. The annual costs of establishment in both the short and long run are then presented. Finally, we compare the costs of eradication to the costs of establishment to determine if the expected benefits of eradication would be greater than the costs.

Eradication Program

Government Urban Eradication Program
Because of inspection and monitoring requirements, the minimum cost to eradicate citrus canker in an urban area would be $90,060 even if only one tree were infested (Table 9.3). This is for the policy option of a 125-foot eradication zone around the infested tree and no compensation paid to the homeowner. When the eradication zone is 125 feet, on average only the infested tree would need to be removed.

However, the 125-foot boundary would result in a very low probability of successful eradication. The minimum estimated cost of an eradication program with a 95 percent probability of success would be $102,000 when no compensation is paid. Under this policy option, the

boundary of the eradication zone is 1,900 feet and 199 trees are removed.

Should policymakers decide to increase the probability of successful eradication to 99 percent, the boundary of the eradication zone would need to be 3,000 feet. This results in an eradication zone of one square mile. Total costs would now be $120,000 (Table 9.3). Therefore, raising the probability of successful eradication by 4 percentage points would raise costs by about 17.5 percent.

As the size of the infestation increases, eradication costs would also rise. However costs do not rise in proportion to the increase in size of the infestation as measured by the radius of the area (Table 9.4). For example, as the infestation spreads by 2,500 feet from 5,000 feet to 7,500 feet, costs would increase by $421,000. However, as the infestation spreads by 2,500 feet, from 15,000 feet to 17,500 feet, costs would increase by $1,100,000.

Therefore, costs increase exponentially as the infestation spreads (Figure 9.7). This highlights the importance of early detection when an exotic pest enters, as well as the importance of containing and eradicating the infestation quickly.

When compensation is paid, costs incurred by government agencies are higher. These costs depend on the compensation policy pursued. Vouchers would increase agency costs of the urban eradication program by 41 percent for all infestation levels. Compensation according to the appraised value of a citrus tree would increase costs by 164 percent over the no-compensation policy. Clearly, whether paid or not, compensation costs are real, and ignoring them ignores the full cost of an infestation. Furthermore, while regulatory agencies have eminent domain and can enter and remove trees as necessary to successfully eradicate an exotic pest, compensation to homeowners may also be

Table 9.3 Minimum Eradication Costs for Citrus Canker in an Urban Setting

Radius (feet)	Square miles	Number of trees	Costs with no compensation ($)	Costs with compensation[a] Voucher ($20/tree)	Appraised value ($400/tree)
125	0.002	1	60	160	460
1,900	0.41	199	12,000	32,000	92,000
3,000	1	497	30,000	80,600	228,700

[a]Plus 90,000 per square mile in inspection and monitoring costs.

Table 9.4 Costs for Citrus Canker in an Urban Setting as Size of Infestation Increases

| Radius (feet) | Square miles | Number of trees | Total Cost - Tree Removal | | | Inspection costs[a] |
			No compensation ($)	Voucher ($100/tree)	Appraised value ($400/tree)	
5,000	3	1.4	83	221	635	254
7,500	6	3.1	186	497	1,430	571
10,000	11	5.5	331	884	2,541	1,015
12,500	18	8.6	518	1,381	3,971	1,585
13,500	21	10.1	604	1,611	4,632	1,849
15,000	25	12.4	746	1,989	5,718	2,283
17,500	35	16.9	1,015	2,707	7,783	3,107
20,000	45	22.1	1,326	3,536	10,165	4,058

[a]Inspection Costs are $90,000 per square mile

a cost-effective policy if the compensation increases homeowner cooperation. With more cooperation, regulatory agencies may eradicate the disease more quickly and lower direct eradication costs.

Government Commercial Grove Eradication Program Taxpayer costs for an eradication program in commercial groves rise in proportion to the number of acres eradicated (Table 9.5). For a 100-acre infestation, total costs to inspect, monitor, and remove trees without compensation is $44,000. Inspection and monitoring costs are $14,000, and tree removal costs are $30,000. Because infested groves are not necessarily contiguous, calculating the costs based on the spread of the infestation, as measured by the radius of a circle, is not appropriate. Consequently, the results show that as the size of the infestation doubles, costs also double. However, citrus canker would also

spread in a circular fashion within groves. Therefore, costs will rise exponentially as the radius of the infestation increases at a constant rate.

Paying compensation recognizes the true costs of eradication and increases government outlays. Under the federal crop insurance policy option, government outlays will increase by 660 percent from the no-compensation policy. This assumes that every producer purchases crop insurance. If some producers do not purchase insurance, the cost of compensation does not disappear. The cost shifts from the government to the producer.

Federal crop insurance reimburses producers according to the predetermined actuarial value of a citrus tree. An alternative would be to pay the remaining investment value of a grove. Compensating producers based on the investment value would increase government outlays by 1,220 percent over the no-compensation pol-

Table 9.5 Costs for Citrus Canker Eradication in a Commercial Setting as Size of Infestation Increases

| Acres | Tree removal with no compensation ($300/acre) | Tree removal with compensation | | Inspection and monitoring costs ($140/acre) |
		Insurance ($2,916+ $300/acre)	Capital value of orchard ($5,300+ $300/acre)	
100	30	322	560	14
500	150	1,608	2,800	70
1,000	300	3,216	5,600	140
5,000	1,500	16,080	28,000	700
10,000	3,000	32,160	56,000	1,400
25,000	7,500	80,400	140,000	3,500

Costs

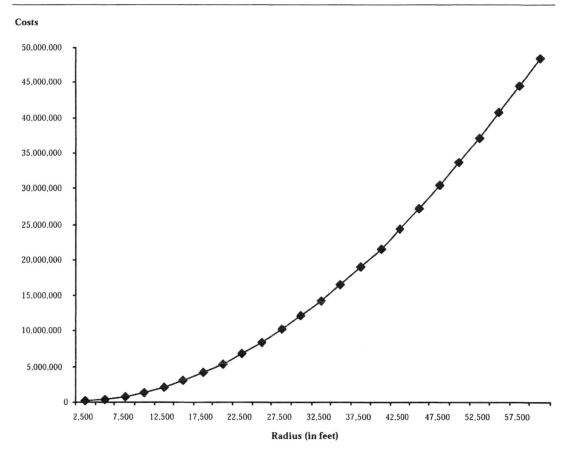

Radius (in feet)

Figure 9.7 Eradication costs as size of infestation increases.

icy, but only 71 percent over the crop insurance option for all infestation levels.

Welfare Effects of a Commercial Grove Eradication Program

If the infestation is identified before it has time to spread throughout many groves, the destruction of groves will not have any measurable market effects. However, if many acres need to be eradicated, the reduction in quantity supplied could cause prices to increase and affect the welfare of consumers and producers outside the eradication area. The reduced supply and higher prices may make producers in the eradication zone worse off, producers outside of the eradication zone better off, and consumers worse off.

The Effects on Market Quantity and Price

The percentage change in prices and quantities

as a result of a commercial eradication program is estimated for oranges, lemons, and grapefruit (Table 9.6). Eradicating citrus trees in commercial production would decrease market supplies. The decrease in quantity supplied would cause prices to increase. In response to the increase in prices, producers divert more production to the fresh market and plant more acres. Therefore, the net change in U.S. market quantities is less than the reduction in production due to eradication (Table 9.6). For example, for fresh oranges when an eradication program causes two percent of production to be removed, market supplies only fall by 1.3 percent. The net decrease in market supply is only 35 percent of the amount of production lost.

Similar patterns are observed for lemons and grapefruit. Given our assumed supply response elasticity of 0.5, final quantity supplied for

Table 9.6 Production and price effects of eradicating citrus canker from commercial groves

	Scenario Attributes			Results		
Acres eradicated	Percentage reduction in supply due to eradication	Elasticity of demand	Elasticity of supply	Quantity produced by growers not subject to eradication	U.S. market quantity	U.S. market price
	- - - (%) - - -			- - - - - - - - - - - - - - - - (%) - - - - - - - - - - - - -		
Oranges						
5,000	2	−0.85	0.5	0.7	−1.3	1.5
10,000	4	−0.85	0.5	1.5	−2.5	2.9
25,000	10	−0.85	0.5	3.6	−6.1	7.1
Lemons						
5,000	9	−0.5	0.5	4.3	−4.3	8.6
10,000	17	−0.5	0.5	7.8	−7.8	15.7
25,000	43	−0.5	0.5	17.7	−17.7	35.4
Grapefruit[a]						
5,000	5	−0.45	0.5	2.6	−2.3	5.1
10,000	11	−0.45	0.5	5.5	−4.9	11.0

[a]California does not have 25,000 acres planted in grapefruit.

lemons is about 47 percent of the lost production and about 50 percent for grapefruit.

The difference in the percentage of net change in market supply is due to differences in the elasticity of demand. The more responsive the quantity demanded is to changes in market price, the smaller the changes in market supply. Oranges have the highest elasticity of demand and grapefruit the lowest in absolute value. Consequently, the net percentage change in market supply is lowest for oranges and highest for grapefruit.

Changes in Welfare for Producers and Consumers The scope of the eradication program also influences the final annual change in welfare to producers and consumers (Table 9.7). For each crop and eradication level, total producer welfare decreases even though prices increase. Not all producers would be worse off, however. Producers who do not have an infestation on their land would benefit from the higher prices. For example, the decline in producer welfare for orange growers who have trees destroyed is $18 million when 2 percent of U.S. production is lost. However, producer welfare increases by $5 million for California growers who continue production and by $1 million for producers in the rest of the United States. Due to large losses to growers with destroyed trees, welfare for all growers would decline by $12 million.

If citrus canker was to become established, consumers in California and the United States are worse off due to lower market supplies and higher prices. For an eradication program of oranges that lowers production by 2 percent, annual consumer welfare declines by $4 million in California and by $37 million for the United States as a whole (Table 9.7).

Whereas the magnitude of the welfare losses is different for lemon and grapefruit eradication programs, the losses to growers who have trees destroyed are greater than the gains to the remaining producers, and the net change in producer welfare is negative for those crops. With lower market supplies and higher prices, the change in consumer welfare is also negative. The longer it takes to identify the infestation, more trees become infested and the losses are greater to both consumers and producers. Over time, as growers replant, market supply would increase, price would fall, and the losses to consumer and producer welfare would decline.

The Welfare Effects of Citrus Canker Becoming Established

The scenarios developed to estimate the effects of citrus canker becoming established in California look at the effects of rising costs of production in both the short and long run (Table 9.8).

Table 9.7 Producer and consumer welfare effects of a citrus canker eradication program

Scenario Attributes[a]			Results												
			Producer welfare								Consumer welfare			All	
Acres removed	U.S. prod. reduced	Elast. of demand	Calif.[c] in the erad. area	Calif. not in the erad. area	Total Calif.[c]	Share of prod. rev.	Rest of U.S.	Share of prod. rev.	Total U.S.[c]	Share of prod. rev	Calif.	U.S.	Share of cons. costs	Calif.[c]	U.S.[c]
	(%)		($mil)	($mil)	(%)	($mil)	(%)	($mil)	(%)	($mil)	(%)	($mil)	(%)	($mil)	($mil)
Oranges															
5,000	2	−0.85	−18	5	−13	−4	1	1	−12	−3	−4	−37	−1	−17	−49
10,000	4	−0.85	−36	10	−26	−8	2	3	−24	−6	−9	−73	−3	−35	−97
25,000	10	−0.85	−91	25	−66	−19	4	7	−62	−16	−21	−175	−7	−87	−237
Lemons															
5,000	9	−0.5	−38	15	−23	−13	2	9	−20	−10	−9	−74	−8	−32	−94
10,000	17	−0.5	−73	28	−45	−26	5	16	−40	−20	−16	−132	−15	−61	−173
25,000	43	−0.5	−194	67	−127	−73	11	38	−116	−57	−34	−284	−32	−161	−400
Grapefr.[b]															
5,000	5	−0.45	−31	3	−28	−54	7	5	−22	−12	−5	−43	−5	−33	−65
10,000	11	−0.45	−69	6	−63	−120	14	11	−49	−27	−11	−91	−11	−74	−140

[a]The elasticity of supply is 0.5.
[b]California does not have 25,000 acres planted in grapefruit.
[c]Unadjusted for any compensation paid to growers whose groves were infested with citrus canker.

The increase in costs of production per short ton is 5 percent for oranges in the short run and 6.5 percent in the long run. Because it costs more per ton to produce lemons than oranges, the increase in costs is 2 percent for lemons in the short run and 3 percent in the long run. For grapefruit the increase is 4 percent in the short run and 5 percent in the long run.

Changes in Market Quantity and Price Increased costs of production would cause the supply of California fresh citrus to go down, putting upward pressure on U.S. prices. The higher prices cause growers in the rest of the U.S. to increase the quantity supplied of fresh citrus and consumers to demand less, putting downward pressure on U.S. prices. Consequently, the net change in prices is not enough to offset the rise in costs for California producers (Table 9.9).

For example, the fresh market production of oranges in California would decrease by 1.8 percent. Fresh orange quantity supplied by the rest of the U.S. would increase by 0.7 percent. Market prices would rise by 1.4 percent in the short run while costs would increase for California growers by 5 percent (Table 9.9). Consequently, California growers would decrease production in the long run and growers in the rest of the United States would increase production of fresh market oranges.

Because California has a large share of the U.S. market, the decrease in production by California is greater than the increase in production by other states, and the net change in market supply is negative. In response, prices would rise further in the long run (Table 9.9). The fresh market production of oranges by California would decrease by 3.8 percent when the long-run elasticity of supply is 1.0 and by 9.4 percent when it is 4.0. Production by the rest of the United States would increase in the long run by 2.7 percent when the elasticity of supply is 1.0 and by 16.7 percent when it is 4.0. The corresponding increase in market prices would be 2.7 percent for an elasticity of 1.0 and 4.2 percent when it is 4.0.

Similar results are observed for lemons and grapefruit; however, the final percentage changes in market price and quantity are smaller than they are for oranges (Table 9.9). The differences between the crops depend on the elasticity of demand. The less responsive demand (i.e., the lower the elasticity of demand) is to changes in prices, the lower is the percentage change in market supply and the lower is the percentage change in price.

Change in Producer Welfare Annual losses to California growers are greatest in the short run and the long run when the elasticity of supply is 1.0. For example, the decline in producer welfare for orange growers is $12.1 million in

Table 9.8 Scenario Attributes to Measure the Welfare Effects of the Establishment of Citrus Canker

Time period	Number of fungicide applications	Production cost increase	Elasticity of demand	Elasticity of supply
		(%)		
Oranges				
Short run	4	5	−0.85	0.5
Long run	4	6.5	−0.85	1
Long run	4	6.5	−0.85	4
Lemons				
Short run	4	2	−0.5	0.5
Long run	4	3	−0.5	1
Long run	4	3	−0.5	4
Grapefruit				
Short run	4	4	−0.45	0.5
Long run	4	5	−0.45	1
Long run	4	5	−0.45	4

Table 9.9 Production and price changes from the establishment of citrus canker in California

	Scenario Attributes			Results			
Time Period	Production cost increase	Elasticity of demand	Elasticity of supply	Calif. production	Rest of U.S. production	U.S. market quantity	U.S. market price
	(%)						
Oranges							
Short run	5	−0.85	0.5	−1.8	0.7	−1.2	1.4
Long run	6.5	−0.85	1	−3.8	2.7	−2.3	2.7
Long run	6.5	−0.85	4	−9.4	16.7	−3.5	4.2
Lemons							
Short run	2	−0.5	0.5	−0.6	0.4	−0.4	0.8
Long run	3	−0.5	1	−1.3	1.7	−0.8	1.7
Long run	3	−0.5	4	−3.0	9.0	−1.1	2.2
Grapefruit							
Short run	4	−0.45	0.5	−1.8	0.2	−0.2	0.4
Long run	5	−0.45	1	−4.4	0.6	−0.3	0.6
Long run	5	−0.45	4	−16.9	3.1	−0.4	0.8

the short run and $12.7 million in the long run when the elasticity of supply is 1.0 (Table 9.10). Losses increase in the long run for California growers when the elasticity of supply is 1.0 because the industry cost to produce fresh citrus would increase in the long run due to growers needing to treat new orchards while the orchards are being established.

When growers have more flexibility in responding to an infestation (i.e., the elasticity of supply is higher), then losses are lower in the long run, even though costs are higher. For oranges, the decline in producer welfare is only $7.6 million in the long run when the elasticity of supply is 4.0.

Higher market prices would cause growers in the rest of the United States to increase production. With higher prices and greater fresh citrus production, benefits to producers in the rest of the United States increase. For oranges, annual producer welfare for the rest of the United States increases from $0.8 million in the short run to $1.5 million when the elasticity of supply is 1.0 and to $2.4 million when it is 4.0.

For all growers in the United States, the net loss in welfare declines over time and as the elasticity of supply increases. In the short run the annual net loss in producer welfare for orange growers is $11.3 million. In the long run the net loss declines slightly to $11.2 million when the elasticity of supply is 1.0. However,

when it is 4.0, the net loss is only $5.2 million.

The results are similar for lemons. The decline in producer welfare for California growers would increase in the long run, when the elasticity of supply is 1.0 due to the higher costs of production. However, annual losses would decline when it is 4.0 (Table 9.10). For all growers in the United States, the net loss in welfare is about the same in the short run and in the long run when the elasticity of supply is 1.0. When the long run elasticity of supply is 4.0, the net loss in producer welfare is smaller. For grapefruit, the net loss in welfare for all U.S. growers would increase in the long run, compared with the short-run level. Losses increase slightly from $1.4 million to $1.5 million. When the elasticity of supply is 4.0, however, net losses are also smaller for grapefruit growers (Table 9.10).

Change in Consumer Welfare Whereas the losses to producer welfare would, in general, decrease over time, the annual losses to consumer welfare would increase. Over time, the decrease in quantity supplied and increase in prices would increase losses to consumers. For oranges, the loss in consumer welfare increases from $36.2 million in the short run to $68.3 million when the elasticity of supply is 1.0 and $103.6 million when it is 4.0 (Table 9.10). Similar losses in consumer welfare for lemons and grapefruit would also occur.

Table 9.10 Producer and consumer welfare effects for California citrus from the establishment of citrus canker

	Scenario Attributes			Results										
	Production cost increase	Elasticity of demand	Elasticity of supply	Producer welfare						Consumer welfare			All	
				Calif.	Share of prod. rev.	Rest of U.S.	Share of prod. rev.	U.S.	Share of prod. rev.	Calif.	U.S.	Share of con. costs.	Calif.	U.S.
	($mil)			($mil)	(%)	($mil)	(%)	($mil)	(%)	($mil)		(%)	($mil)	
Oranges														
Short run	5	−0.85	0.5	−12.1	−3.5	0.8	1.4	−11.3	−2.9	−4.3	−36.2	−1.4	−16.5	−47.5
Long run	6.5	−0.85	1	−12.7	−3.7	1.5	2.8	−11.2	−2.8	−8.2	−68.3	−2.7	−20.9	−79.5
Long run	6.5	−0.85	4	−7.6	−2.2	2.4	4.5	−5.2	−1.3	−12.4	−103.6	−4.1	−20.1	−108.9
Lemons														
Short run	2	−0.5	0.5	−2.0	−1.2	0.2	0.8	−1.8	−0.9	−0.9	−7.4	−0.8	−2.9	−9.2
Long run	3	−0.5	1	−2.3	−1.3	0.5	1.7	−1.8	−0.9	−1.8	−14.7	−1.7	−4.0	−16.5
Long run	3	−0.5	4	−1.3	−0.7	0.7	2.3	−0.6	−0.3	−2.4	−19.6	−2.2	−3.7	−20.2
Grapefruit														
Short run	4	−0.45	0.5	−1.9	−3.6	0.5	0.4	−1.4	−0.8	−0.4	−3.1	−0.4	−2.3	−4.5
Long run	5	−0.45	1	−2.3	−4.3	0.8	0.6	−1.5	−0.8	−0.6	−5.1	−0.6	−2.9	−6.6
Long run	5	−0.45	4	−2.0	−3.9	1.0	0.8	−1.0	−0.6	−0.8	−6.6	−0.8	−2.8	−7.6

Total Welfare Effects For oranges and lemons, welfare losses to California producers are greater than the losses to California consumers in the short run and in the long run when the elasticity of supply is 1.0. As grower responsiveness increases (i.e., the elasticity of supply is higher), consumers incur a larger share of the total welfare losses. For example, annual losses for orange growers account for 61 percent of total welfare losses when the long run elasticity of supply is 1.0 (Table 9.10). When it is 4.0, producer losses account for only 38 percent of total losses. For grapefruit, due to the relatively smaller share of California in the fresh market, California growers always incur a larger share of the total welfare losses. Welfare losses to grapefruit growers account for over 70 percent of total welfare losses in the long run (Table 9.10).

For the entire United States, consumers would incur most of the total welfare losses associated with the establishment of citrus canker. For oranges, in the short run losses to consumers account for 76 percent of total welfare loss. The consumer share of losses is greatest for lemons when the long-run elasticity of supply is 4.0. In this scenario, consumer losses account for 97 percent of total losses.

Full Assessment of Costs and Benefits

Eradication Costs The total costs to eradicate citrus canker include the costs to regulatory agencies plus the net present value of the losses in welfare to producers and consumers. A small eradication program consisting of 25 square miles of an urban infestation and 100 acres of commercial groves and a large eradication program consisting of 400 square miles and 5,000 acres are both included for comparison because it is unknown how large an infesta-

tion would be when initially discovered (Table 9.11). As stated previously, to ensure a 95 percent probability of successfully eradicating citrus canker, the urban eradication boundary needs to extend 1,900 feet around the infestation. For a 99 percent probability the zone needs to be 3,000 feet. Homeowners are paid $100 per tree removed and growers are compensated in an amount equal to the capital value of the destroyed grove.

Total government outlays for a small eradication program that has a 95 percent probability of success would be $5.1 million (Table 9.11). California would contribute $1.6 million, or 31 percent to total outlays. Because the U.S. government pays compensation, its outlays would be higher and account for 69 percent of total outlays. Increasing the probability of success to 99 percent would increase outlays by $0.6 million, or 12 percent.

For the large infestation total outlays would be $101.4 million to achieve a 95 percent probability of success. California would contribute about 26 percent to total government outlays. Total government expenditures would need to increase by only 2.6 percent, or $2.7 million, to increase the probability of success to 99 percent.

The eradication program associated with a large area would also result in changes in commercial production to the extent that market prices would be affected. Therefore, all producers and consumers of citrus in California and in the United States would be affected. The net present value of welfare changes to all producers and consumers in California would range from $43 million for oranges to $85 million for grapefruit (Table 9.12). Losses to producers account for the majority of total losses for all crops among California consumers and producers.

Table 9.11 Government Costs to Eradicate Citrus Canker from California

Initial infestation	Eradication boundary	Urban square miles	Commercial acreage	Costs Urban	Costs Commercial	Costs Total	Cost sharing California	Cost sharing US
	(feet)					*($mil)*		
Small	1,900	27	100	4.5	0.57	5.1	1.6	3.5
	3,000	31	100	5.2	0.57	5.7	1.9	3.9
Large	1,900	432	5,000	72.7	28.7	101.4	26.9	74.6
	3,000	447	5,000	75.4	28.7	104.1	27.8	76.2

Table 9.12 Present value of changes in producer and consumer welfare for a commercial eradication program[a]

Crop	U.S. production reduced	Elasticity of Demand	California			U.S.		
			Producer	Consumer	All	Producer	Consumer	All
	(%)		- ($mil) - - - - - - - - - - - - - - - - - -					
Oranges	2	−0.85	−28	−15	−43	−25	−123	−149
Lemons	9	−0.5	−49	−27	−76	−41	−228	−269
Grapefruit	5	−0.45	−70	−16	−85	−50	−131	−181

[a]The discount rate is 7% and 5,000 acres are eradicated.

Total welfare changes to all U.S. producers and consumers of citrus would range from $149 million for oranges to $269 million for lemons (Table 9.12). For the United States as a whole, losses to consumers would account for the majority of total welfare losses for all crops.

Establishment Costs The costs of eradication now need to be compared to the costs of establishment. If citrus canker were to become established, the net present value of total welfare losses to California producers and consumers of fresh oranges would range from 393 million when the long-run elasticity of supply is 1.0 to 378 million when it is 4.0 (Table 9.13). The present value of total welfare losses for California declines as the elasticity of supply increases because losses to growers will decline when they have more flexibility in adjusting to sudden increases in costs.

The effects on California lemon and grapefruit producers and consumers show similar results (Table 9.13). Total losses would range from a high of $74 million for lemons and $53

million for grapefruit to a low of $67 million for lemons and $53 million for grapefruit.

The present value of U.S. producer and consumer welfare losses when the long-run elasticity of supply is 1.0 would be $1.44 billion for oranges, $292 million for lemons and $119 million for grapefruit. When the long-run elasticity of supply is 4.0, the present value of the losses in welfare increases to 1.95 billion for oranges, 357 million for lemons and 139 million for grapefruit. Losses increase when the elasticity of supply increases because consumer losses are increasing.

Adding the losses in welfare for all citrus crops, the total losses to producers and consumers within California would be $520 million when the elasticity of supply is 1.0 and $498 million when the elasticity of supply is 4.0 (Table 9.14). For the United States as a whole, the losses in welfare increase as the elasticity of supply increases. Losses increase from $1.85 billion when the elasticity of supply is 1.0 to $2.45 billion when the elasticity of supply is 4.0.

Table 9.13 Present value of changes in producer and consumer welfare if citrus canker becomes established[a]

Crop	Production cost increase	Long-run elasticity of demand	Elasticity of supply	California			U.S.		
				Producers	Consumers	All	Producers	Consumers	All
	(%)			- - - - - - - - - - - - - - - - - ($mil) - - - - - - - - - - - - - - - -					
Oranges	6.5	−0.85	1	−247	−146	−393	−221	−1,215	−1,435
	6.5	−0.85	4	−158	−220	−378	−116	−1,835	−1,951
Lemons	3	−0.5	1	−43	−31	−74	−35	−257	−292
	3	−0.5	4	−26	−41	−67	−15	−342	−357
Grapefruit	5	−0.45	1	−42	−11	−53	−29	−90	−119
	5	−0.45	4	−39	−14	−53	−21	−118	−139

[a]The discount rate is 7%.

Table 9.14 Total welfare changes when citrus canker becomes established

Long-run elasticity of supply	California			U.S.		
	95%	99%	Total	95%	99%	Total
			($mil)			
1	−494	−515	−520	−1,754	−1,828	−1,847
4	−473	−493	−498	−2,323	−2,421	−2,446

A Comparison of the Costs and Benefits of Eradicating Citrus Canker The expected value of the establishment costs is calculated for a 95 percent probability of success and a 99 percent probability to properly compare the benefits of eradication to the costs (Table 9.14). Total costs of eradication include government outlays plus the changes in producer and consumer welfare during a large eradication program. Because it is unknown which crops would be affected by an invasion and eradication program, the effects of each crop on producers and consumers should not be added together. Also, when aggregating the producer and consumer losses with government outlays, the compensation to producers needs to be deducted from the total losses to producers to avoid double counting. For purposes of comparison to the costs of establishment, the losses in welfare to producers and consumers from an eradication program for lemons are included because lemons are more susceptible to citrus canker than oranges or grapefruit.

The expected benefits of eradicating citrus canker would be greater than the costs of eradication (Table 9.15). This holds even when the size of the infestation is relatively large to begin with, and for both 95 percent and 99 percent probabilities of success. For a large eradication program that would achieve either a 95 or 99 percent probability of success, the benefits within California would be about six times greater than the costs. The net benefits (expected benefits less costs) would be about $400 million for the 95 percent probability of success and just over $415 million for the 99 percent probability.

The benefits of eradication would also be greater than the costs for the United States as a whole, even for a large eradication program. The costs of establishment would be five to six times greater than the costs of eradication at the

95 percent probability level. However, the absolute value of the net benefits increases to over $1 billion when the elasticity of supply is 1.0 and by almost $2 billion when it is 4.0.

The additional expected benefits of increasing the probability of success to 99 percent would also be greater than the additional costs both within California and for the United States. For the case of the large eradication program the additional costs would be $1 million for California government outlays and losses in welfare. The additional expected benefits would be about $20 million.

The gains for the entire United States would be even greater. Increasing the probability of success to 99 percent would cause total U.S. outlays to increase by $3 million. Benefits would increase by $74 million when the elasticity of supply is 1.0 and by $98 million when the elasticity of supply is 4.0.

One option not included in this study is the cost and benefit of a second eradication program should the first one fail. The costs and benefits of a second program would be highly variable depending upon the reason the first one failed. If the first one failed because a few trees were missed and the infestation was discovered quickly, then the costs of a second eradication

Table 9.15 Comparison of the costs and benefits of eradicating citrus canker

		California		U.S.	
		95%	99%	95%	99%
			($mil)		
Costs	Small	1.6	1.9	3.5	3.9
	Large	76.8	77.7	370	373
Benefits	Long-run elasticity of supply				
	1	−494	−515	−1,754	−1,828
	4	−473	−493	−2,323	−2,421

program would be similar to the costs of a small eradication program. Even using just the 95 percent probability of success, the probability of failing a second time would be 0.05×0.05 or 0.0025, which is approximately equal to zero. If, however, the eradication program fails because the disease spread faster than trees could be destroyed, costs would be even greater than in the first program. In this case, a whole new program would have to be evaluated and the costs and benefits estimated for a 95 percent probability of success and a 99 percent probability of success.

General Discussion and Implications

The most epidemiologically significant way that citrus canker is likely to be introduced into California is through infected planting stock and/or other propagative material (except seeds). Clinically or subclinically, Xcc-infected fruit with lesions or asymptomatic Xcc-infested fruit would not be likely sources for establishment of the pathogen in California. California's environmental conditions are generally conducive to Xcc infection and citrus canker development; however, the likely rate of spread of any citrus canker infestation and the extent of infection in California are largely unknown. Introduction and establishment of the citrus leaf miner in California could exacerbate leaf infection. Timely, decisive, and resolute measures would have to be taken. Along with adequate resources (i.e., funds, personnel) and broad-based industry and public support, sound biologically based regulatory policies would be needed to implement effective measures to minimize the effects of any citrus canker infestation in California. Regulatory policies and actions would also have to be harmonious with current policies of the USDA and international plant protection organizations for protecting citrus production in the United States and in other citrus-growing areas around the world.

Other timely policy decisions regarding eradication versus disease management (e.g., copper-containing sprays, planting windbreaks, establishment of citrus canker-free areas) would be needed. The choice of policy depends upon the relative costs and benefits of the alternative disease management choices. The relative costs and benefits of eradication as opposed to establishment would depend to a large extent on how soon the infestation is identified and how quickly regulatory agencies cooperated with homeowners and producers to destroy affected trees. Therefore, there is a need to effectively engage the public in developing and implementing Xcc exclusion strategies, as well as disease eradication methods.

References

Agricultural Issues Center. 1999. *Improving the Level of Accuracy Reported in State-by-State Agricultural Export Statistics.* University of California.

Alston, Julian M., George W. Norton, and Phillip G. Pardey. 1995. *Science Under Scarcity: Principles and Practice for Agricultural Research Evaluation and Priority Setting.* New York: Cornell University Press.

California Chapter of the American Society of Farm Managers and Rural Appraisers. 1999. *Trends in Agricultural Land and Lease Values.* California Chapter of the American Society of Farm Managers and Rural Appraisers.

California Department of Food and Agriculture. 1994. *Final Programmatic Environmental Impact Report: The Exotic Fruit Fly Eradication Program Using Aerial Application of Malathion and Bait.* State Clearinghouse Number 91043018.

California Department of Food and Agriculture. 2000. Press Release CDFA00-012.

California Department of Food and Agriculture. 2002. California Agricultural Statistics Service. "County Agricultural Commission Report Data," http://www.nass.usda.gov/ca/bul/agcom/indexcac.htm.

Civerolo, E.L. 1984. "Bacterial Canker Disease of Citrus." *Journal of the Rio Grande Valley Horticultural Society.* 37:127–146.

Civerolo, E.L. 1994. "Citrus Bacterial Canker Disease in Tropical Regions." In M. Lemattre, S. Freigoun, K. Rudolph, and J.G. Swings, Eds., *Proceedings of the 8th International Conference on Plant Pathogenic Bacteria.* Paris: ORSTOM/INRA. pp. 45–50.

Commonwealth Mycological Institute. 1996. *Distribution Maps of Plant Diseases,* Xanthomonas campestris *pv. citri.* Map No. 11, 6th Ed. Wallingford: CAB International.

Elitzak, Howard. "Food Cost Review, 1950-1997." Agricultural Economic Report Number 780. Economic Research Service Report. Washington, D.C.: U.S. Dept. of Agriculture.

Goto, M. 1992a. *Fundamentals of Bacterial Plant Pathology.* San Diego: Academic Press. pp. 316–319.

Goto, M. 1992b. "Citrus Canker." In J. Kumar, H.S. Chaube. U.S. Singh, and A.N. Mukhopadhyay,

Eds., *Plant Diseases of International Importance,* Vol. III. Englewood Cliffs: Prentice Hall. pp. 170–208.

Gottwald, T.R., J.H. Graham, and T.S. Schubert. 1997. "An Epidemiological Analysis of the Spread of Citrus Canker in Urban Miami, Florida, and Synergistic Interaction with the Asian Citrus Leafminer." *Fruits.* 52:383–390.

Haab, L. 1999. Florida Department of Agriculture and Consumer Services. Personal communication to K.M. Jetter, December 1999.

Haylock, Doug. 1999. Florida Department of Agriculture and Consumer Services. Personal communication to K.M. Jetter, December 1999.

Huang, Kuo S. 1993. "A Complete System of U.S. Demand for Food." Technical Bulletin Number 1821. Washington, D.C.: Economic Research Service, United States Department of Agriculture.

ISA (International Society of Arboriculture). 1992. *Guide for Plant Appraisal.* Urbana, IL: International Society of Arboriculture. 150 pp.

Kinney, William, Hoy Carman, Richard Green, and John O'Connell. 1987. "An Analysis of Economic Adjustments in the California Arizona Lemon Industry." Giannini Foundation Research Report No. 337. Division of Agricultural and Natural Resources, University of California.

O'Connell, Neil, Karen Klonsky, Mark Freeman, Craig Kallsen, and Pete Livingston. 1999. *Sample Costs To Establish an Orange Grove and Produce Oranges—Low Volume Irrigation in the San Joaquin Valley.* Department of Agricultural and Resource Economics, University of California, Davis.

Pruvost, O., C. Verniere, J. Hartung, T. Gottwald, and H. Quetelard. 1997. "Towards an Improvement of Citrus Canker Control in Reunion Island." *Fruits.* 52:375–382.

Schubert, T.S. 1999. Chief Plant Pathologist, Florida Department of Agriculture and Consumer Services. Personal communication, December.

Schubert, T.S., and J.W. Miller. 1999. *Bacterial Citrus Canker.* Gainesville. Florida Department of Agriculture and Consumer Services.

Stall, R.D., J.S. Miller, G.M. Marco, and B.I.C. Canteros. 1982. "Pathogenicity of the Three Strains of Citrus Canker Organism on Grapefruit." In C. Lozano and P. Gwin, Eds., *Proceedings of the 5th International Conference on Plant Pathogenic Bacteria.* Cali, Colombia: Centro Internacional de Agricultura Tropical. pp. 334-340.

Stall, R.E., and E.L. Civerolo. 1991. "Research Relating to the Recent Outbreak of Citrus Canker in Florida." *Annual Review of Phytopathology.* 29:339–420.

Stall, R.E., and E.L. Civerolo. 1993. "*Xanthomonas campestris* pv. citri: Cause of Citrus Canker." In J.G. Swings and E.L. Civerolo, Eds., *Xanthomonas.* London: Chapman & Hall. pp. 48–51.

Stutsman, Walter. 1999. Supervisor, Lincove. Personal communication.

Takele, Etaferahu, Nicholas Sakovich, and Delos Walton. 1997a. *Establishment and Production Costs, Grapefruit, Riverside County.* Riverside: University of California Cooperative Extension, Division of Agricultural and Natural Resources.

Takele, Etaferahu, Nicholas Sakovich, and Delos Walton. 1997b. *Establishment and Production Costs, Lemons, Ventura County.* Riverside: University of California Cooperative Extension, Division of Agricultural and Natural Resources.

U.S. Department of Agriculture, Federal Crop Insurance Corporation. 1999. *Florida Fruit Tree Pilot Crop Insurance Provisions.* Revised 12/99. Washington, D.C.: U.S. Dept. of Agriculture.

U.S. Department of Agriculture. 2002. National Agricultural Statistics Service. http://www.usda.gov/nass/.

U.S. Department of Commerce, Bureau of the Census, Economics and Statistics Administration. 1991. *1990 Census of Population and Housing, Summary Population and Housing Characteristics, California.* CPH 1-6. Washington, D.C.: U.S. Dept. of Commerce.

Vauterin, L., B. Hoste, K. Kersters, and J. Swings. 1995. "Reclassification of *Xanthomonas.*" *International Journal of Systematic Bacteriology.* 45:472–489.

Verniere, C., J.S. Hartung, O.P. Pruvost, E.L. Civerolo, A.M. Alvarez, P. Maestri, and J. Luisetti. 1998. "Characterization of Phenotypically Distinct Strains of *Xanthomonas axonopodis* pv. citri from Southwest Asia." *European Journal of Plant Pathology.* 104:477–487.

Young, J.M., G.S. Saddler, Y. Takikawa, S.H. DeBoer, L. Vauterin, L. Gardan, R.I. Gvozdyak, and D.E. Stead. 1996. "Names of Plant Pathogenic Bacteria 1864–1995." *Review of Plant Pathology.* 75:721–763.

Appendix 9.1

Estimating Losses in Producer and Consumer Welfare

For both the eradication and establishment scenarios, producers and consumers may be affected through changes in supply and production costs. Models were developed to reflect how these shocks affected total quantities supplied and demanded, and the price for those commodities.

The basic approach used in these models sets out supply and demand conditions in log-differential form such that $d \ln X = (X_1 - X_0)/X_0$, where subscript 1 indexes the new level and subscript 0 indexes the original level of variable X. We use the model to show how the equilibrium quantities, prices, and other variables respond to shocks to the system, such as the removal of acreage during an eradication program, or increases in production costs when an exotic pest becomes established. The model is parameterized with market and biological data.

The Single Output Model

The model has three parts. The first part describes the demand side of the market, the second describes the supply side of the market, and the third describes the equilibrium conditions between quantities and prices.

Consumer Demand

Consumers are separated into two groups, consumers in the U.S. market, D_{US}, and consumers in the foreign market, D_T. Equations 9.1 and 9.2 state that the quantity demanded in each market depends upon the price, P, of the commodity. The log-differential form states that the percentage of change in the quantity demanded is equal to the elasticity of demand for each region, η, times the percentage of change in price (Equations 9.1a and 9.2a). The elasticity of demand is negative to reflect the fact that as price increases, consumers will buy less of the commodity.

(9.1) $D_{US} = d_{US}(P)$

(9.1a) $d \ln D_{US} = \eta_{US} d \ln P$

(9.2) $D_T = d_T(P)$

(9.2a) $d \ln D_T = \eta_T d \ln P$

Equation 9.3 states that the total quantity demanded, D, is equal to the quantity demanded in each region. The log-differential form states that the total percentage of change in demand is equal to the percentage of change in demand in each region, weighted by the original share, α, of the market consumed by each region (Equation 9.3a).

(9.3) $D = D_{US} + D_T$

(9.3a) $d \ln D = \alpha d \ln D_{US} + (1-\alpha) d \ln D_T$

Agricultural Supply

The agricultural supply equations are generalized to account for both the eradication and establishment scenarios. Total quantity supplied, S, is equal to quantity supplied from uninfested producers, Q_u, quantity supplied from acreage that will be destroyed during an eradication program, Q_e, and quantity supplied from producers with citrus canker infestations, Q_i (Equation 9.4).

(9.4) $S = Q_U + Q_i + Q_e$

(9.4a) $d \ln S = \lambda_U d \ln Q_U + \lambda_i d \ln Q_i + \lambda_e d \ln Q_e$

When the eradication scenario is estimated, Q_i is equal to zero and when the establishment scenario is estimated, Q_e is equal to zero. The percentage of change in quantity supplied is equal to the sum of the share weighted proportional changes in quantity supplied from each area (Equation 9.4a).

Producers in citrus canker-free areas will produce where the market price, P, they receive for crop Q_u is equal to the marginal costs of producing the commodity (Equation 9.5). Equation 9.5a relates the percentage of changes in prices to the percentage of changes in production. The percentage of change in price is equal to one divided by the elasticity of supply, ε_u, for growers in unaffected regions times the proportional change in quantity produced (Equation 9.5a).

$$(9.5) \quad P = \frac{\partial C_U(Q_U)}{\partial Q_i}$$

$$(9.5a) \quad d\ln P = \left(\frac{1}{\varepsilon_U}\right) d\ln Q_U$$

When an eradication program is undertaken, the production from the acreage on which trees would be destroyed is equal to the supply shifter ω (Equation 9.6). Because Q_e is defined as the region where the commercial eradication program will take place, the change in supply is a decrease of 100 percent. Therefore, the proportional change in production is equal to -1 (Equation 9.6a).

$$(9.6) \quad Q_e = \omega$$

$$(9.6a) \quad d\ln Q_e = -1$$

When citrus canker becomes established, production costs increase. Producers in the infested region will produce where price is equal to marginal costs (Equation 9.7). However, because growers now need to treat for citrus canker, the proportional change in price is equal to 1 divided by the elasticity of supply, ε_i, for growers in the affected regions multiplied by the proportional change in quantity produced plus the proportional change in the costs of production (Equation 9.7a).

$$(9.7) \quad P = \frac{\partial C_i(\mu_i, Q_i)}{\partial Q_i}$$

$$(9.7a) \quad d\ln P = \left(\frac{1}{\varepsilon_i}\right) d\ln Q_i + d\ln \mu_i$$

10

An Insect Pest of Agricultural, Urban, and Wildlife Areas: The Red Imported Fire Ant

John H. Klotz, Karen M. Jetter, Les Greenberg, Jay Hamilton, John Kabashima, and David F. Williams

Introduction

The red imported fire ant (RIFA), *Solenopsis invicta* (Buren), is an insect pest of particular importance in California due to its potential impact on public health, agriculture, and wildlife. In 1997, RIFAs hitchhiked to the Central Valley on honeybee hives brought in from Texas for pollination of an almond orchard (Dowell et al. 1997). There has been local spread from these locations to surrounding irrigated areas. In 1998 the ants were detected in several other locations, including an area covering at least 50 square miles of Orange County. As a consequence, all of Orange County, parts of Riverside County between Palm Springs and Indio, and one square mile of the Moreno Valley were quarantined. The size and distribution of the infestations indicate that the RIFA has been established and spreading for several years in southern California.

The RIFA has both beneficial and detrimental effects on our environment. In a few cases they are predators of agricultural pests, but mostly they have a negative impact. Their large mounded nests, which can be 35 cm (1.1 feet) high, damage mowing and harvesting equipment. When people or animals disturb their nest, the highly aggressive ants swarm out and attack and sting the unwary intruder. In some cases people hypersensitive to their venom have died.

They are attracted to irrigation lines during times of drought, plugging sprinkler heads and chewing holes in drip systems (Vinson 1997). Their aggregation near electrical fields (Slowik et al. 1996) can result in short circuits or interference with switches and mechanical equipment such as water pumps, computers, and air conditioners. More serious problems can arise when they infest traffic signals and airport runway lights (Lofgren et al. 1975).

Biology and Ecology

The RIFA is the most thoroughly studied ant. It has been the focus of research and control efforts for more than four decades (Williams 1994). Comprehensive reviews on their biology and ecology can be found in Vinson and Greenberg (1986), Vinson (1997), Taber (2000), and for California, Greenberg et al. (1999, 2001).

Fire ants undergo complete metamorphosis in their life cycle, which consists of four stages: egg, larva, pupa, and adult. The queen lays hundreds of eggs each day. After 7 to 10 days the eggs hatch into larvae. In another one to two weeks the larvae molt into a quiescent pupal stage. Pupae resemble curled-up adults and cannot move. Over the next one to two weeks the pupae acquire the reddish-brown pigmentation of adults. In the final molt, female pupae become either adult workers or reproductives. Mature colonies of RIFAs have 200,000 to 300,000 workers, and either one queen (monogyne) or many queens (polygyne).

Monogyne colonies are territorial and reproduce by mating flights. The males die after copulating, while the newly mated queens seek out nest sites. Fire ants are not strong fliers, but can fly several miles before landing. They are attracted to reflective surfaces such as pools and truck beds where they will land, and in the latter case, sometimes be transported for hundreds of miles. In the more typical case a newly mated queen lands on the ground, removes her wings, and then searches for moist, soft soil where she digs a small hole. Inside the hole, she seals the entrance and begins laying eggs. After one or two years the colony matures, and large numbers of winged reproductives (alates) are produced in preparation for mating flights in spring. These nuptial flights can occur at other times if conditions are favorable. Alates prefer to fly after it rains, on warm, clear days with no wind.

Polygyne colonies are not territorial and may consist of many mounds. As a result, they are larger than monogyne colonies and have higher mound densities. Polygyne infestations have hundreds of mounds per acre, whereas monogynes have 30-40. In addition to mating flights, polygyne colonies can also spread by fission or budding (Vargo and Porter 1989), an adaptation that may allow them to invade areas where conditions are not favorable for mating flights.

RIFAs can, and do, fly almost any time of the year in California (Les Greenberg, personal observation). Instead of rain being the triggering event for a flight, water from sprinklers is adequate. To be successful, though, mating flights must be coordinated over large areas so that males and females from different colonies can form large mating swarms hundreds of feet above the ground. In addition, whether an infestation is monogyne or polygyne is useful information, because the latter with larger and more numerous colonies will have more frequent and intense interactions with people.

The RIFA has an omnivorous diet and opportunistic feeding habits. They will feed on any plant or animal they encounter (Lofgren et al. 1975). Their primary diet, however, is insects and other small invertebrates (Vinson and Greenberg 1986), including some that are pests of important agricultural crops such as the cotton boll weevil (Sterling 1978), sugar cane borer (Reagan 1981), and tobacco budworm (McDaniel and Sterling 1979, 1982). They are also scavengers and feed on carrion.

In heavy infestations RIFAs saturate the environment and become the dominant ecological force. As a consequence, coexisting species of ants, other invertebrates (Porter and Savignano 1990), and vertebrates (Lofgren 1986) suffer and are sometimes eliminated. The negative effects of RIFAs on invertebrate and vertebrate biodiversity in the South are extensive (Wojcik et al. 2001).

Their notoriety, of course, is due mainly to their aggressive defense of the nest accompanied by their painful sting, which they are able to inflict in unison after crawling up the legs of an unwitting victim. In order to sting, they must first grab the skin with their mandibles for leverage, and then curl their abdomens to insert the stinger. The venom contains piperidines, which cause a burning sensation, and proteins, which can cause life-threatening anaphylactic shock in a small percentage (< 1 percent) of the population. Their sting causes a white pustule to form on the skin.

Introduction and Spread

The RIFA originates in lowland areas of South America and was most likely introduced into the United States between 1933 and 1945 (Lennartz 1973). The initial colonization in Mobile, Alabama probably occurred as a result of infested soil from South America used as ship ballast or dunnage, and dumped at the port. At that time several native fire ant species thrived in the southeast and the presence of another exotic one created little concern. But by the 1950s their rapid spread and aggressive nature alarmed the public. Now they inhabit all of the southern states from Florida to Texas and as far north as southern Oklahoma, Arkansas, Virginia, and Tennessee.

Since their first documented interception at a border station in California in 1984 (Lewis et al. 1992), RIFAs have been found in several counties. The first outbreak was discovered in Carpinteria in Santa Barbara County in 1988 and was eradicated (Knight and Rust 1990). Recent outbreaks are more serious because they are not confined to a single location and may have gone undetected for three to five years, giving the ants time to spread. Outbreaks are associated with commerce, with the ants arriving on trucks, trains, or other vehicles. A partial list of likely sources includes the root balls of nursery stock, sod, dirt attached to honeybee hives and encrusted on land-moving equipment, and produce brought into the state. New housing developments, with their inflow of building materials, trees and plants, and dirt-moving tractors, are especially vulnerable.

Since the 1997 outbreak in Kern County, more extensive infestations have been found in Orange and Riverside counties, but it is not known how they were brought in. Additional isolated infestations have been found in San Diego, San Bernardino, Los Angeles, Fresno, Madera, and Stanislaus counties. Commerce from infested states will continue to bring imported fire ants into California.

There is no way of predicting how far the RIFAs will spread in California, but if their history in the South is any indication, their future distribution in California could be extensive. Two factors are critical to their survival: temperature and moisture. A map of the expected distribution of the RIFA in the United States based on a 0° minimum temperature shows them inhabiting the entire West Coast from southern California to northern Washington (Killion and Grant 1995). Water, however, is a limiting factor in many areas in southern California.

The arid climate of southern California's inland deserts is inhospitable to RIFAs. But due to irrigation the RIFA became established on golf courses, nurseries, horse facilities, and turf farms in the Coachella Valley. Flood irrigation can even spread the RIFA because they form rafts of living ants that are carried by the water to new locations. The queen and brood are within these rafts, so a new mound can spring up instantly wherever they touch land. As soil conditions become dry, the RIFA will move its nest to an area with more moisture, such as around homes, irrigated farmlands, watering holes on rangelands, and near lakes, ponds, and streams.

Another factor that may limit or slow down its spread in California is competition with other species of ants. In southern California, for example, there are reports of intense interspecific competition between the RIFA and the Argentine ant, *Linepithema humile*. In the South, reinfestation of treated areas by the RIFA is common because control measures often eliminate other species of ants that are competitors (Williams 1986).

Intervention Strategies

There are three levels of policy action that address the RIFA threat: (1) prevention of their entry into the state, (2) quick eradication of outbreaks, and (3) containment and management if they become established. Policy options designed to prevent their entry range from government inspections and monitoring to quarantines on the importation of agricultural commodities that may harbor stowaway RIFAs.

Once an outbreak of RIFAs is discovered, eradication should be attempted as soon as possible. The longer the time lag between surveys to map infestations and the initial treatment, the more time the ants have to spread. As the RIFA spreads in all directions into surrounding areas, survey and treatment costs increase exponentially with the elapsed time between infestation, detection, and eradication. Eradication efforts in the South have failed due to reinfestation by RIFAs from surrounding untreated areas.

The situation is different in California because the outbreaks are localized and surrounded by inhospitable nonirrigated land. Consequently, eradication is a realistic policy choice for controlling the RIFA in California. Small, discrete infestations in California have been successfully eradicated. In addition, new, highly effective insecticides such as fipronil will soon be available in California for use against the RIFA (Chris Olsen, Aventis, personal communication 2002). Registration has been approved by EPA and is pending in California. Currently, eradication can only be completed with chemical treatments, including baits and contact insecticides.

To address the fire ant crisis, the California Department of Food and Agriculture (CDFA) developed a short-term interim plan to deal with the immediate problem and a long-term control plan to prevent future infestations if current eradication efforts are successful. Both plans were developed with the aid of the RIFA Science Advisory Panel, a group of university and U.S. Department of Agriculture (USDA) fire ant experts.

The interim plan was announced in March 1999 and called for treatment to begin in April 1999. Beginning in July 1999, treatment programs are coordinated by CDFA through contract agreements with local agencies. Funding is through $40 million in budget commitments by the state legislature and California Governor Gray Davis. The money is available over a five-year period. In 2004, the eradication program will be reevaluated for feasibility. Objectives of the interim plan include limiting the local spread of the RIFA and training personnel in local agencies on proper identification and treatment of fire ants. To coordinate eradication efforts, the CDFA developed a treatment protocol for county administrators. The protocols include: (1) pest identification; (2) detailed location of RIFA mounds; (3) surveys of local areas to find additional mounds; (4) application of a metabolic inhibitor (hydramethylnon) and an insect growth regulator (pyriproxyfen or fenoxycarb) in granular bait form when soil temperature is between 65° and 90°F and free of rain or irrigation for 36 hours (the protocol allows for the use of insecticide drenches if reproductives are found); and (5) a visual and bait survey of treated mounds six weeks after the insect growth regulator application. If RIFA mounds are found on private property, the protocol requires the owner's permission before a treatment can be applied.

The interim plan also contains a protocol for surveys in areas where an infestation is suspected and one for monitoring to assess efficacy of a treatment. It specifies how long monitoring should continue and how visual monitoring of bait stations should be conducted in different areas such as orchards, golf courses, and parks. Treatments may be undertaken by city, county, state, or federal agencies, but should be reported to the CDFA. In conjunction with the interim plan the California Environmental Protection Agency's Department of Pesticide Regulation will be monitoring the impact of the insecticides on the environment.

In addition, the CDFA has developed the California Action Plan for RIFA. This comprehensive plan supplements the interim plan with public outreach efforts to inform and train local agencies on the protocols described in the interim plan. The state will coordinate multicity programs, but actual treatment will be administered by local agencies. The action plan also calls for monitoring industries that have a high risk of transporting the RIFA to new locations. Quarantines will be used to slow the spread of RIFAs when new infestations are found.

Surveillance for RIFAs at California's inspection stations will be strengthened. The exterior quarantine improvements include an additional inspector for each work shift at southern border inspection stations to improve the detection of RIFAs on high-risk vehicles, new inspection stations and 10 new inspectors, and research into rapid identification techniques for RIFAs.

The state will also employ biologists to survey high-risk areas for RIFA infestation. Research funds will be made available for studies on the optimal treatment of the RIFA under California's unique conditions. The goal of the California Action Plan is to eradicate or control the spread of RIFAs in 5 to 10 years. If eradication is successful, surrounding areas will need to be surveyed for at least 1 year. Since newly mated females can travel several miles, the monitoring and survey area should be at least three miles around the eradication zone. If eradication efforts fail, the current plan would form the foundation of future management programs and another set of policy decisions would need to be made regarding the scope of public expenditure for containment and management of the RIFA. Another option would be to stop public management measures, allowing the RIFA to spread and establish itself throughout its climatic range in California.

Parties Affected

The RIFA is unique among California's exotic pests because of its potential impact on so many aspects of the state's economy. They pose a threat to homeowners, growers, and wildlife with their sting, their direct damage to crops and livestock, their interference with electrical and irrigation equipment, and their ability to displace native species.

The RIFA prefers to nest in soil in open, sunny areas, but it can be a serious household pest (Klotz et al. 1995). For homeowners the potential problems include medical treatment for stings, interference with communications and electrical equipment, direct and indirect costs (such as environmental degradation) of increased pesticide use, and reduced use of recreational facilities. In infested areas, picnics and recreation involving ground contact are avoided, especially around lakes. Many homeowners become frustrated by their inability to keep their lawns free of fire ant mounds. Children avoid going barefoot or playing in yards that are infested with RIFAs, and gardening activities are curtailed. The fear of being stung has even led to liability considerations and reduced property values (Vinson 1997).

Agriculture in southern states has been significantly damaged by fire ants both directly through lost production and indirectly through economic losses from quarantines. The RIFA feeds on many crops, including the seeds or seedlings of corn, peanuts, beans, Irish potatoes, cabbage, and young citrus. Their mounds often interfere with harvesting equipment and reduce usable pasture. They cultivate and defend plant lice from predators, thereby interfering with biological control. They can cause blindness and death in livestock as well as diminish the overall quality of livestock. The painful stings are a nuisance to farm laborers, and RIFAs can cause automatic feeding and irrigation systems to malfunction. Quarantines impose additional costs, because hay, equipment, beehives, and nursery products must all undergo special treatments to meet regulations.

In the South, RIFAs reduce invertebrate and vertebrate biodiversity and threaten endangered species (Wojcik et al. 2001). They inflict damage on ground-nesting reptiles, birds, and mammals, especially their newborns. Their foraging efficiency is such that other species of ants, invertebrates (Porter and Savignano 1990), and vertebrates (Lofgren 1986) are eliminated. In addition, many chemical control measures for RIFAs adversely affect wildlife. In California, similar negative effects may occur in lowland and coastal wilderness areas if the RIFA becomes established.

Policy Scenarios

The policy options for managing RIFAs are either eradication or allowing it to become established and then imposing private controls and quarantines. The expected costs to taxpayers of a public eradication program will be compared to the expected benefits to households and agricultural industries if establishment is avoided.

The CDFA eradication program has been funded for 5 years, with the possibility of another 5 years, depending on progress, for a maximum of 10 years. Taxpayer funding for the RIFA eradication program is fixed for the 5-year period and has not changed in response to the discovery of new infestations. Because a biological risk assessment has not been done, the probability of success for the eradication program has not been estimated. Therefore, the cost/benefit analysis will determine the probability of success needed for the expected benefits to be at least as great as the expected costs. This probability will be estimated for a 5-year

eradication program, a 10-year eradication program, and two 5-year programs.

While containment is another policy option, the lack of knowledge on how the RIFA interacts with the California environment prevents us from making any meaningful biological or economic risk assessments of possible strategies. However, a policy of containment to slow the spread of RIFAs would be important to consider should eradication efforts fail.

Economic Analysis

Eradication Costs

Eradication costs are incurred by taxpayers and nurseries within quarantined areas. Taxpayers pay the regulatory agency costs of implementing the interim and long-term action plans. As part of any eradication program all nurseries and infested golf courses within the quarantine area must treat their premises for RIFAs, earth-moving equipment must be free of soil, and other restrictions met. The total cost of the eradication program in this study will be the cost to taxpayers of the public project, plus the costs to nurseries and other businesses to comply with quarantine regulations. Insufficient data are available on the number of golf courses in the affected areas. Consequently, those quarantine compliance costs are excluded from the analysis. Treatment on land around private residences is done through the public project.

The current 5-year public funding level of the RIFA eradication program is $40 million. This includes $8.4 million for the first year, $7.4 million a year for the remaining 4 years and an additional $2 million general allocation. We assume that the annual funding level for the next 5-year period is also $7.4 million a year, with no other allocations or increases in funding.

The cost to nurseries is calculated as the amount of acreage affected times the treatment and monitoring costs per acre. The amount of acreage that is affected in the quarantined areas is equal to the total nursery acreage in Orange County, plus 10 percent of the acreage in Los Angeles and Riverside counties. Total affected acreage is 2,300. At per acre treatment costs of $650, total private costs are $1.5 million a year. Total annual private and taxpayer costs are $8.9 million.

The present value of the initial 5-year project is $39.4 million when discounted at a long-term interest rate of 7 percent. Should the eradication project require an additional 5-year period, the present value of taxpayer and private costs for the second 5-year period is $26 million. In total, the present value of the 10-year project is $65.4 million.

Establishment Costs

Households, agriculture, and wildlife are all affected by RIFA. However, the costs and benefits of the RIFA spreading throughout California would not be evenly distributed among these groups. Some households, farms, or ranches may suffer from large infestations and costs, while nearby homes and agricultural operations may have little or no damage. The costs and benefits estimated in this chapter are based on average costs per acre from studies of damage by RIFAs in the southeastern United States. Actual costs incurred by individual households and agricultural producers can vary substantially from these average costs. Because of its drier climate, costs in California may also deviate from the wetter, southeastern United States.

Costs to Urban Households
Urban households incur costs to treat mounds, repair damage to electrical equipment, and for medical and veterinary expenses. In a survey of South Carolina households, the average total cost per household due to RIFAs was $80 (Dukes et al. 1999). Costs, however, were not the same across regions. In lower risk regions average costs were only $33, while in higher risk regions they were $104.

Given the wide range in costs and climatic conditions in California, three methods were used to estimate the economic effects of RIFA infestations on urban households. The first was to multiply the number of households in counties susceptible to RIFA infestations by the average cost per household for all households. The second method was to multiply the number of households in the low-risk counties by the average low-risk cost, and the number of households in the high-risk counties by the average high-risk cost, and then add the two together. The third method was to multiply the number of households in susceptible counties by the average costs per low-risk household.

In 1999 the total number of households in susceptible counties was 10,363,432 (Department of Finance 2000). In the low-risk counties there were 2,711,036 households, and in the high-risk counties 7,652,396. Total estimated cost of RIFAs to urban households would then be $829 million when average costs for all households are used to calculate total cost, $885 million when cost is calculated by region, and $342 million when the average low-risk cost is used for all susceptible households.

Costs to Agriculture
TREE CROPS AND VINEYARDS Tree crops and vineyards use hand labor throughout the year. Tasks requiring hand labor include pruning, raking, and harvesting. In fields infested with RIFAs, crews may not be able to enter to complete these tasks because of the aggressive nature of the ant and the painful stings, or may request a higher fee to compensate for the additional health risks. Alternatively, producers could treat fields with insecticides and control RIFAs before crews enter. In our analysis we assume that producers would treat twice a year to control RIFAs with the growth regulator Extinguish, which is registered for use on all tree crops and vineyards in California. Total application costs for both treatments are $55 per acre.

The extent to which the RIFA would establish in groves, orchards, and vineyards may vary depending on previous treatments and agro-climatic conditions. Therefore, a range of acreage is used to estimate the additional costs to tree fruit, nut, and vine industries in California. A low-impact level of 10 percent of total acreage affected, a medium level of 25 percent, and a high level of 40 percent are used based on conversations with scientists familiar with RIFA problems in Florida and Arkansas (Thompson 2000).

Absolute increases in costs would range from $81,000 for figs at low-impact levels to $16.45 million for grapes at high levels (Table 10.1). Total increases in costs for all crops would range from $12 million at low-infestation levels to $48 million at high levels. While the dollar amount is substantial, as a percentage of total farm receipts it is less than 1 percent, even when 40 percent of acreage is affected. Costs as a percentage of farm receipts are greatest for figs, walnuts, and prunes, and lowest for lemons, nectarines and peaches, pears, apples, and plums.

ADDITIONAL EFFECTS ON CITRUS The RIFAs may also damage young citrus when they build their nests around or near the base of trees one to four years old. The ants feed on the bark and cambium to obtain sap, often girdling and killing the young trees. They also chew off new growth at the tips of branches and feed on flowers of developing fruit. Dead trees must be removed and replanted, raising the costs to establish an orchard. Based on field experiments in Florida, nursery stock mortality in untreated groves increased three- to fivefold per hectare, and total loss of newly planted groves due to RIFA feeding occurred in a few instances (Banks et al. 1991, Knapp 2000).

To prevent tree mortality, growers may choose to treat groves with insecticides. Groves should be treated for two to three years until young trees develop woody bark that RIFAs cannot chew through (Knapp 2000). RIFA control undertaken during grove establishment would increase investment costs and must be depreciated over the life of the grove. Establishment costs increase to $110 per acre if the grove is only treated the first two years and to $165 per acre for three years when groves are treated with two applications of Extinguish at an annual cost of $55 per acre. Depreciation of the additional investment costs to establish the grove would increase annual cash costs by $9 per acre when treatments last two years, and by $13 per acre for three years. This increase in costs is less than 0.5 percent of the total annual cash costs based on University of California Cooperative Extensive farm budgets for citrus.

VEGETABLES AND MELONS The RIFA builds nests around the edges of fields planted in vegetable crops because frequent discing in the fields disrupts nests in the interior. From the edges they can enter fields and damage crops. Most damage is from consumption of developing fruit, seeds, roots, or tubers. Documented losses from RIFAs include a 50 percent yield loss on eggplants and a 2.4 to 4 percent plant loss on sunflowers (Adams 1983; Stewart and

Table 10.1 RIFA effects on selected tree and vine crops

Crop	Acres	Farm receipts	Additional costs to industry[a] Acreage affected			Percent of farm receipts Acreage affected		
			10%	25%	40%	10%	25%	40%
	(000)	---------- *($000s)* ----------				----- *(%)* -----		
Almonds	456	1,165,150	2,509	6,273	10,037	0.22	0.54	0.86
Apples	39	207,151	216	541	865	0.10	0.26	0.42
Apricots	21	57,309	114	286	457	0.20	0.50	0.80
Avocados	58	272,406	321	802	1,283	0.12	0.29	0.47
Cherries	18	79,103	96	241	386	0.12	0.30	0.49
Figs	15	18,149	81	203	325	0.45	1.12	1.79
Grapefruit	17	73,794	93	232	371	0.13	0.31	0.50
Grapes	747	3,178,940	4,111	10,277	16,444	0.13	0.32	0.52
Lemons	49	347,329	271	677	1,083	0.08	0.19	0.31
Nectarines & peaches	110	556,535	604	1,511	2,417	0.11	0.27	0.43
Olives	34	73,677	185	463	741	0.25	0.63	1.01
Oranges	205	906,317	1,125	2,813	4,500	0.12	0.31	0.50
Pears	19	90,479	105	264	422	0.12	0.29	0.47
Pistachios	65	181,678	358	895	1,431	0.20	0.49	0.79
Plums	43	199,801	238	595	952	0.12	0.30	0.48
Prunes	86	151,822	471	1,176	1,882	0.31	0.77	1.24
Walnuts	202	344,848	1,109	2,774	4,438	0.32	0.80	1.29
Total	2,183	7,904,486	12,009	30,022	48,035	0.15	0.38	0.61

[a]Treatment costs are $55 per acre.

Vinson 1991). In the sunflower field no further damage was observed after a treatment with insecticides. It is often the case though, that crop damage will not be significant enough to make it economically justifiable to treat (Lofgren 1986).

While losses from crop damage may not always be greater than the costs to treat the RIFA, many vegetable and melon crops are hand-harvested. Therefore, growers may need to treat fields for worker protection, even though direct damage by RIFAs may be minor. To control RIFAs in vegetable and melon fields two applications of Extinguish would be applied per year at a total cost of $55 per acre. Because ant pressures will vary from year to year, a range of acreage is again used to determine the potential range in costs. Thus, industry costs were calculated for infestation levels of 10, 25, and 40 percent.

Total potential costs to the vegetable and melon industries would range from $3.7 million when only 10 percent of acreage is infested to $9.2 million when the infestation level is 25 percent, and to $14.8 million when the level is 40 percent (Table 10.2).

While the dollar figures would be large, as a percentage of farm receipts they would be less than 1 percent in all cases, and under 0.5 percent in most, even when up to 40 percent of acreage is affected.

ROW AND FIELD CROPS   The large nest mounds of RIFAs interfere with cultivation and mowing. In mowing weeds or cutting alfalfa, farm operators must either raise the cutting bar to prevent damage, switch from sickle bar to disc type cutters, repair equipment damaged by the mounds, or use insecticides to destroy colonies (Thompson et al. 1995).

Nonyield damages to row crops such as wheat, rice, and cotton include downtime to repair combines, electrical problems with pumps and machinery, other equipment damage, building damage, and medical expenses. In a survey of Arkansas row crop producers, nonyield costs of RIFAs per farm were $1,478. Over half of these costs were due to combine breakage and downtime for repairing cutter blades. Most damage to combines occurs from harvesting soybeans, a crop not grown in Cali-

Table 10.2   RIFA effects on selected vegetable and melon crops

Crop	Acres	Farm receipts	Additional costs to industry[a]			Percent of farm receipts		
			Acreage affected			Acreage affected		
			10%	25%	40%	10%	25%	40%
	(000)	---------- ($000s) ----------			----- (%) -----			
Artichokes	10	68,405	55	138	220	0.08	0.20	0.32
Asparagus	31	109,624	171	428	685	0.16	0.39	0.63
Beans, fresh	5	25,758	25	63	101	0.10	0.25	0.39
Broccoli	120	467,088	660	1,650	2,640	0.14	0.35	0.57
Brussels sprouts	3	21,715	18	44	70	0.08	0.20	0.32
Cabbage	14	74,401	76	191	306	0.10	0.26	0.41
Cantaloupe	63	240,525	345	861	1,378	0.14	0.36	0.57
Cauliflower	39	189,263	213	533	853	0.11	0.28	0.45
Celery	24	227,443	133	333	534	0.06	0.15	0.23
Cucumbers	6	52,676	35	87	139	0.07	0.16	0.26
Garlic	34	220,199	184	461	737	0.08	0.21	0.33
Honeydew	21	71,720	113	282	451	0.16	0.39	0.63
Lettuce, head	142	868,571	778	1,946	3,113	0.09	0.22	0.36
Lettuce, leaf	42	261,755	231	578	924	0.09	0.22	0.35
Lettuce, Romaine	27	156,520	149	371	594	0.09	0.24	0.38
Onions	39	169,254	214	534	855	0.13	0.32	0.50
Peppers, bell	22	162,707	118	296	473	0.07	0.18	0.29
Spinach, fresh	15	84,816	83	208	332	0.10	0.24	0.39
Watermelon	17	84,216	93	233	373	0.11	0.28	0.44
Total	672	3,556,651	3,694	9,236	14,777	0.10	0.26	0.42

[a]Treatment costs are $55 per acre.

fornia. The next highest cost was for repair of electrical equipment. When costs were calculated on a per acre basis, the costs for all yield and nonyield damage were $1 for rice, $0.25 for wheat, and $1.35 for hay. In general, it was not cost effective to treat for RIFAs in field crops.

RIFAs are predators of many agricultural pests. Among the pests that are present in cotton grown in California, the tobacco budworm and pink and cotton bollworms would all be preyed upon by the RIFA. Field experiments in Texas show that the presence of RIFAs significantly decreases bollworms in cotton fields and increases yields (Brinkley 1991). However, because the RIFA also damages electrical machinery and clogs sprinklers and irrigation equipment, the net result on profits is ambiguous. Surveys from Arkansas show a net profit in some cases and net losses in others (Semevski 1995). Therefore, no losses or benefits are estimated for cotton.

The total number of susceptible farms in California, based on the 1997 Census of Agriculture, is 5,526 (U.S. Department of Agriculture 2000). This includes grain, oilseeds, and hay enterprises. The cost per farm is set at the average level incurred per farm by Arkansas growers. As in the case of tree and vine crops, all field crops would not be affected. Costs are again calculated assuming 10, 25, and 40 percent of acreage would be affected. Total estimated costs are $817 thousand when 10 percent is infested, $2.0 million when 25 percent is infested and $3.3 million when 40 percent is infested.

Hay growers may have additional costs due to quarantine regulations. Hay stored on the ground may not be moved out of a quarantined area. How this affects growers would depend on the amount of production that would leave the area and the cost of alternative storage methods. Even if hay is not transported out of the region, growers would need to take precautionary measures against RIFAs because horses, cattle and other livestock would not consume ant infested hay.

NURSERY INDUSTRY All nurseries within a quarantine area would need to meet quarantine regulations in order to ship plants outside of the quarantined region. Open land on which nursery stock is grown would need to be treated once every three months with either fenoxycarb or hydramethylnon, alternating between the two insecticides. In addition, growers would need to treat the individual containers in which the plants are grown. Acceptable treatments include either a drench with chlorpyrifos, diazinon, or bifenthrin, 30 days before shipping, or incorporating a granular formulation of bifenthrin into the soil every six months. Because of environmental regulations concerning pesticide runoff and the need to treat frequently with chlorpyrifos, bifenthrin is more commonly used than chlorpyrifos.

Annual costs to treat nurseries for RIFAs would be about $650 per acre. The applications of fenoxycarb and hydramethylnon are $60 per acre, with the use of bifenthrin accounting for the remaining costs. According to the American Nursery and Landscape Association, the treatment cost per plant per container is 2¢. Only open nursery acreage that produces container plants would be affected by the quarantine regulations. Based on the 1997 Census of Agriculture, 28,000 acres were devoted to open-field nursery production of bedding and flower plants, foliage, potted flowers, and other nursery stock. Because nurseries within the quarantined regions must treat in order to ship outside of the quarantine, even if the nursery does not have RIFAs, almost all nurseries would be affected by the regulations. Total costs to the nursery industry are thus calculated on all open-field acreage and are equal to $18.2 million. In addition, nurseries would need to be inspected for RIFAs by placing bait out quarterly and observing the presence or absence of RIFAs on the bait at a cost of $38 per acre. Additional costs for inspection and certification are about $1.40 per acre.

Sod growers are also affected by quarantine regulations. Insecticide treatment for sod would be an application of chlorpyrifos. Materials and application costs are $330 per acre. Based on the 1997 Census of Agriculture, a total of 13,665 acres would be affected. Total costs are equal to $4.5 million.

Greenhouses that use containers placed on benches are exempt from the quarantine regulations. However, greenhouse operations would still need to treat if infested with RIFAs for worker safety and to protect electrical and irrigation equipment and machinery. These expenses would increase the costs to the nursery industry.

ANIMAL INDUSTRIES   The RIFA stings cattle and other livestock, infests hay and other food sources, and damages electrical and irrigation equipment (Barr and Drees 1994). The ants are attracted to mucous membranes located in the eyes and nostrils. Fire ant stings cause blindness and swelling and may end in suffocation. Immobilized animals, such as penned or newborn livestock are at the greatest risk. A survey of Texas veterinarians indicated that the most common livestock problem was skin inflammations from RIFA stings (49.6 percent of all cases). The next most common problem was blindness (20.1 percent) followed by secondary infections (14.4 percent) and injury to convalescent animals (12.3 percent).

Over 50 percent of the cases seen by the veterinarians were to treat pets and small animals. While pets and small animals were treated more often, mortality associated with the RIFA was greatest for cattle. However, it was often difficult to determine if RIFAs caused cattle death or if the ants were observed on animals after death. As a percentage of all cases seen by veterinarians, cases involving RIFA-related problems account for less than 1 percent.

In avoiding ants, livestock may also become malnourished or dehydrated when the ants invade their food and water. Cattle would not consume hay, nor would poultry eat feed, infested with RIFAs. The agitation caused by RIFAs invading poultry houses can decrease egg production. Extra expenses would be incurred to purchase RIFA-free hay or to treat around the perimeter of buildings to prevent RIFA invasions of calving pens, dairy and hog barns, and poultry houses.

Since the RIFA preys on insects, it may provide a benefit to the cattle industry from predation on ticks and horn flies in their immature stages. Because ticks and flies are disease vectors, the RIFA may potentially decrease the incidence of animal diseases carried by them.

RANGELAND EFFECTS   Losses to ranchers from the RIFA include damage to electrical equipment, hay-harvesting equipment, and cattle injury and loss. In a survey of Texas ranchers, 71 percent of respondents reported some type of economic loss (Teal et al. 1998). The largest damage levels were estimated at $28.06 per acre, but many counties in the drier, western regions had damages of less than $2 per acre. Even though damages are estimated on a per acre basis, about 95 percent of the total costs occur on about 5 percent of the land.

Most costs would be from damages around buildings, electrical equipment, and water sources. Also, as in the case of households and cropland, costs would vary widely. Some ranchers would experience large infestations and, consequently, large costs while nearby ranchers may have little damage.

Because California's climate differs markedly from that of Texas, costs in California are more likely to resemble costs incurred by ranchers located in Texas's western counties than for all counties in Texas. Furthermore, a significant proportion of rangeland in California is in counties too cold or dry to support RIFAs. These rangelands are located in northern California, along the Sierra Nevada mountain range, and in southern California.

Excluding rangelands in counties not susceptible to RIFAs results in a potential 15,759 acres at risk (U.S. Department of Agriculture 2000; FRRAP 1988). This acreage includes private rangelands, Bureau of Land Management land, and land grazed in National Forests. As in the case of agricultural crops, different impact levels are used to determine the potential range in costs. RIFAs will not be a problem on all susceptible acreage, however. Because a higher proportion of ranchers reported economic losses from the RIFA than were reported by growers, a higher range of acreage is used. Infestation levels of 25 percent, 40 percent, and 65 percent of all susceptible acres are used to determine the range in costs. Per acre costs are $1.50. Total annual potential costs are $5.9 million for the low-impact level of 25 percent affected, $9.5 million for 40 percent, and $15.4 million for 65 percent.

OTHER EFFECTS   Quarantine regulations would require that farm machinery and soil must be treated before leaving a quarantine area. Other agricultural activities, such as beekeeping, would also have to meet quarantine restrictions before being moved from one field or orchard to another.

Not included in our analysis are the costs to repair and replace irrigation equipment. Because the RIFA has previously established in areas with rain-fed agriculture, costs involving damage to irrigation equipment are not available.

Wildlife   Many claims have been made that imported fire ants affect wildlife and reduce biodiversity (Allen et al. 1994). When imported fire ants move into an area, they often displace native organisms. Due to their enormous population size and foraging efficiency, they become formidable competitors and predators within their territory. Thus, biodiversity in many coastal and low-altitude wilderness areas of California may be at risk. Imported fire ants displace other ants and invertebrates and also inflict damage on ground-nesting birds and mammals. The displacement of native ants and other animals may also disrupt native plant communities. Native ants assist the propagation of native plants by spreading seeds. As the ants decline, native plant species may also decline in fragile areas, and in turn threaten the animals that feed on those plants.

The RIFA appears to primarily affect bird and reptilian populations by destroying the eggs and the young. One study in Texas found that RIFA predation caused a 92 percent reduction in the number of waterbird offspring when natural habitants were not treated for infestations. Of special significance to California are studies that have documented ant predation on tortoise and reptile hatchlings. Fire ants may also prey on quail, but biologists have yet to definitively answer this question. In addition, many past chemical control measures for fire ants adversely affected wildlife. The newer products, however, do not adversely affect wildlife.

Many endangered species are among the wildlife threatened (Table 10.3). Either directly as a source of food or indirectly from predation on a food source, 58 out of California's 79 endangered animal species are susceptible to RIFAs. Insects, young rodents, reptiles, amphibians, and ground-nesting birds are directly susceptible through RIFA feeding. In addition several endangered birds, such as the northern spotted owl and bald eagle, may be at risk through a reduction in food sources. While no exact value has been estimated for the increased risk of extinction of specific endangered species, most people value preservation of endangered species and their potential increased risk

represents an additional cost of RIFA establishment.

Discussion of the Consequences of the Establishment of the RIFA   The spread of RIFAs throughout California will result in the establishment of a major nuisance pest. The greatest costs will be from the repair of electrical and irrigation equipment, insecticide treatments to prevent harm to human and animal health, and treatments to meet quarantine restrictions. Annual aggregate losses are estimated to be between $387 million at the low-impact level and $989 million at the high (Table 10.4). Costs to households account for about 89 percent of the total estimated costs.

Other significant costs would accrue from the disruption of ecosystems, which in turn would threaten California's native plant and animal biodiversity. It is also possible that dozens of endangered species in California will face a greater risk of extinction.

Cost/Benefit Analysis

The cost/benefit analysis will compare the expected costs of eradication to the expected benefits of preventing establishment. The cost/benefit analysis takes into account uncertainty over the success of the eradication program and differences in the number of years during which the costs and benefits accrue. Eradication costs are incurred for one 5-year period, two 5-year periods and one 10-year program. Eradication benefits will continue into perpetuity.

Uncertainty is incorporated into the cost/benefit analysis by estimating an expected value. An expected value is equal to the probability of an event happening times the value of the event. For a one-period model, the expected costs are equal to the total discounted program costs because it is known with certainty that those costs will be incurred. The expected benefits are equal to the probability of success times the present value of the benefits of preventing establishment.

For the two 5-year programs it is uncertain if the costs will be incurred during the second period. The expected costs are equal to the actual discounted costs that will be incurred during the first period plus the expected additional costs.

Table 10.3 Endangered species susceptible to a RIFA invasion

Endangered species	Reason	Endangered species	Reason
Beetle, delta green ground	Yes—insect	Fairy shrimp, vernal pool	Yes—eggs in soil of dried pools
Butterfly, bay checkerspot	Yes—insect	Tadpole shrimp, vernal pool	Yes—eggs in soil of dried pools
Butterfly, El Segundo blue	Yes—insect	Lizard, blunt-nosed leopard	Yes—reptile
Butterfly, Lange's metalmark	Yes—insect	Lizard, Coachella Valley fringe-toed	Yes—reptile
Butterfly, lotis blue	Yes—insect	Lizard, Island night	Yes—reptile
Butterfly, mission blue	Yes—insect	Snake, giant garter	Yes—reptile
Butterfly, Myrtle's silverspot	Yes—insect	Snake, San Francisco garter	Yes—reptile
Butterfly, Oregon silverspot	Yes—insect	Tortoise, desert	Yes—reptile
Butterfly, Palos Verdes blue	Yes—insect	Turtle, green sea	Yes—reptile
Butterfly, San Bruno elfin	Yes—insect	Turtle, leatherback sea	Yes—reptile
Butterfly, Smith's blue	Yes—insect	Turtle, loggerhead sea	Yes—reptile
Fly, Delhi Sands flower-loving	Yes—insect	Turtle, olive (=Pacific) Ridley sea	Yes—reptile
Flycatcher, Southwestern willow	Yes—insect	Snail, Morro shoulderband	Yes—mollusk
Gnatcatcher, coastal California	Yes—insect	Kangaroo rat, Fresno	Yes—rodent young
Moth, Kern primrose sphinx	Yes—insect	Kangaroo rat, giant	Yes—rodent young
Beetle, valley elderberry longhorn	Yes—insect	Kangaroo rat, Morro Bay	Yes—rodent young
Goose, Aleutian Canada	Yes—ground-nesting bird	Kangaroo rat, Stephens'	Yes—rodent young
Plover, western snowy	Yes—ground-nesting bird	Kangaroo rat, Tipton	Yes—rodent young
Rail, California clapper	Yes—ground-nesting bird	Mouse, Pacific pocket	Yes—rodent young
Rail, light-footed clapper	Yes—ground-nesting bird	Mouse, salt marsh harvest	Yes—rodent young
Rail, Yuma clapper	Yes—ground-nesting bird	Vole, Amargosa	Yes—rodent young
Shrike, San Clemente loggerhead	Yes—ground-nesting bird	Mountain beaver, Point Arena	Yes—habitat disruption
Tern, California least	Yes—ground-nesting bird	Condor, California	Possible—reduction in food source
Towhee, Inyo California	Yes—ground-nesting bird	Eagle, bald	Possible—reduction in food source
Pelican, brown	Yes—ground and tree nesting	Falcon, American peregrine	Possible—reduction in food source
Frog, California red-legged	Yes—soft-shelled eggs	Owl, northern spotted	Possible—reduction in food source
Salamander, desert slender	Yes—soft-shelled eggs	Sparrow, San Clemente sage	Possible—reduction in food source
Salamander, Santa Cruz long-toed	Yes—soft-shelled eggs	Murrelet, marbled	Possible—low tree-nesting bird
Toad, arroyo southwestern	Yes—soft-shelled eggs	Vireo, least Bell's	Possible—low tree-nesting bird

The additional costs are calculated as the probability that additional costs will be needed the second period, times the actual discounted costs for the second period.

Total expected costs = $C_1 + (1-P_1)*C_2$

where subscripts denote the period, C is total discounted costs, and P is the probability of success for the first period.

The expected benefits are equal to the probability of receiving them during the first period times the benefit amount, plus the probability

Table 10.4 Total annual costs of RIFA establishment in California

Category	Impact		
	Low	Medium	High
	----- ($ million) -----		
Tree and vine crops	12.0	30.0	48.0
Vegetable crops	3.7	9.2	14.8
Field crops	0.8	2.0	3.3
Nursery	18.2	18.2	18.2
Sod	4.5	4.5	4.5
Rangelands	5.9	9.5	15.4
Total agricultural	45.1	73.5	104.2
Total household	342.0	829.0	885.0
Total	387.1	902.5	989.2

that they will not, times the probability that they will be received during the second period, times the benefit amount. With two unknown probabilities, the probability of success in period one is set at 0.1 percent, which reflects the qualitative assessment that success during the first 5 years is unlikely.

Expected benefits $= P_1*B+(1-P_1)*P_2*B$

where B is equal to the present value of total benefits.

The annual costs of establishment shown in Table 10.4 are the estimated losses once the RIFA has spread completely throughout its susceptible range in California. We assume that this level would be achieved in 10 years if all public control activities cease based on infestation rates in the southeastern United States. The costs for years 1–10 depend on the rate of spread of the pest. For an exotic species such as the RIFA, the rate of spread will be relatively slow at first. It increases exponentially as the size of the infestation increases and then tapers off as the ant spreads into the last few susceptible areas.

For this analysis the rate of spread is expressed as a percentage, or share, of the total susceptible area and is given by the expression

$$Share(t) = \frac{Share(max)}{1+e^{-(a+\beta *t)}}$$

where

$$\alpha = ln\left(\frac{Share(t_1)}{Share(max)-Share(t_1)}\right)-\beta *Share(t)$$

and

$$b=\frac{ln\left(\frac{5*Share(max)}{(Share[max]-.5*Share[max])}\right) - ln\left(\frac{Share(t_1)}{Share(max)-Share(t_1)}\right)}{t_{50\%}-t_1}$$

Share(max) is equal to 100 percent and represents the share of total annual costs incurred once the RIFA is fully established. Share(t) is the share incurred at time t while the ant is spreading and becoming established. $T_{50\%}$ is the time period when the ant has spread 50 percent.

To estimate the rate of spread, two pieces of information are needed: the initial share at t_1 and the time period at which the ant has achieved a share of 50 percent. We assume that the initial share is 1 percent and that the RIFA has spread throughout 50 percent of its range by year 6. The present value of the benefits is calculated as the sum of the discounted annual cost of establishment multiplied by the share infested from year 1 to year 10, plus the sum of the discounted values of the total annual costs from year 11 into perpetuity.

If the probabilities were known, then the expected costs and benefits can be calculated directly and compared. For the RIFA eradication program, these probabilities are not known. From the expected cost and benefits equations, however, the probability at which the expected benefits equal at least the expected costs may be calculated and then compared to a qualitative assessment to determine feasibility. The qualitative assessment may rank the probability of success anywhere from very high to very low. As the value of the breakeven probability increases, the likelihood that it will be greater than the qualitative assessment decreases.

Discussion of Cost/Benefit Results

The three cost scenarios included in the analysis and breakeven probabilities are calculated for the one-period program of 5 years, the one-period program of 10 years, and the two 5-year periods at the low-, medium-, and high-benefit level.

As shown in the table, the higher the costs of establishment, the lower the probability needed for the breakeven value to be reached. In all cases the breakeven probability of success is relatively low. When the length of the eradication program increases from 5 to 10 years, eradication costs increase, causing the breakeven probability of success to also increase. The ab-

Table 10.5. Cost/benefit analysis

Benefits		Breakeven probability		
Level	Amount	One 5-year period	Two 5-year periods[a]	One 10-year period
	($ billion)	---------- *(%)* -----------		
Low	3.8	1.04	1.72	1.73
Medium	8.8	0.45	0.73	0.74
High	9.9	0.41	0.67	0.68

[a]When the probability of success in year 1 is 0.1%.

solute increase in percentage points is relatively small, however. Between the 5-year program and the 10-year program, the increase in percentage points is only 0.26 for the high economic impact level to 0.68 for the low-impact level. While low, this represents an approximate increase of 64 percent over the 5-year program.

When the eradication program increases from one 5-year period to two 5-year periods, the probability of success again must increase. However, the probabilities increase by slightly less than one 10-year program. At the high-impact level, the probability of success increases to 1.73 percent for the 10-year program, but only to 1.72 percent for the two 5-year programs. Even though the probability of success is 0.1 percent for the first 5 years of the two 5-year programs, having a nonzero probability of success lowers the probability of success needed for the expected benefits of an additional 5-year program when compared with the 10-year program.

While the estimated probabilities are very low, it is possible that they may not be low enough. At the start of the public eradication program expert opinion was solicited, and a consensus emerged that a nonzero probability existed that the RIFA could be eradicated given the size of the infestation at that time and the amount of resources available. Since the start of the eradication program new discrete infestations have been identified; however, no increase in resources has been provided to increase the scope of the eradication program. Consequently, updating qualitative assessments of the biological feasibility of eradicating the RIFA is important.

otic pest problems. This approach has worked well in Texas, and the fire ant program there should serve as a model for California. The Texas Agricultural Experiment Station and Extension Service, Texas Department of Agriculture, Texas Park and Wildlife Department, Texas Technological University, and the University of Texas are all collaborating in a coordinated effort to address their fire ant problem through research, education, and regulatory programs. Basic and applied research is designed to improve methods of control. Community-based education provides training on control. Regulatory programs through surveys determine distribution and abundance of fire ants and provide effective quarantine programs to prevent their spread.

In California, a close collaboration between CDFA and the University of California would bring together two complementary organizations, each bringing their own strengths and talents to bear on the current fire ant crisis. CDFA, as a regulatory agency, is in charge of survey and detection, as well as quarantine. The University of California with its Experiment Station and Extension Service is ideally suited for research and education. The University of California's Exotic Pest Center is a consortium of University of California scientists who are experts on a variety of pests. The Exotic Pest Center is uniquely qualified to offer its expertise to help find solutions to urgent problems such as the one California is now facing with fire ants. In order to succeed, these two organizations must be dedicated to working together quickly and efficiently, before fire ants become permanently entrenched in California.

Conclusion

Ideally, regulatory agencies and academic institutions would collaborate closely to address ex-

References

Adams, C.T. 1983. "Destruction of Eggplants in Marion County, Florida by Red Imported Fire

Ants (Hymenoptera: Formicidae)." *Florida Entomologist*. 66:518–520.

Allen, C.R., S. Demarais, and R.S. Lutz. 1994. "Red Imported Fire Ant Impact on Wildlife: An Overview." *Texas Journal of Science*. 46:51-59.

Banks, W.A., C.T. Adams, and C.S. Lofgren. 1991. "Damage to Young Citrus Trees by the Red Imported Fire Ant (Hymenoptera: Formicidae)." *Journal of Economic Entomology*. 84:241–246.

Barr, C.L., and B.M. Drees. 1994. "Results from the Texas Veterinarian Survey: Impact of Red Imported Fire Ants on Animal Health." In *Proceedings of the 1994 Imported Fire Ant Conference,* Mobile, AL, May 9-11.

Brinkley, C.K., R.T. Ervin, and W.L. Sterling. 1991. "Potential Beneficial Impact of Red Imported Fire Ant to Texas Cotton Production." *Biological Agriculture and Horticulture*. 8:145–152.

Department of Finance, State of California. 2000. City/County Population and Housing Estimates, 1991-2000 with 1990 Census Counts. Sacramento, California. http://www.dof.ca.gov/html/Demograp/E-5.xls.

Dowell, R.V., A. Gilbert, and J. Sorensen. 1997. "Red Imported Fire Ant Found in California." *California Plant Pest & Disease Report*. 16:50–55.

Dukes, F.R., S.E. Miller, M.S. Henry, B.J. Vandermey, and P.M. Horton. 1999. "Household Experiences with the Red Imported Fire Ant in South Carolina." Research Report. Department of Agricultural and Applied Economics. Clemson University. 17 pp.

Forest and Rangeland Resources Assessment Program (FRRAP). 1988. "California's Forests and Rangelands: Growing Conflict Over Changing Uses—An Assessment." California Department of Forestry and Fire Protection. Sacramento, California. 348 pp.

Greenberg, L., J. Kabashima, J. Klotz, and C. Wilen. 1999. "The Red Imported Fire Ant in California." *Pacific Coast Nurseryman*. 58:8.69–73.

Greenberg, L., J. Klotz, and J. Kabashima. 2001. "Red Imported Fire Ant." *Pest Notes*, publication 7487. UC ANR publication. Davis, California. 3 pp.

Killion, M.J., and W.E. Grant. 1995. "A Colony-Growth Model for the Imported Fire Ant: Potential Geographic Range of an Invading Species." *Ecological Models*. 77:73–84.

Klotz, J.H., J.R. Mangold, K.M. Vail, L.R. Davis, Jr., and R.S. Patterson. 1995. "A Survey of the Urban Pest Ants (Hymenoptera: Formicidae) of Peninsular Florida." *Florida Entomologist*. 78:109–118.

Knapp, Joseph. Professor of Entomology. University of Florida. November 2000. Personal communication.

Knight, R.L., and M.K. Rust. 1990. "The Urban Ants of California with Distribution Notes of Imported Species." *Southwestern Entomologist*. 15:167–178.

Lennartz, F.E. 1973. "Modes of Dispersal of *Solenopsis invicta* from Brazil into the Continental United States—A Study in Spatial Diffusion." MS Thesis, University of Florida, Gainesville. 242 pp.

Lewis, V.R., L.D. Merrill, T.H. Atkinson, and J.S. Wasbauer. 1992. "Imported Fire Ants: Potential Risk to California." *California Agriculture* 46:29-31.

Lofgren, C.S. 1986. "The Economic Importance and Control of Imported Fire Ants in the United States." In S.B. Vinson, Ed., *Economic Impact and Control of Social Insects*. New York: Praeger Publishers. pp. 227–256.

Lofgren, C.S., W.A. Banks, and B.M. Glancey. 1975. "Biology and Control of Imported Fire Ants." *Annual Review of Entomology*. 20:1-30.

McDaniel, S.G., and W.L. Sterling. 1979. "Predator Determination and Efficiency on *Heliothis virescens* Eggs in Cotton Using ^{32}P." *Environmental Entomology*. 8:1083–1087.

McDaniel, S.G., and W.L. Sterling. 1982. "Predation of *Heliothis virescens* (F.) Eggs on Cotton in East Texas." *Environmental Entomology*. 11:60–66.

Olsen, Chris. 2002. Aventis Environmental Science. Personal communication to J. Klotz, May 15, 2002.

Porter, S.D. and D.A. Savignano. 1990. "Invasion of Polygyne Fire Ants Decimates Native Ants and Disrupts Arthropod Community." *Ecology*. 71:2095–2106.

Reagan, T.E. 1981. "Sugarcane Borer Pest Management in Louisiana: Leading to a More Permanent System." In *Proceedings of the 2nd Inter-American Sugarcane Seminar: Insect and Rodent Pests, 1981*. Florida International University, Miami, FL, Oct. pp. 100–110.

Semevski, F., L. Thompson, and S. Semenov. 1996. "Economic Impact of Imported Fire Ants and Selected Crops: A Synthesis from the Literature." In *Proceedings of the 1996 Imported Fire Ant Research Conference*. New Orleans, LA, April 16-18. pp. 93-80.

Slowik, T.J., H.G. Thorvilson, and B.L. Green. 1996. "Red Imported Fire Ant (Hymenoptera: Formicidae) Response to Current and Conductive Material of Active Electrical Equipment." *Journal of Economic Entomology*. 89:347–352.

Sterling, W.L. 1978. "Fortuitous Biological Suppression of the Boll Weevil by the Red Imported Fire Ant." *Environmental Entomology*. 7:564–568.

Stewart, J.S., and S.B. Vinson. 1991. "Red Imported Fire Ant Damage to Commercial Cucumber and Sunflower Plants." *Southwestern Entomologist*. 16:168–170.

Taber, S.W. 2000. *Fire Ants*. College Station, TX: Texas A&M University Press. 308 pp.

Teal, S., S. Segarra, and K. Moates. 1998. "Spatial Economic Impacts of the Red Imported Fire Ant on the Texas Cattle Industry." Technical Research Report No. T-1-484. Department of Agricultural and Natural Resources, Texas Tech University. 38 pps.

Thompson, L.C., D.B. Jones, F.N. Semevski, and S.M. Semenov. 1995. "Fire Ant Economic Impact: Extending Arkansas' Survey Results over the South." In *Proceedings of the Fifth International Pest Ant Symposia and the 1995 Annual Imported Fire Ant Conference*, San Antonio, TX, May 2-4.

Thompson, Lynne. 2000. Professor, School of Forest Resources, University of Arkansas. Personal communication.

U.S. Department of Agriculture. 2000. *1997 Census of Agriculture*, Vol. 1. Geographic Area Series. Part 5, California. AC97-A-5. Washington, D.C.: U.S. Dept. of Agriculture.

Vargo, E.L., and S.D. Porter. 1989. "Colony Reproduction by Budding in the Polygyne Form of *Solenopsis invicta* (Hymenoptera: Formicidae)." *Annals of the Entomological Society of America*. 82:307–313.

Vinson, S.B. 1997. "Invasion of the Red Imported Fire Ant: Spread, Biology, and Impact." *American Entomologist*. 43:23–39.

Vinson, S.B., and L. Greenberg. 1986. "The Biology, Physiology, and Ecology of Imported Fire Ants."

In S.B. Vinson, Ed., *Economic Impact and Control of Social Insects*. New York: Praeger Publishers. pp. 193–226.

Williams, D.F. 1986. "Chemical Baits: Specificity and Effects on Other Ant Species." In C.S. Lofgren and R.K. Vander Meer, Eds., *Fire Ants and Leaf-cutting Ants: Biology and Management*. Boulder, Colorado.: Westview Press. pp. 378–386.

Williams, D.F. 1994. "Control of the Introduced Pest *Solenopsis invicta* in the United States." In D.F. Williams, Ed., *Exotic Ants: Biology, Impact, and Control Of Introduced Species*. Boulder, CO: Westview Press. pp. 282–292.

Wojcik, D.P., C.R. Allen, R.J. Brenner, E.A. Forys, D.P. Jouvenaz, and R.S. Lutz. 2001. "Red Imported Fire Ants: Impact On Biodiversity." *American Entomologist*. 47:16–23.

11

A Rational Regulatory Policy: The Case of Karnal Bunt

Joseph W. Glauber and Clare Narrod

Introduction

The U.S. Department of Agriculture (USDA) has had responsibility for implementing plant quarantines since 1912 (Palm 1999). Under the Federal Plant Pest Act and the Plant Quarantine Act, the USDA has the authority to impose restrictions on the interstate movement of any article believed to be infested with exotic pests or diseases. There are currently 17 federal quarantines in place, ranging from restrictions affecting peach orchards in Pennsylvania infected by the plum pox virus to hardwood forests in the eastern United States infested with gypsy moths (Table 11.1). The range of the combined quarantines covers most of the United States and affects most crops produced there. The federal cost to maintain these quarantines is estimated to be almost $50 million in 2000 (U.S. Department of Agriculture 2000).

The costs attributable to plant pests and diseases in the United States in lost productivity and expenses for protection and control have been estimated to be as much as $41 billion annually (U.S. General Accounting Office 1997). Although these loss estimates are controversial, the threat of foreign pests and diseases to U.S. crop production has long been used to argue for strict import regulations and broad domestic quarantine authorities.[1] Aside from benefits, however, quarantines can impose substantial costs on producers, handlers, and others affected directly by regulations as well as potentially adversely affecting consumers and others through restrictions in supply (James and Anderson 1998). Federal quarantine policy has generally followed guidelines developed by the National Plant Board in 1931.[2] These guidelines state that (1) the pest concerned must be of such nature as to offer actual or expected threat to substantial interests; (2) the proposed quarantine must represent a necessary or desirable measure for which no other substitute, involving less interference with normal activities, is available; (3) the objective of the quarantine, either for preventing introduction or for limiting spread, must be reasonable of expectation; (4) the economic gains expected must outweigh the cost of administration and the interference of normal activities (Sim 1998).

Assessing the economic effects of quarantines is oftentimes difficult because of the uncertainty surrounding the risks that the quarantine policy seeks to mitigate (James and Anderson 1998). Yet, even when probabilistic risk assessments exist, regulators often consider the costs and benefits separately. Ignoring the underlying distribution of costs and benefits not only overstates the certainty of the analysis, but it can potentially lead to regulatory actions where the expected costs exceed the expected benefits.

This chapter examines the federal quarantine established by the USDA in 1996 to prevent the spread of Karnal bunt, a minor disease of wheat. During the early stages of establishing its regulatory strategy, the USDA made extensive use of probabilistic risk assessments to determine the efficacy of various quarantine protocols. There was less careful consideration given to the costs and benefits of the actions, however. In early press releases and *Federal Register* notices, the benefits were expressed largely in terms of the value of the U.S. wheat market believed to be at risk (e.g., 61 FR 12058, Docket No. 96-016-1). Likewise, when the regulatory impact analysis for the final rule was published on May 6, 1997, the costs and benefits of the regulations were discussed without

Table 11.1 Federal domestic quarantines

Plant pest	Year initiated[a]	Crops potentially affected	Regulated area
Pink bollworm	1967	Cotton, kenaf, okra	AZ, AR, CA, NM, OK, TX
Witchweed	1970	Corn, sorghum, sugarcane, rice	NC, SC
Golden nematode	1972	Potatoes	NY
Japanese beetle	1979	Ornamentals, tree fruits, row crops, turf	AL, CT, DE, DC, GA, IL, IN, KY, ME, MD, MA, MI, MN, MO, NH, NJ, NY, NC,OH, PA, RI, SC, TN, VT, VA, WV, WI
Sugarcane diseases	1983	Sugarcane	HI, PR
Mexican fruit fly	1983	Tree fruits	CA, TX
European larch canker	1984	Larch trees	ME
Citrus canker	1985	Citrus fruit	FL
Black stem rust	1989	Wheat and small grains	48 conterminous states and DC
Mediterranean fruit fly	1991	Fruit, vegetables	CA, FL
Pine shoot beetle	1992	Pine trees	IL, IN, MD, MI, NY, OH, PA, WV, WI
Imported fire ant	1992	Impedes harvest and cultivation	AL, AR, CA, FL, GA, LA, MS, NM, NC, OK, PR, SC, TN, TX
Gypsy moth	1993	Hardwood forests	CT, DE, DC, IN, ME, MD, MA, MI, NH, NJ, NY, NC, OH, PA, RI, VT, VA, WV, WI
Oriental fruit fly	1993	Fruits, vegetables	CA
Karnal bunt	1996	Wheat, rye, triticale	AZ, CA, TX, NM
Asian longhorn beetle	1997	Hardwoods	IL, NY
Plum pox	2000	Stone fruit	PA

[a]Reflects year that current regulatory policy was implemented.

consideration of the distribution of potential outcomes. If risk had been incorporated directly into the cost/benefit analysis, it is likely that different conclusions would have been drawn about the expected impact of the regulations.

The chapter is organized as follows. First a brief history of Karnal bunt and the events leading to the establishment of the federal quarantine in 1996 is presented. Next, the probabilistic risk assessments undertaken in 1996 to assess how proposed regulatory actions mitigated the risks of Karnal bunt are discussed. The potential benefits and costs of the regulations are considered and the expected costs and benefits of regulations incorporating information on the distribution of potential outcomes given various regulatory actions are examined. Conclusions are presented in the last section.

Regulatory History

Karnal bunt is a disease affecting wheat, rye, and triticale (a hybrid of wheat and rye) caused by the fungus *Tilletia indica Mitra* (Bonde et al. 1997). Karnal bunt can cause production losses to wheat in the form of reduced yields due to

the infestation of kernels and reduction in the quality of the wheat flour. Generally, wheat containing more than 3 percent bunted kernels is considered unsatisfactory for human consumption because of a fishy odor that makes wheat products unpalatable (Warham 1986), but it poses no risk to human health.

Karnal bunt was first reported in 1931 in the Indian state of Haryana in wheat-growing areas near the city of Karnal, from which the disease gets its name. From that time through the early 1970s the disease went largely unnoticed and was believed to be limited in its distribution to similar environments in Pakistan, Iraq, Afghanistan, Nepal, and Iran (Singh et al. 1998). In 1970 Karnal bunt appeared in Mexico but caused little economic loss until the early 1980s, when disease incidence increased sharply. Initially found in Sonora, the disease spread south into the neighboring states of Sinaloa and Baja California Sur (Brennan and Warham 1990).

In 1982 diseased wheat kernels were intercepted in wheat imported from Mexico. Following confirmation of Karnal bunt in Mexico, the USDA took action to prevent the importation of host plant material (including seed and grain) and any other articles that might spread the disease (Poe 1997). These actions were made permanent in October 1983 by adding Mexico and other countries where Karnal bunt was known to occur to the list of countries in the Wheat Disease subpart of the Foreign Quarantine Notices (7 Code of Federal Regulations 319.59). All of the major wheat exporting countries followed suit. In 1982 only four countries had phytosanitary trade restrictions involving Karnal bunt. Following the U.S. action against Mexico, that number jumped to 22 (Beattie and Bickerstaff 1999).

A risk assessment of Karnal bunt completed by the USDA in 1988 concluded that because of the close proximity of wheat growing areas of Arizona and California to infested areas in northwestern Mexico and the flow of prevailing winds, "transport of the Karnal bunt pathogen is extremely likely" (Schall 1988). A subsequent pest risk analysis conducted in 1991 concluded that Karnal bunt was a high risk pest, primarily because "wheat from infested areas would probably be denied or restricted access in the export market"[3] (Schall 1991). Because of its potential adverse effects on exports, the analysis recommended that in the event of introduction of the Karnal bunt pathogen the USDA should establish and maintain quarantines to restrict distribution.

On March 8, 1996, Karnal bunt was detected in Arizona during a seed certification inspection done by the Arizona Department of Agriculture.[4] On March 20, 1996, the Secretary of Agriculture signed a "Declaration of Extraordinary Emergency" authorizing the USDA to take emergency action under 7 U.S.C. 150dd with regard to Karnal bunt within the states of Arizona, New Mexico, and Texas. The quarantine was extended to Imperial and Riverside counties in California on April 12, 1996. In an interim rule effective March 25, 1996, and published in the *Federal Register* on March 28, 1996, the Animal and Plant Health Inspection Service (APHIS) established the Karnal bunt regulations and quarantined all of Arizona and portions of New Mexico and Texas because of Karnal bunt. The regulations defined regulated articles and restricted the movement of these regulated articles from the quarantined areas.

The imposition of federal quarantine and emergency actions was seen by the USDA as a "necessary, short-run measure taken to prevent the interstate spread of the disease to other wheat producing areas in the outbreak area, so that eradication could be eventually achieved" (62 *Federal Register* 24754–24755). The USDA described its objectives as threefold: (1) to protect U.S. wheat producers in Karnal bunt–free areas, (2) to protect U.S. export markets, and (3) to provide the best possible options for producers in quarantined areas who are affected by the Karnal bunt detections (U.S. Department of Agriculture, Animal and Plant Health Inspection Service 1997).

The USDA's initial actions were to require producers in New Mexico and Texas who had planted fields with infected seed to plow down their crop immediately. Because crop development was further along in Arizona and California, plowing down crops was not considered viable. Instead, a number of regulations were implemented that affected persons or entities that produced wheat in the regulated area and/or moved certain articles associated with wheat out of a regulated area (Table 11.2). These articles were subject to regulatory actions to minimize the risk of spreading the pathogen to other, uninfected areas. Regulated articles itemized in the Karnal bunt protocols included:

1. farm machinery and equipment used to produce wheat;
2. conveyances from field to handler, such as farm trucks and wagons;
3. grain elevators, equipment, and structures at facilities that store and handle grain;
4. conveyances from handler to other marketing channels, such as railroad cars;
5. plant and plant parts, such as grain for milling, grain for seed, and straw;
6. flour and milling by-products;
7. manure from animals fed wheat/wheat by-products from quarantine area;
8. used sacks;
9. seed-conditioning equipment;
10. by-products of seed cleaning;
11. soil-moving equipment;
12. root crops with soil; and
13. soil.

All wheat fields within the regulated areas of Arizona, California, New Mexico, and Texas were sampled at harvest for Karnal bunt

Table 11.2 Impact of Karnal bunt quarantine actions

Action	Regulated article	Affected entities	Numbers affected	Types of impacts due to KB[b] and quarantine actions
Plow-down & seed plot destruction	· Fields planted with infected seed at preboot stage · Tools and farm equipment · Harvesters	· Certain producers in Texas and New Mexico · Wheat producers in RA[a] · Farmer-owned and custom combines	· 4,100 acres · 73 producers · 145 growers · 389 combines	· Loss in value of wheat crop destroyed · Cost of cleaning · Cost of cleaning
Cleaning/disinfection	· Grain trucks · Grain storage and load-out facilities · Harvesters · Harvesters · Harvesters · Railcars	· Grain haulers from field to grain elevators · Grain-handling firms · Combine harvester owners · Combines involved in preharvest sampling · Custom combine companies · Grain-handling firms	· 976 trucks · 17 elevators · 36 to 40 combines · 5 to 10 combines · 5 companies · 10,880 cars (511 for positive grain)	· Cost of cleaning · Cost of cleaning · Excess wear and tear on equipment · Downtime on harvesters due to field testing · Loss of income due to termination of contracts outside the RA · Cost of cleaning
Restriction on use or marketing	· KB-positive milling wheat · KB-negative milling wheat · Millfeed · Movement restrictions on wheat seed · Straw, manure, millfeed	· Producers · Grain-handling firms · Producers in RA · Handlers in RA · Millers, millfeed processors Seed producers, researchers, and companies · Straw producers and handlers · Users of straw · Livestock producers using wheat or straw produced in the RA · Flour millers · Millfeed processors/users	· 145 growers · 6 handlers · 664 producers · 26.7 million bushels · 108 mills · 45,644 tons · 15 producers · 9 research firms · 20 seed marketers · 25 growers · 3 contractors · 1 straw user, making of straw mats for erosion control · 7 millers in 5 states · 2 millfeed processors	· Loss in value of KB-positive wheat · Loss in value of KB-negative wheat in RA · Millers' reluctance to mill KB-negative wheat from RA · Loss in premiums · Loss in market value · Loss in royalties · Loss in income · Increased cost of production

(Table 11.2 continued)

Action	Regulated Article	Affected Entities	Numbers Affected	Types of Impacts Due to KB[b] and Quarantine Actions
	· Moratorium on wheat production on KB-positive fields	· Producers with KB-positive properties	· 109 growers · 13,674 acres	Loss in income from wheat
	· Soil on root crops grown on infected properties	· Vegetable producers on KB-positive properties	· Unknown number	· Increased cost of production
	· Used seed sacks · Seed-conditioning equipment · By-products of seed	· Seed research and marketing companies	· 9 research firms · 20 seed marketers	· Increased cost of production

[a]RA, regulated area.
[b]KB, Karnal bunt.
Source: Karnal Bunt Regulatory Flexibility Analysis and Regulatory Impact Analysis published in the *Federal Register*, May 6, 1997.

teliospores. Any wheat shipped outside of the regulated area was again tested for Karnal bunt teliospores. Grain that tested positive for Karnal bunt was prohibited from moving out of the regulated areas, but could be milled or fed to cattle within the regulated area. Other contaminated articles were required to be cleaned and sanitized before movement out of the regulated area. To determine whether Karnal bunt was present in areas outside the quarantined areas, a comprehensive national survey of wheat elevators was planned for the fall of 1996.

Commercial seed intended for planting or for breeding and seed development purposes was prohibited from moving outside the regulated areas. Wheat seed could be planted within the quarantine areas, but only if it tested negative for Karnal bunt teliospores and was treated prior to planting. Grain that tested negative was permitted to move outside the regulated areas under limited permit. Grain was required to be shipped in sealed railcars, and the railcars had to be sanitized after the grain was delivered to its destination. Grain that was exported received a phytosanitary certificate from USDA certifying that the grain had been tested twice and found negative for Karnal bunt.[5]

Negative-testing grain was permitted to move to approved domestic flour mills. Due to the grinding process and intended use, the risk of spread of the disease through movement of the flour was viewed by the USDA as negligible. In the milling process, however, a considerable amount of by-product or millfeed is produced. The millfeed is typically sold as cattle feed, which represents about 10 percent of the value of the milled wheat. Because of the risk that manure from the cattle could be deposited on wheat fields and thus potentially be a pathway for spread of Karnal bunt, the USDA required that mills heat the millfeed to 130°F for 30 minutes or steam treat to 170°F.

As will be seen in a later section, the protocols imposed large costs on the southwestern wheat industry. As the full extent of the quarantine became understood, opposition within the quarantine area grew, and many questioned whether an eradication strategy was appropriate.[6] The USDA maintained that the principal rationale for the quarantine was to assure foreign wheat importers that they could import wheat from the United States that was from areas where Karnal bunt was not known to occur.

This discussion revises the original analyses (both risk assessment and the economic analysis) to assess this view. In order to assess whether the expected benefits of the quarantine exceed the costs, a model of quarantine policy must first be developed.

Assessing the Probability of Outbreak

To estimate the effects of various quarantine protocols on the likelihood of outbreaks of Karnal bunt in areas outside the quarantined area, the USDA relied on a number of probabilistic risk assessments conducted prior to discovery of Karnal bunt in Arizona (Schall 1988, 1991; Podleckis 1995) and in the first two months following the outbreak (Podleckis and Firko 1996a, 1996b, 1996c, 1996d). Probabilities of outbreak were estimated for a variety of potential pathways including millfeed, export elevators, seed originating in the quarantined area, railcars transporting grain from the quarantined area to domestic mills and export elevators, grain storage facilities, and combines and other harvesting machinery.

The risk assessment presented here is based on the USDA risk assessments. However, unlike the USDA analysis, which focused on measuring risk of individual pathways, this risk assessment focuses on the overall level of risk of outbreak from any source.[7] The probability of an outbreak of Karnal bunt occurring outside the quarantined area, $p*$, can be written as:

$$(11.1) \quad p* = 1 - (1-p_1)(1-p_2)(1-p_3)(1-p_4)(1-p_5)$$

where p_1 = probability of an outbreak of Karnal bunt outside the quarantined area from millfeed

p_2 = probability of an outbreak of Karnal bunt in host fields outside the quarantined area from grain in transit to mills or export elevators

p_3 = probability of an outbreak of Karnal bunt outside the quarantined area from combines or other harvesting machinery

p_4 = probability of an outbreak of Karnal bunt outside the quarantined area from railcars after grain is unloaded at mills or export elevators

p_5 = probability of an outbreak of Karnal bunt outside the quarantined area from seed

In general, the probability of outbreak via a given pathway is positively related to the number of railcars or other conveyances transporting grain or seed outside the quarantine areas. The number of railcars leaving the quarantined area is, in part, determined by the incidence of infested fields within the quarantined area. The higher the infestation of Karnal bunt within the quarantined area the less negative-testing wheat is available for export or domestic milling purposes and the lower the probability of outbreak outside of the quarantined area.[8]

The overall level of risk tends to be influenced by the riskier pathways. Changes in the probability of outbreak in a given pathway may be large in absolute terms, but have little effect on the overall level of risk. By focusing on individual pathways, the risk-reducing potential of the protocol may be overestimated. For example, in the initial analysis the controversial requirement to heat treat millfeed was justified by the USDA on the basis of the relatively sharp reduction in the risk of outbreak from contaminated millfeed. Yet when we separate this out, the results indicate that while the millfeed treatment requirement reduced the mean risk of Karnal bunt outbreak from contaminated millfeed from 1 in 15,175 to 1 in 60 million, the effect of the protocol was negligible in reducing the overall level of risk (Table 11.3).

Likewise, restrictions on the movement of negative-testing seed also had a relatively small effect on the overall risk of outbreak. One of the pathways with the highest probability of outbreak was p_4, the probability of outbreak of Karnal bunt in elevators that received grain that

had been transported in contaminated railcars. The mean risk of outbreak from this pathway, assuming that railcars were not required to be cleaned after delivery, was 1 in 35. This risk was significant since a contaminated elevator would potentially be identified when sampled in the national survey of wheat elevators.

The USDA analysis also ignored the level of ambient risk that had existed prior to the discovery of Karnal bunt in Arizona. Podleckis (1995) had estimated that the probability of outbreak in the United States from contaminated Mexican boxcars was as high as 2.59×10^{-3} (1 in 386). This ambient risk was higher than the risks of outbreak from contaminated railcars from the regulated areas, millfeed, or negative-testing seed and potentially reduced the effect any such protocols might have in mitigating the overall risks of outbreak.

In the analysis that follows, eight quarantine options were considered. The options were based on the following protocols: (1) the restriction on the movement of negative-testing seed outside of the quarantine area; (2) the requirement that railcars be cleaned after delivery of wheat from the quarantined area; and (3) the requirement to heat treat millfeed. These protocols were chosen because they imposed large costs on the wheat industry in the Southwest and, as a result, were controversial. Option 1 reflects the least-restrictive option where the quarantine protocols were limited to restrictions on the movement of positive-testing grain. Grain and seed that twice tested negative for Karnal bunt teliospores would be free to move to export and domestic locations with no addi-

Table 11.3 The effects of various protocols on the risk of Karnal bunt outbreak

Protocol	Probability of an outbreak[a]	
	For that pathway	Overall
Railcar cleaning:		
- with	6.43×10^{-4}	2.14×10^{-3}
- without	5.18×10^{-2}	5.67×10^{-2}
Restrictions on the movement of negative-testing seed:		
- with	0	5.53×10^{-2}
- without	1.40×10^{-3}	5.67×10^{-2}
Millfeed treatment:		
- with	1.66×10^{-8}	5.66×10^{-2}
- without	6.59×10^{-5}	5.67×10^{-2}

[a]Evaluated at mean.

tional restrictions. Railcars would not be required to be cleaned. Option 8 reflects protocols put in place by APHIS in March of 1996 following the discovery of Karnal bunt in Arizona. The other options reflect various combinations of the three protocols, plus the baseline option.

The effects of the options on the risk of outbreak are presented in Table 11.4. The probabilistic risk assessments provide estimates of the probability of outbreak with an estimated mean and distribution. The table presents two measures of central tendency (median and mean) and the 95th-percentile value. Current APHIS policy uses the 95th-percentile value in making regulatory decisions (Firko et al. 1996). Viscusi (1998) discusses the potential for a "conservatism" bias when the 95th-percentile value is used for every component of the estimate. In the risk assessment presented here, the 95th-percentile value was drawn from the joint distribution $p*$, not from a combination of the 95th-percentile values for the individual p_i.

Of the individual protocols considered, railcar cleaning had the largest effect on the overall level of risk of outbreak because of the relatively high risk of contamination through railcars. Restrictions on the movement of nega-

tive-testing seed and millfeed treatment requirements had minimal effects on the overall level of risk. Taken together, the three protocols reduced the level of risk by almost 99 percent relative to the baseline level.

Estimated Benefits and Costs of the Federal Quarantine Program

To assess the welfare effects of the quarantine actions, we must first calculate the welfare effects in the event of an outbreak of Karnal bunt outside the regulated area. From the initial detection of Karnal bunt in Arizona and the USDA's subsequent announcement of a declaration of extraordinary emergency, protection of U.S. export markets was articulated as a primary goal of the USDA's regulatory efforts (Glickman 1996). The United States typically exports about 1.2 billion bushels of wheat annually, with an estimated value of about $3 to $4 billion. About half of U.S. wheat exports were to countries that (at the time Karnal bunt was discovered in Arizona) maintained restrictions against wheat imports from countries where Karnal bunt was known to occur. The

Table 11.4 Probability of an outbreak of Karnal bunt under alternative quarantine options

Quarantine option	Probability of outbreak[a]		
	Median	Mean	95th percentile
Option 1: Baseline[b]	2.92E-02	5.67E-02	1.93E-01
	(—)	(—)	(—)
Option 2: Railcar cleaning	1.11E-03	2.14E-03	7.43E-03
	(0.038)	(0.038)	(0.038)
Option 3: Restrictions on seed	2.78E-02	5.53E-02	1.92E-01
movement	(0.951)	(0.976)	(0.994)
Option 4: Millfeed treatment	2.91E-02	5.66E-02	1.93E-01
	(0.997)	(0.999)	(1.000)
Option 5: Railcar cleaning;	2.32E-04	7.08E-04	2.45E-03
restrictions on seed movement	(0.008)	(0.013)	(0.013)
Option 6: Railcar cleaning;	1.05E-03	2.07E-03	7.35E-03
millfeed treatment	(0.036)	(0.037)	(0.038)
Option 7: Restrictions on seed	2.77E-02	5.53E-02	1.92E-01
movement; millfeed treatment	(0.949)	(0.975)	(0.994)
Option 8: Railcar cleaning;	1.91E-04	6.40E-04	2.29E-03
restrictions on seed	(0.007)	(0.011)	(0.012)
movement; millfeed treatment			

[a]Expressed in scientific notation; e.g., 2.92E-02 = 2.92×10^{-2} = 0.0292.
[b]Includes prohibition of movement of positive-testing grain and seed from quarantined area; all negative-testing grain and seed moved in sealed hopper cars; all combines disinfected before leaving quarantined area.
( ) Denotes level of risk relative to baseline.

USDA argued that failure to implement the quarantine would jeopardize trade with those countries. Benefits of federal quarantine, therefore, were regarded largely as the avoided losses in the export market.

In its Regulatory Impact Analysis published on May 6, 1997, the USDA estimated that a 50 percent reduction in U.S. wheat exports would likely reduce U.S. wheat prices by 30 percent and lower net sector income by $2.7 billion. This estimate takes into account the dampening effect on domestic wheat prices, because wheat for export is diverted into the domestic consumption market, animal feed outlets, and ending stocks.

The reduction in U.S. wheat exports, however, would likely be less than 50 percent. Not all countries that have restrictions against Karnal bunt would, in practice, strictly prohibit wheat imports from the United States. (Italy and Germany currently import wheat from countries where Karnal bunt is known to occur despite European Union regulations to the contrary). Second, while some markets would be captured by wheat from exporting countries that are free of Karnal bunt, U.S. wheat exports to countries that have no restrictions against Karnal bunt would likely increase. In the long

run, the effects could be minimal, depending on whether the market were to treat Karnal bunt as a quality issue and develop discounts for Karnal bunt.

In the impact analysis, the USDA estimated that the impact of Karnal bunt on exports, because of substitution effects, would likely result in a 10 percent reduction in U.S. wheat exports. A decrease of 10 percent in exports would cause a 22¢ per bushel drop in the wheat prices and a drop in annual wheat sector income of $545 million. The effects of decreases in wheat exports of various percentages are presented in Table 11.5.

While the effect on prices and incomes would likely affect all producers of wheat, it is noteworthy to point out that the majority of benefits from federal quarantine actions were received by producers outside the regulated areas who produce over 95 percent of the wheat grown in the United States. Beattie and Bickerstaff (1999) have recently argued that the regulations were largely the result of rent-seeking behavior on the part of wheat producers outside the regulated areas. It is certainly true that wheat producers outside the quarantine area were strong supporters of the USDA quarantine actions.[9]

Table 11.5 Estimated net welfare effects of reduced exports due to an outbreak of Karnal bunt outside of the regulated area[a]

Item	Unit	Reduction in exports			
		0%	10%	25%	50%
Exports	mil. bu.	1,200	1,080	900	600
Total use	mil. bu.	2,462	2,394	2,295	2,138
Price	$/bu	3.85	3.63	3.29	2.68
Value of production	mil. dol.	9,543	8,998	8,146	6,637
Government payments[b]	mil. dol	1,815	1,815	1,815	1,943
Gross income	mil. dol.	11,358	10,813	9,961	8,580
Variable expenses	mil. dol.	4,823	4,823	4,823	4,823
Net cash income	mil. dol.	6,536	5,990	5,138	3,758
Welfare effects:					
Producer losses	mil. dol.	—	− 545	− 1,397	− 2,778
Consumer gains	mil. dol.	—	284	747	1,674
Change in government payments	mil. dol.	—	0	0	128
Net welfare	mil. dol.	—	− 261	− 650	− 976
Over 10 years[c]	mil. dol.	—	− 2,098	− 5,214	− 7,830

[a]Estimates based on 1997/1998 marketing year.
[b]Includes AMTA payments ($1,815 million) plus loan deficiency payments.
[c]Discounted at 7 percent annually.
  Source: Adapted from Karnal Bunt Regulatory Flexibility Analysis and Regulatory Impact Analysis (*Federal Register*, 62:24755, May 6, 1997).

The impact analysis failed to consider changes in consumer welfare. Based on the price and domestic demand levels in Table 11.5 and an implied domestic demand elasticity of −0.7, consumer surplus effects were estimated. Subtracting consumer gains and any additional government price support payments due to low prices, annual net welfare effects ranged from $261 million for a 10 percent loss in exports to $976 million assuming a 50 percent reduction in exports.

Since the potential adverse effects of an outbreak of Karnal bunt on export markets may last longer than a year, we calculated the net present value of benefits assuming losses over a 10-year period using a 7 percent discount rate. Based on the annual net welfare losses in Table 11.5, the discounted welfare effects ranged from $2.1 billion to $7.8 billion. This should be viewed as a conservative assumption. In the long run, if export losses due to Karnal bunt remained large and prices depressed, many wheat producers would likely switch to alternative crops, mitigating sector losses. Because of the factors mentioned above, it is likely the long-term losses would be less than $2 billion.

In its regulatory impact analysis, the USDA estimated that the costs of the Karnal bunt regulations in 1996 incurred by producers, handlers, and other affected parties was $44 million (Table 11.6). It was estimated that about 8 percent of the 1996 crop wheat produced in the regulated area tested positive for Karnal bunt. This wheat was largely diverted to feed use in the regulated area resulting in an estimated loss to producers and handlers of $4.2 million.

Regulatory requirements to treat millfeed caused many domestic mills to drop contracts with producers and handlers of grain from the quarantined areas, resulting in a decline in prices for negative-testing wheat within the regulated areas. In the absence of the regulatory requirement on millfeed, domestic wheat millers would have likely purchased negative-testing grain from the infected areas. Although some millers were reluctant, the high quality of the durum wheat produced within this area would have helped counter their reluctance to the purchase of uninfected grain. The requirement, however, that millfeed be treated and railcars sanitized increased the costs of milling wheat from the regulated area and prompted many contracts with grain producers and handlers to

Table 11.6 Estimated costs due to Karnal bunt regulations, 1996 crop year

Item	Estimated costs (mil. dollars)
Plowdown of NM and TX fields planted with infected seed	1.2
KB-positive grain diverted to animal feed market	4.2
Cleaning and disinfecting railcars	0.6
Loss in value of seed	6.0
KB-negative grain that experienced loss in value	28.0
Other[a]	4.1
Total	44.1

[a]Includes losses related to cleaning and disinfecting combine harvesters, sanitizing storage facilities, and loss in value of straw.

Source: Adapted from Karnal Bunt Regulatory Flexibility Analysis and Regulatory Impact Analysis (*Federal Register*, 62:24755, May 6, 1997).

be canceled. The estimated loss in value to producers and handlers of negative-testing wheat was estimated to be $28 million.

Under the 1996 quarantine and emergency actions, wheat seed produced in the regulated areas was prohibited from sale outside the regulated areas. Wheat seed intended for planting within the regulated areas had to be sampled and tested for Karnal bunt, and for seed originating in a regulated area, treated prior to planting. These restrictions were estimated to have a significant impact on the seed industry, largely due to the high value that is commanded by wheat sold for seed relative to grain. It is estimated that 1.5 million bushels of wheat seed sustained loss in value of $5 to 6 million. Seed developers, who earn returns on their investment in research and development of wheat varieties, also claim potential long-term losses in royalties; by receiving plant variety protection (or patent rights), seed developers then obtain royalties on future sales of wheat that are developed and sold for propagative purposes. Other economic losses suffered by the seed industry are difficult to quantify; they include additional handling, storage, and finance costs on seed that could no longer be sold outside the regulated areas and costs to relocate wheat-breeding operations outside the regulated areas.

In a report submitted as an exhibit in a lawsuit brought by the Arizona Wheat Growers Association against the USDA, Beattie (1996) argued that the quarantine had adverse effects on wheat seed development. He estimates that the loss in productivity due to the quarantine likely cost producers and consumers between $177 and $357 million on a net present value basis.

The USDA impact analysis also enumerated losses to other parties such as wheat straw producers, custom harvesters, and producers who were required to destroy their crops prior to harvest because of the regulations. These losses were estimated to total approximately $5 to 6 million in 1996.

Estimated Expected Costs and Benefits

In the Regulatory Impact Analysis accompanying the final Karnal bunt regulations on compensation, USDA concluded that:

> . . . our quarantine measures were appropriate and justifiable when compared with the magnitude of the benefits achieved. Even a 10 percent reduction in wheat exports would have a significant effect on wheat sector income. It is estimated that a 10 percent decline in wheat exports would cause a decline in wheat sector of over $500 million. (62 FR 24765)

But can these conclusions be justified if one examines the *expected* costs and benefits of the regulations?

Cost/benefit analysis for alternative quarantine options can be completed under the assumptions given above (Table 11.7). For the baseline (option 1), the cost of diverting positive-tested wheat to feed markets and destroying any crops planted with contaminated seed is $5.4 million ($4.2 million plus $1.2 million). The probability of an outbreak outside the quarantine area was reduced from certainty with no protocol to 0.0567. For a 10 percent diversion of exports with present value of costs of $2.098 billion, the expected loss due to an outbreak of Karnal bunt outside the quarantined area is $119 million (0.0567 * 2.098), and the welfare gain from using the baseline option is $1,979 million (i.e., $2,098 million − $119 million). Each of the other options also shows a large expected cost/benefit ratio when considered individually.

Table 11.8 presents the marginal benefits and costs of options 2, 5, and 8, assuming various levels of export market effects due to an outbreak of Karnal bunt. Under the baseline option, a minimal quarantine is put into place to regulate positive-testing grain, but the marginal benefits are large relative to the costs. Likewise, the addition of option 2—railcar cleaning—provides from $115 to $427 million in additional benefits for additional costs less than $1 million. The addition of protocols restricting the movement of negative-testing seed (option

Table 11.7 Expected costs and benefits of alternative quarantine actions assuming as 10 percent loss in annual exports (million dollars)

Quarantine option	Expected net present value of benefits	Expected costs	Net
Option 1: Baseline[a]	1,978.8	5.4	1,973.4
Option 2: Railcar cleaning	2,093.2	6.0	2,087.3
Option 3: Restrictions on seed movement	1,981.7	11.4	1,970.3
Option 4: Millfeed treatment	1,979.0	33.4	1,945.6
Option 5: Railcar cleaning; restrictions on seed movement	2,096.2	12.0	2,084.3
Option 6: Railcar cleaning; millfeed treatment	2,093.4	34.0	2,059.4
Option 7: Restrictions on seed movement; millfeed treatment	1,981.7	39.4	1,942.3
Option 8: Railcar cleaning; restrictions on seed movement; millfeed treatment	2,096.4	40.0	2,056.4

[a]Includes prohibition of movement of positive-testing grain and seed from quarantined area; all negative-testing grain and seed moved in sealed hopper cars; all combines disinfected before leaving quarantined area.

5) imposed direct costs of an additional $6 million, while the reduction in expected welfare loss was only $3 million, assuming a 10 percent loss in exports over 10 years and when evaluated at the mean probability estimates. If export losses were as high as 50 percent annually over 10 years, the expected marginal benefit rises to $11 million. The seed protocol is likewise marginally cost effective when evaluated using the more conservative 95th-percentile value for the risk of outbreak. However, when one includes the potential loss in productivity as estimated by Beattie, the seed protocol costs far exceed its benefits at any measure of risk. The costs of the millfeed treatment requirement (option 8) exceed the expected benefits even under the most conservative assumptions (i.e., 50 percent loss in exports over 10 years evaluated at the 95th percentile of risk of outbreak).

Conclusions

While the USDA continues to regulate for Karnal bunt, many of the original areas placed under quarantine have been deregulated. During a national survey of elevators in the fall of 1996, the USDA detected Karnal bunt-like spores in a number of grain facilities in the Southeast. It was determined that the teliospores were those of a fungus that infects ryegrass but not wheat. Because the spores were indistinguishable from Karnal bunt teliospores, the USDA did not impose a quarantine. In 1997, the USDA changed the standard for defining regulated areas based on the presence of bunted kernels rather than Karnal bunt teliospores. The immediate effect of the regulatory change was to remove the millfeed treatment requirement. In 1998, the USDA relaxed the quarantine to allow commercial seed to move outside the regulated area. These changes have allowed much of the original regulated area to return to more normal marketing; losses in recent years have been small and confined to positive-testing grain. While the number of countries requiring phytocertificates on U.S. wheat has increased to 54 countries, importing countries have generally accepted the changes.

The cost imposed by the quarantine has been controversial since the quarantine was established in March 1996. To increase cooperation, the USDA agreed to pay producers, grain handlers, and other affected parties compensation for losses suffered due to the federal quarantine action. Compensation payments have totaled more than $40 million since 1996.

Table 11.8 Marginal costs and benefits of alternative quarantine options (million dollars)

Quarantine option	Marginal cost	Marginal benefit assuming that an outbreak of karnal bunt outside regulated area will cause annual wheat export losses of:		
		10%	25%	50%
Probability of outbreak evaluated at the mean:				
Option 2: Railcar cleaning	0.6	114.5	284.5	427.2
Option 5: Railcar cleaning; restrictions on seed movement	6.0	3.0	7.5	11.2
Option 8: Railcar cleaning; restrictions on seed movement; millfeed treatment	28.0	0.1	0.4	0.5
Probability of outbreak evaluated at the 95th percentile:				
Option 2: Railcar cleaning	0.6	389.3	967.5	1,453.1
Option 5: Railcar cleaning; restrictions on seed movement	6.0	10.4	26.0	39.0
Option 8: Railcar cleaning; restrictions on seed movement; millfeed treatment	28.0	0.31	0.8	1.3

A larger issue has been the regulatory status of Karnal bunt as a plant disease. Even at the time Karnal bunt was discovered in Arizona in 1996, many scientific bodies (e.g., American Phytopathological Society) considered Karnal bunt to be a minor plant pest that could be controlled much like other wheat pests, i.e., without the use of quarantine measures. In 1997, the USDA convened an international symposium on Karnal bunt with the intent of convincing other nations to deregulate Karnal bunt. To date, no countries have agreed to change their phytosanitary restrictions on wheat imports containing Karnal bunt.

From the analysis presented here, a number of conclusions can be drawn concerning the USDA's Karnal bunt quarantine policy. From the late 1980s, the USDA has made extensive use of probabilistic risk assessments to guide regulatory decisions. In the case of Karnal bunt, the risk assessments have been comprehensive in their analysis of the effects of various quarantine policies on the probability of outbreak along potential pathways. However, in their analysis of risks associated with Karnal bunt, the USDA tended to focus on risk mitigation for individual pathways, seemingly without regard to the effect on the overall level of risk. As a result, the effects of individual protocols were arguably overstated.

In their regulatory impact analyses, the USDA ignored the effects of the quarantine policies on consumers that tended to overestimate the benefits of the quarantine. Their analysis also failed to look at the expected marginal benefits and costs of various quarantine alternatives. Had they considered the expected marginal effects in their decisions, it is likely that at least two of the more controversial and costly protocols—seed restrictions and the millfeed requirement—would have received closer scrutiny and possibly been rejected as viable options.

Since the establishment of the Karnal bunt quarantine in 1996, the USDA has established new quarantines to control the Asian Longhorn beetle and plum pox, and it has increased the scope of the quarantine to control citrus canker in Florida. Like Karnal bunt, these quarantines have been justified on the basis of the potential liability worth billions of dollars. Yet, like Karnal bunt, these quarantines also impose large costs on those who are regulated as well as consumers and taxpayers more indirectly affected by the quarantine actions.

Bridging the gap between regulatory analysis and risk assessment has become increasingly important in public policy due to the complex array of supporting documents that regulatory decision makers must consider during the decision-making process. The method used here departs from most USDA analyses, which historically have separated the risk assessment from the economic analysis. We offer this method as a potential way that future analysis, when appropriate, can be combined to improve the analysis and aid in the regulatory rule-making process.

Notes

[1]For example, estimates of the costs of invasive species to the United States range from $1.1 billion annually (Office of Technology Assessment 1993) to $137 billion (Pimentel et al. 2000). See also Pinstrup-Anderson (1999) and Orke et al. (1994).

[2]The National Plant Board is an organization of state plant pest regulatory agencies created in 1925 to promote efficiency and uniformity in the promulgation and enforcement of plant quarantines and plant inspection policies (Sim 1998).

[3]An economic analysis conducted by USDA in 1994 indicated that annual crop losses due to Karnal bunt in Arizona, Texas, New Mexico and California would total between $406 thousand and $1 million per year and that annual losses in export markets could total over $57 million for Arizona and Texas alone (cited in Podleckis 1995).

[4]Checks of seed lots dating back to 1993 from the same area in Arizona revealed the presence of Karnal bunt teliospores at low levels (Nelson 1996).

[5]Grain originating from outside the regulated areas received phytosanitary certificates certifying that the grain was from areas where "Karnal bunt was not known to occur."

[6]In a position statement released in August 1996, the American Phytopathological Society questioned the "zero tolerance" requirement for teliospores in seed lots and concluded that "experience from countries where this disease has occurred would suggest further that it is a minor disease, and what little risk does exist can be effectively managed without the use of quarantines."

[7]A more detailed description of the risk assessment model is summarized in Appendix 11.1.

[8]This assumes that the probability of teliospores surviving shipment outside the quarantined area is uncorrelated with the incidence of infection within the quarantined area.

[9]A number of agricultural commissioners from wheat-producing states were concerned, however,

that the quarantine actions themselves were having an adverse impact on trade (Sim 1998). Indeed, a number of wheat-importing countries that had no prohibitions on Karnal bunt prior to the Declaration of Extraordinary Emergency, soon afterward adopted the requirement that U.S. wheat have an additional phytosanitary certificate certifying that the wheat was from an area where Karnal bunt was known not to occur.

References

Beattie, Bruce R. 1996. "Economic Impact on U.S. Wheat Producers and Consumers Due to Karnal Bunt Quarantine Restrictions on Wheat Seed Breeding in the Desert Southwest." University of Arizona, Department of Agricultural and Resource Economics Working Paper, November. 23 pp.

Beattie, Bruce R., and Dan R. Bickerstaff. 1999. "Karnal Bunt: A Wimp of a Disease But an Irresistible Political Opportunity." *Choices.* Second Quarter:4-8.

Bonde, M.R., G.L. Peterson, N.W. Schaad, and J.L. Smilanick. 1997. "Karnal Bunt of Wheat." *Plant Disease.* 81:1370–1377.

Brennan, John P., and Elizabeth J. Warham. 1990. "Economic Losses from Karnal Bunt of Wheat in Mexico." CIMMYT Economics Working Paper 90/02. CIMMYT, Mexico. 56 pp.

Firko, Michael J., Edward V. Podeleckis, and Thomas Perring. 1996. "Comparison of Karnal Bunt Risk Assessments." U.S. Department of Agriculture, Animal and Plant Health Inspection Service memorandum. June 6. 2 pp.

Glickman, Dan. 1996. "Statement at Karnal Bunt Press Conference." USDA Press Release No. 0137.96. March 21. 1 p.

James, Sallie, and Kym Anderson. 1998. "On the Need for More Economic Assessment of Quarantine Policies." *Australian Journal of Agricultural and Resource Economics.* 42(4):425–444.

Nelson, Merritt. 1996. "Karnal Bunt, an Arizona Perspective." Paper posted on the online symposium hosted by the American Phytopathological Society. Available at http://www.scisoc.org on June 5, 2000. 4 pp.

Office of Technology Assessment. 1993. "Harmful Non-Indigenous Species in the United States." Washington, D.C.: Office of Technology Assessment.

Orke, E.C., H.W. Dehne, F. Schonbeck, and A. Weber. 1994. *Crop Production and Crop Protection: Estimated Losses in Major Food and Cash Crops.* Amsterdam: Elsevier.

Palm, Mary E. 1999. "Mycology and World Trade: A View from the Front Line." *Mycologia.* 91(1):1-12.

Pimentel, David, Lori Lach, Rodolfo Zuniga, and Doug Morrison. 2000. "Environmental and Economic Costs of Nonindigenous Species in the United States." *BioScience.* 50(1):53–65.

Pinstrup-Anderson, Per. 1999. "The Future World Food Situation and the Role of Plant Diseases." The Glenn Anderson Lecture presented at the joint meeting of the American Phytopathological Society and the Canadian Phytopathological Society, Montreal, Canada. August 8. Available at http://www.scisoc.org on June 5, 2000.

Podleckis, Edward V. 1995. "Karnal Bunt Introduction via Wheat Contaminants in Conveyances: Mexican Boxcars." Washington, D.C.: U.S. Department of Agriculture, Animal and Plant Health Inspection Service.

Podleckis, Edward V., and Michael J. Firko. 1996a. "Karnal Bunt: Likelihood of Spread via Conveyances, Harvest Equipment and Wheat Shipments." Washington, D.C.: U.S. Department of Agriculture, Animal and Plant Health Inspection Service. May 8.

Podleckis, Edward V., and Michael J. Firko. 1996b. "Karnal Bunt: Special Risk Assessment Addendum." Washington, D.C.: U.S. Department of Agriculture, Animal and Plant Health Inspection Service. May 14.

Podleckis, Edward V., and Michael J. Firko. 1996c. "Karnal Bunt: Special Risk Assessment Addendum II." Washington, D.C.: U.S. Department of Agriculture, Animal and Plant Health Inspection Service. May 22.

Podleckis, Edward V., and Michael J. Firko. 1996d. "Karnal Bunt: Special Risk Assessment Addendum III." Washington, D.C.: U.S. Department of Agriculture, Animal and Plant Health Inspection Service. May 28.

Poe, Stephen. 1997. "APHIS Response to Karnal Bunt Prior to March 1996." In V.S. Malik and D.E. Mathre, Eds., *Bunts and Smuts of Wheat: An International Symposium.* Ottawa: North American Plant Protection Organization. pp. 108-111.

Schall, R. 1988. "Karnal Bunt: The Risk to the American Wheat Crop." Washington, D.C.: U.S. Department of Agriculture, Animal and Plant Health Inspection Service.

Schall, R. 1991. "Pest Risk Analysis on Karnal Bunt." Washington, D.C.: U.S. Department of Agriculture, Animal and Plant Health Inspection Service.

Sim IV, Thomas. 1998. "Plant Pest Quarantines: Their Role and Use." In V.S. Malik and D.E. Mathre, Eds. *Bunts and Smuts of Wheat: An International Symposium.* Ottawa: North American Plant Protection Organization. pp. 367-382.

Singh, D.V., K.D. Srivastava, and R. Aggarwal. 1998. "Karnal Bunt: Constraints to Wheat Production and Management." In V.S. Malik and D.E. Mathre, Eds. *Bunts and Smuts of Wheat: An International Symposium.* Ottawa: North American Plant Protection Organization. pp. 201-222.

U.S. Department of Agriculture. Animal and Plant Health Inspection Service. 1996. "1996 National Karnal Bunt Survey." November 26. 18 pp.

U.S. Department of Agriculture. Animal and Plant Health Inspection Service. 1997. "Biology of Kar-

nal Bunt." January 2. Available at http://www.
aphis.usda.gov/oa/bunt/kbbiol.html on July 7,
1999.

U.S. Department of Agriculture. 2000. *2001 Budget
Summary*. February.

U.S. General Accounting Office. 1997. "Agricultural
Inspection: Improvements Needed to Minimize
Threat of Foreign Pests and Diseases."
GAO/RCED-97-102. May.

Viscusi, W. Kip. 1998. *Rational Risk Policy*. Oxford:
Clarendon Press.

Warham, E.J. 1986. "Karnal Bunt Disease of Wheat:
A Literature Review." *Tropical Pest Management*.
34:229-242.

Appendix 11.1

Karnal Bunt Risk Assessment Procedure

Joseph W. Glauber and Clare Narrod

In this analysis we tried to be true to the original analysis (Podleckis and Firko 1996a, 1996c) upon which regulatory assumptions were based. Below we describe how the approach used in this paper differs from the original model.

The probability of at least one outbreak of Karnal bunt occurring outside the quarantined area is modeled through a series of multiplicative steps. This probability is modeled as a function of the quarantine protocols and the number of railcars or other conveyances transporting grain or seed outside the quarantined areas. Furthermore, the number of infected railcars of grain shipped out of the quarantined area is modeled as a function of the amount of wheat testing positive for Karnal bunt in fields, railcars, or elevators in the quarantined area.

The exact pathways by which contamination can occur are detailed in Figure 2. This analysis departs from the original analysis, however, in calculating some of the probabilities. In the original model (P8), the probability that grain going to storage was infected with Karnal bunt was considered an additive function of the probability that the harvested grain was infected/contaminated with Karnal bunt (P3), the probability that the grain was contaminated by equipment (P6), and the probability that local conveyances were contaminated (P7). Technically this is not correct. The system of protocols must be considered together when assessing the probability of a positive find. This analysis departs from the original analysis by computing this probability as $p_8 = [1 - (1 - p_3)(1 - p_6)(1 - p_7)]$. Similarly, in the original analysis the probability of a shipment having Karnal bunt (P12) is modeled as an additive function of the probability that the grain going to storage had Karnal bunt (P8) and the probability that grain picked up Karnal bunt in local storage (P11). In this analysis this probability was changed to $p12 = [1 - (1 - p_8)(1 - p_{11})]$.

Monte Carlo simulation is used to compute the probability of at least one outbreak of Karnal bunt outside the quarantine area. In each iteration of the model, this value is determined by the multiplicative contribution of a series of steps raised to the frequency with which either railroad cars were shipped or combines moved out of the quarantine area.

Typically, these steps include the probability that a shipment had Karnal bunt P12, the probability that the Karnal bunt was in the shipment and detected (P13), the probability that viable Karnal bunt survived the shipment (P15), the probability that Karnal bunt reached a suitable host (P16) and the probability that Karnal bunt was able to become established (P17).

For each scenario, the following formula is used to calculate the probability of an outbreak:

$$F_3 = 1 - (1 - p_{12}*p_{13}*p_{14}*p_{15}*p_{16}*p_{17})^\wedge F_1$$

In most scenarios F_1 is the frequency of railroad cars shipped to the mill. When combine movement is being considered F_1 is replaced by F_2, which is the frequency of combines moved out of the quarantine area. F_3 is the frequency of Karnal bunt outbreaks.

Probabilities were estimated for a variety of potential pathways including millfeed, export elevators, seed originating in the quarantined area, railcars transporting grain from the quarantined area to domestic mills and export elevators, grain storage facilities, and combines and other harvesting machinery. From the scenarios originally used by Podleckis and Firko (1996a), it was determined that there were nine different

Table A11.1 Option used and changes to scenarios included

	Baseline Option 1	Rail Option 2	Seed Option 3	Mill Option 4	Rail/Seed Option 5	Rail/Mill Option 6	Seed/Mill Option 7	Rail/Seed/Mill Option 8
Millfeed	2*	2	2	1	2	1	1	1
Transit/elevator	3 & 4	3 & 4	3 & 4	3 & 4	3 & 4	3 & 4	3 & 4	3 & 4
Combine	5	5	5	5	5	5	5	5
Railroad car	6, P15=1 7, P14=1 8, P13=1	normal	6, P15=1 7, P14=1 8, P13=1	6, P15=1 7, P14=1 8, P13=1	normal	normal	6, P15=1 7, P14=1 8, P13=1	normal
Seed	9	9	–	9	–	9	–	–

* Numbers represent scenarios included under each option; P13, P14, P15 defined in figure.

scenarios that would lead to the probability that at least one outbreak of Karnal bunt would occur outside the regulated area. These scenarios included:

1. grain to the mill, risk of Karnal bunt outbreak in mill state, millfeed untreated;

2. grain to mill, risk of Karnal bunt outbreak in mill state, millfeed treated;

3. grain to mill, risk of Karnal bunt outbreak in transited states, millfeed treated;

4. grain to export elevator, risk of Karnal bunt outbreak in transited states, millfeed treated;

5. combine/harvest equipment moved out of quarantine area, risk of Karnal bunt outbreak in states receiving equipment;

6. grain to mill, risk of Karnal bunt outbreak in secondary state (state receiving railcar after grain is unloaded at mill);

7. grain to mill, risk of Karnal bunt contamination in storage facility in secondary state;

8. grain to export elevator, Karnal bunt contamination in storage facility in secondary state; and

9. risk of outbreak via seed harvested and planted in Arizona.

To capture the effect of various combinations of options eight potential combinations of options were developed as seen in Table A11.1.

Monte Carlo analysis was performed using the @Risk Software. Each option was run for 10,000 iterations, and the random seed numbers generated were fixed at 2. The specific values used for the probabilities in the model are summarized in Table A11.2. The values include an unspecified mix of the variability and uncertainty that can occur under each event.

Table A11.2 Parameters used

F1	Frequency of rail cars shipped per year	Triangle	4500	5530	6500
a	Frequency of railroad cars shipped to the mill per year (45% of F1)	Triangle	2025	2488.5	2925
b	Frequency of railroad cars exported per year (55% of F1)	Triangle	2475	3041.5	3575
c	Frequency of railroad cars shipped to seed per year (10% of F1)	Triangle	450	553	650
F2	Frequency of combines shipped per year	Triangle	50	100	200
P1	Probability that wheat in field infected/contaminated with KB				
a		Beta	1.2	10	
b		Beta	4	20	
P2	Probability that KB not detected in field				
a		Lognormal	0.01	0.025	
b		Beta	2	20	
P3	Probability that harvested grain infected/contaminated with KB	P1xP2			
P4	Probability that farm equipment is contaminated with KB				
a		Lognormal	0.05	0.05	
b		Beta	4	20	
P5	Probability that decontamination of farm equipment fails	Lognormal	0.01	0.025	
P6	Probability that grain is contaminated by equipment	P4xP5			

(Table A11.2 continues)

(Table A11.2 continued)

P7	Probability that local conveyances (trucks) get contaminated			
a		Lognormal	0.001	0.0025
b		Beta	4	20
P8	Probability that grain going to storage has KB	1−(1−P3)(1−P6)(1−P7)		
P9	Probability that local storage gets contaminated with KB			
a		Lognormal	0.01	0.025
b		Lognormal	0.0001	0.0001
P10	Probability that KB is in local elevator and not detected			
a		Lognormal	0.01	0.025
b		Constant	1	
P11	Probability that grain picks up KB in local storage	P9xP10		
P12	Probability that shipment has KB	1−(1−P8)(1−P11)		
P13	Probability that KB in shipment is not detected			
a		Lognormal	0.01	0.025
b		Constant	1	
P14	Probability that grain is transported to a suitable habitat			
a		Beta	2	4
b		Constant	1	
P15	Probability that KB survives shipment (viable KB)			
a		Beta	4	2
b		Lognormal	0.01	0.01
c d		Beta	5	15
		Constant	1	
P16	Probability that KB reaches a suitable host			
a		Lognormal	0.001	0.001
b		Beta	1.75	25
c		Lognormal	0.0001	0.0001
d		Beta	4	2
e		Constant	1	
P17	Probability that KB is able to become established			
a		Lognormal	0.001	0.001
b		Beta	1.75	25
c		Lognormal	0.0001	0.0001
P18	Probability that decontamination of rail car fails - Scenario 8, 9	Lognormal	0.01	0.01
P19	Probability that KB remains with grain - Scenario 8, 9	Beta	4	2
P20	Probability that KB is transferred to storage facility - Scenario 8, 9	Beta	4	2
P21	Probability that combines harvest bunted kernels	Lognormal	0.1	0.1
P22	Probability that bunted kernels with viable spores remain after decontamination	Lognormal	0.01	0.01
P23	Probability that kernels are transported to suitable habitats outside quarantine area	Beta	2	4
P24	Probability that decontamination of rail cars fails	Lognormal	0.01	0.01
P25	Probability that KB in pile is not detected	Beta	1.2	20

12

Introduction and Establishment of Exotic Insect and Mite Pests of Avocados in California, Changes in Sanitary and Phytosanitary Policies, and Their Economic and Social Impact

Mark S. Hoddle, Karen M. Jetter, Joseph Morse

Introduction

Since 1996 changes in sanitary and phytosanitary (SPS) regulations and exotic pest introductions have strongly affected the U.S. avocado industry. In 1996 avocado thrips, a previously undescribed pest, was identified in California avocado groves. In 1997 Mexico gained limited access to the U.S. avocado market through the Mexican Hass Avocado Agreement. Even though the recent California thrips infestation was not a result of the Mexican Hass Avocado Importation Agreement, it is rarely the case that an exotic pest outbreak can be traced to the failure of a specific pest protection regulation. However, these two events, relaxation of SPS trade barriers and establishment of an exotic pest, allow us to analyze their simultaneous effects on a specific agricultural industry.

Avocados are a subtropical fruit of New World origin, and three distinguishable ecological races or subspecies of avocado (*Persea americana* Miller [Lauraceae]) are recognized (Bergh and Ellstrand 1986; Popenoe 1915, 1952; Storey et al. 1986). The three subspecies are referred to as Mexican (*P. americana* var. *drymifolia*), Guatemalan (*P. americana* var. *guatemalensis*), and West Indian (*P. americana* var. *americana*) types (Bergh and Ellstrand 1986; Nakasone and Paull 1998). Mexican subspecies are native to dry subtropical plateaus with a Mediterranean type climate. Guatemalan

varieties on the other hand are native to cool high-altitude areas, whereas West Indian types perform best in hot, humid climates. All three races hybridize readily, and over 400 pure race and hybrid varieties are known (Condit 1932). Avocados have been grown in California (Santa Barbara) since 1871.

Commercial avocado production in the United States is limited to California, Hawaii, and Florida. California produces 87 percent of the nation's crop, and 80 percent of this annual harvest is from the Hass cultivar. The year before the Mexican Hass Avocado Agreement was enacted, 6,000 growers in California produced 165,000 tons of fruit on 58,000 acres, and the harvest was worth $259 million (California Avocado Commission 1997).

Historically, pesticide use in avocado orchards has been minimal. Pests like greenhouse thrips (*Heliothrips haemorrhoidalis* (Bouché) [Thysanoptera: Thripidae]) (McMurtry et al. 1991), avocado brown mite (*Oligonychus punicae* (Hirst) [Acari: Tetranychidae]), six spotted mite (*Eotetranychus sexmaculatus* (Riley) [Acari: Tetranychidae]) (Fleschner 1953; Fleschner et al. 1955), the omnivorous looper (*Sabulodes aegrotata* Guenee (Lepidoptera: Tortricidae), the *Amorbia* moth (*Amorbia cuneana* (Walsingham) [Lepidoptera: Tortricidae]) (Bailey et al. 1988), and red banded whitefly (*Tetraleurodes perseae* Nakahara [Homoptera: Aleyrodidae]) (Nakahara 1995) have been kept below economically injurious levels by natural enemies (e.g., predators, pathogens, and parasitoids).

Biological control has succeeded in California avocado orchards because minimal pesticide use has not disrupted natural pest control by indiscriminately killing natural enemies (McMurtry 1992). The luxury afforded to avocado growers by successful biological control has recently been disrupted by two new pests, the persea mite and avocado thrips. These two pests have moved growers from biologically based pest control to insecticide-reliant management strategies.

Entry of Hass avocado fruit into the United States is regulated under 7 CFR 319.56, known as the Fruits and Vegetables Quarantine, or Quarantine 56. Avocado fruit from Mexico and Central America have been prohibited entry since 1914 because of a seed weevil, *Heilipus lauri*, certain tephritid fruit flies, additional seed weevil pests, the avocado stem weevil, and the avocado seed moth. In 1997, the federal Fruits and Vegetables Quarantine was amended to allow the provisional entry of Mexican Hass avocados into the United States. As of June 2001, avocados from Mexican orchards certified pest-free may be imported into 19 northeastern states from November through February. Mexican Hass avocado imports increased total Hass imports by 60 percent after the ruling became effective.

Biology and Ecology of Avocado Thrips and Persea Mite

Avocado Thrips, *Scirtothrips perseae* Nakahara (Thysanoptera: Thripidae)

In June 1996 an unknown species of thrips was discovered damaging foliage and fruit of the Hass avocado in Ventura County, California. By July 1997 the thrips had spread to all coastal avocado-growing areas of California (Hoddle and Morse 1997, 1998). This pest has subsequently been described and named *Scirtothrips perseae* Nakahara (Thysanoptera: Thripidae) (Nakahara 1997). Feeding damage by adult and larval *S. perseae* to young leaves causes distortion and brown scarring along the midrib, and veins on the leaf underside become increasingly visible as leaves mature. Feeding damage to young

leaves can be severe enough to induce premature defoliation. Thrips larvae and adults also feed on developing fruit.

Feeding can scar the entire fruit surface, while localized feeding produces discrete brown scars that elongate as the fruit matures. Economic losses are incurred after harvest when fruit disfigured by thrips feeding is either culled or downgraded in packinghouses.

Avocado thrips females lay eggs directly into plant tissue, and young leaves and fruit are preferred oviposition sites. Developing larvae pass through two immature stages before dropping from leaves and fruit to pupate in leaf duff below the trees. Avocado thrips pass through two pupal stages, both of which are sedentary and nonfeeding, before emerging as winged adults and flying back up into the tree canopy to commence feeding and reproduction. The life cycle is shown in Figure 12.1.

Laboratory data indicate that *S. perseae* survivorship and reproduction are favored at low temperatures. At 20°C, more larvae survive to adulthood, the sex ratio is female biased, significantly more progeny are produced, net reproductive rates (R_o) are three times higher, and the population doubling time (T_d) is 33 percent faster than at 25°C. Intrinsic rate of population increase (r_m) is significantly faster at 20°C and cohort generation time (T_c) is significantly longer at 20°C (Table 12.1).

Laboratory data substantiate field monitoring results in that moderately high temperatures (>25°C for several consecutive days) reduce *S. perseae* population growth and densities even when food is abundant (Figure 12.2).

This result is consistent with the observation that this pest is most problematic in avocado orchards in plant-climate zones where the marine influence has a year round moderating effect on temperature (Kimball and Brooks 1959). Consequently, avocado thrips populations build to their highest levels on young avocado foliage in the winter and spring (which are the coolest time periods in California), and immature and adult thrips damage developing fruit that is set in the spring. Low temperature preferences may help synchronize *S. perseae* population growth when production of young avocado foliage and fruit is maximal. This affinity by avocado thrips for low temperatures is in direct contrast to other *Scirtothrips* pest species. *Scirtothrips citri*, *Scirtothrips auran-*

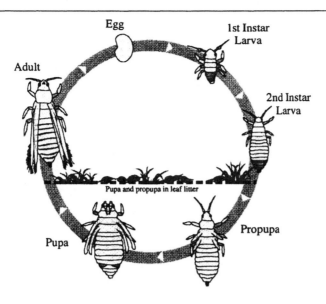

Figure 12.1   A life-cycle diagram showing the sequential development stages of avocado thrips. The entire life cycle (egg to adult) takes 20 days and adults live for 10 days at 25°C (76°F).

tii, and *Scirtothrips dorsalis* all inflict economic damage to crops over summer when temperatures are high, and *S. citri* is typically most damaging to citrus grown in arid interior valleys of California.

Ten species of *Scirtothrips* are formally recognized as economic pests (Mound and Palmer 1981). Of these, *S. citri*, a pest of citrus and mango in California; *S. aurantii*, a pest of citrus and mango in South Africa; and *S. dorsalis*, a pest of tea and chilies in India and grapes in Japan are the best studied (Mound and Teulon 1995; Mound 1997). These three pestiferous species are native to the countries in which they are problematic and use exotic crops in addition to native host plants as food sources. In contrast, *S. perseae* (avocado thrips) is an exotic species in California attacking a host plant (avocado) exotic to California, and to date it has not been recorded from any other host plants in the United States, suggesting that it may have a restricted host range and close evolutionary his-

tory with avocados in Central America (Hoddle and Morse 1997, 1998).

The main source of economic loss attributable to avocado thrips is scarring of immature fruit in late spring and increased pest control costs. Scarred Grade A quality fruit is reduced to standard grade or is culled entirely. With lower prices for standards, grower revenues decrease as a larger share of fruit falls into this category. Fruit that may be downgraded due to an untreated thrips infestation can range from 0 to 95 percent (California Avocado Commission 1998). On average, thrips damage in infested groves reduced revenues by 12 percent in 1998.

Presently, there are no known natural enemies that can effectively control avocado thrips. Work is continuing on the ability of large releases of insectary-reared lacewings and predator thrips for the control of avocado thrips. Growers currently spray orchards with sabadilla (a plant alkaloid), abamectin (a bacterial by-product), and spinosad (a bacterial by-product) for avoca-

Table 12.1   Mean demographic growth parameters (±SE) for *S. perseae*

Temperature	R_o	T_c	r_m	T_d
20°C	15.10 ± 0.06[a]	28.35 ± 0.02[a]	0.10 ± 0.0002[a]	6.88 ± 0.02[a]
25°C	5.27 ± 0.04[b]	24.12 ± 0.02[b]	0.07 ± 0.003[b]	10.34 ± 0.04[b]

Mean followed by different letters across temperatures are significantly different at the 0.05 level.

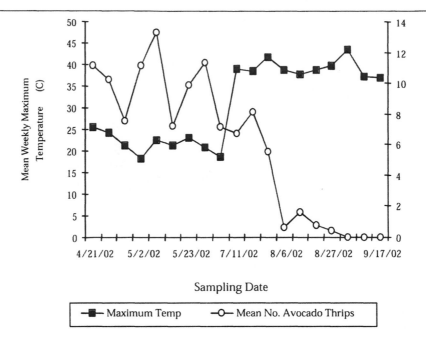

Figure 12.2 Avocado thrips population trends and temperature.

do thrips control. Other species of *Scirtothrips* are efficient virus vectors. Avocado thrips may also be an efficient vector of avocado viral diseases should any establish in California.

Persea Mite, *Oligonychus perseae* Tuttle, Baker, and Abatiello (Acari: Tetranychidae)

Oligonychus perseae was discovered attacking avocados in southern California in 1990 (Bender 1993). Mites feed in colonies beneath protective webbing (i.e., nests) along midribs and veins on the undersides of leaves, and feeding damage produces circular necrotic spots (Aponte and McMurtry 1997). High mite densities (500 per leaf) and subsequent feeding can cause partial or total defoliation of trees (Bender 1993; Aponte and McMurtry 1997; Faber, 1997). Mite-induced defoliation opens the tree canopy, increasing the risk of sunburn to young fruit and exposed tree trunks, and premature fruit drop can occur (Bender 1993). Furthermore, leaf flush following defoliation may exacerbate population growth and feeding damage by avocado thrips that use young leaves for feeding and oviposition.

Persea mite has five developmental stages (egg, larva, protonymph, deutonymph, and adult). All life stages are predominantly found in nests where feeding, mating, reproduction, and development occur. The sex ratio is generally two females to one male. Table 12.2 summarizes important aspects of persea mite biology, and a generalized life cycle for persea mite is shown in Figure 12.3.

Avocado cultivars vary in their susceptibility to persea mite feeding damage. Cultivars can be ranked from least susceptible to most susceptible by calculating the average percentage of leaf area damaged by mite feeding. When cultivars are ordered in this manner, the following ranking from lowest average leaf area damaged by persea mites to highest is attained: Fuerte (13.3%), both Lamb Hass and Reed (16.9%), Esther (29.7%), Pinkerton (30.2%), Gwen (37.4), and then Hass (38.4%). The mechanism responsible for less feeding damage on Fuerte and Lamb Hass is unknown. Host plant resistance may be due to leaf chemistry, which reduces mite survivorship or lowers reproduction rates, leaf hairs that favor natural enemy activity, or some form of repellancy that causes mites to abandon the tree to search for more suitable host plants. Increasing cultivar diversity in orchards should be considered as a strategy to reduce damage and associated yield reductions from persea mite.

Table 12.2 Biology of persea mite on Hass avocado at three different temperatures

Biological attribute	Temperature 20°C (67°F)	Temperature 25°C (77°F)	Temperature 30°C (86°F)
Average adult life span	40 days	27 days	15 days
No. eggs laid per female	37 eggs	46 eggs	21 eggs
Egg to adult development time	17 days	14 days	9 days
No. days for eggs to hatch	7 days	6 days	4 days

In addition to avocados, persea mite can develop on a wide range of fruit, ornamental, and weed plants. This pest has been recorded feeding on leaves of Thompson and flame seedless grapes (*Vitus* spp.), apricots, peaches, plums and nectarines (all *Prunus* spp.), persimmons (*Disopyrus* spp.), milkweed (*Asclepias fusciculares*), sow thistle (*Sonchus* sp.), lamb's quarters (*Chenopodium album*), sumac (*Rhus* sp.), carob (*Ceratonia siliqua*), camphor (*Camphora officinalis*), roses (*Rosa* spp.), acacia (*Acacia* spp.), annatto (*Bixa orellana*), willow (*Salix* spp.), and bamboo (*Bambus* spp.). Good sanitation practices (i.e., elimination of favored weed species) and removal of alternate host plants (i.e., ornamental plants and noncommercial fruit trees in orchards) that act as persea mite reservoirs are useful cultural control practices that should be employed in a persea mite–management program.

Field trials have indicated that inundative releases of predatory mites—either *Neoseiulus californicus* (McGregor) [Acari: Phytoseiidae] or *Galendromus helveous* (Chant) [Acari: Phytoseiidae]—can control persea mite more effectively than chemical sprays (e.g., insecticidal oils), which can cause resurgence of persea mite (resurgence is the phenomenon whereby pest numbers return to the same or higher densities than before application of sprays) (Figure 12.4).

Biological control of persea mite with predator releases is not cost effective currently, and research is continuing to determine the minimum number of predators to release and optimal times to release natural enemies into orchards. Extensive foreign exploration efforts throughout Central America and northern South America have failed to locate effective natural enemies for release into California for permanent control of persea mite.

Introduction and Spread

Avocado Thrips

This pest was first noticed in California in July 1996 when it was discovered damaging fruit in a Saticoy avocado orchard near Port Hueneme in Ventura County. At about the same time, it was observed at the Irvine Ranch in Orange County. In less than a year, the thrips spread north and south of Ventura and Orange counties and was found in San Diego County in May 1997. By July 1997, significant damage attributable to avocado thrips feeding was noticed in orchards in San Diego County, and avocado thrips now infest 95 percent of the avocado-growing acreage in California.

In 1971, a quarantine interception at the Port of San Diego resulted in the collection of two specimens of an undescribed species of *Scirtothrips* on avocado from Oaxaca in southern Mexico. These specimens are very similar to avocado thrips and are considered to be within the acceptable morphological range of *S. perseae*. Avocado thrips are not an avocado-adapted strain of citrus thrips, *Scirtothrips citri*,

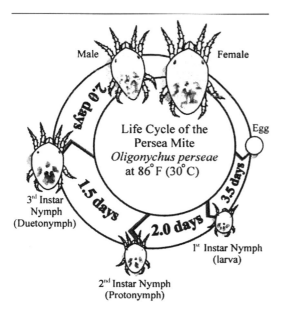

Figure 12.3 A life-cycle diagram showing the sequential developmental stages of persea mite.

which is a citrus pest native to California. Avocado thrips are morphologically more similar to *Scirtothrips aceri,* found on oaks in California and Arizona, and *Scirtothrips abditus,* which inhabits pines and oaks in Mexico and Costa Rica, than it is to *S. citri.*

Foreign exploration efforts for avocado thrips and its natural enemies in Central America indicate that this insect has a range extending from Uruapan in Mexico to Guatemala City in Guatemala. Work in Central America (e.g., Costa Rica) to completely delineate the geographic distribution of avocado thrips, to catalogue other thrips species on avocados, and inventory the natural enemy fauna associated with thrips on avocados in Central America has been completed (Hoddle et al. 2002).

Persea Mite

Oligonychus perseae was first described in 1975 from specimens collected from avocado foliage that were intercepted from Mexico at an El Paso, Texas, quarantine facility. Persea mite is native to Mexico and damages avocados in arid regions, but it is not a major pest in the state of Michoacán, where Hass avocado pro-

duction is greatest, probably because it is controlled by broad-spectrum pesticides used against other pests. Persea mite has also been recorded in Costa Rica but is unknown in Guatemala. Persea mite was discovered attacking avocados in San Diego County in 1990 and was originally misidentified as *Oligonychus peruvianus.* By the summer of 1993, the pest had spread north to Ventura County. Santa Barbara had its first recorded outbreak in spring 1994, and by 1996, persea mite had established in San Luis Obispo County. There are no records of this pest being found in the San Joaquin Valley. Contaminated fruit bins, harvesting equipment, and clothing probably helped disperse persea mite in California. Persea mite currently infests 90 percent of the avocado acreage in California.

An Aside: The Red-Banded Whitefly

The red-banded whitefly, *Tetraleurodes perseae* Nakahara (Homoptera; Aleyrodidae), was first found on avocados in California in 1982 (Rose and Wooley 1984a, 1984b). As with avocado thrips and persea mite, this pest was originally found on avocados growing in the immediate vicinity of a seaport, in this instance, the Port of

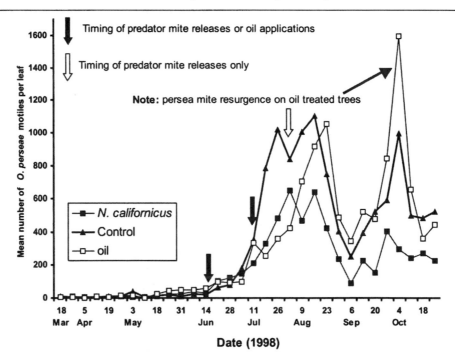

Figure 12.4 Persea mite population trends on avocado trees treated with predators, insecticidal oil, or nothing (control).

San Diego. When this minor pest was discovered in California, it was an undescribed species new to science, and its country of origin was unknown (Nakahara 1995). Red-banded whitefly was officially described in 1995 (Nakahara 1995) and is now found throughout all coastal avocado-growing areas in California. The only known host plant for red-banded whitefly in the United States is avocado (Nakahara 1995). Although this whitefly is a minor pest, honeydew from feeding nymphs promotes sooty mold growth on young leaves. Also, heavy feeding by nymphs and adults can deform young leaves. Whiteflies are efficient virus vectors, and red-banded whitefly is a potential disease vector. At present, this whitefly is under satisfactory control by a parasitoid (*Cales noacki* Howard) that was introduced for control of woolly whitefly (*Aleurothrixus floccosus* [Maskell]). Very little is known about the biology or ecology of red-banded whitefly in California or Central America.

Avocado thrips and persea mite were species unknown to science in their countries of origin until they were intercepted upon entering the United States on smuggled avocados. Similarly, red-banded whitefly was an unknown entity until it became established on California-grown avocados. Border inspections have detected other potentially serious undescribed pest species of avocados on smuggled plants prior to the establishment of these pests in California. However, the small size and cryptic nature of many arthropod pests make detection extremely difficult, even when shipments are known to contain avocados. Importation (legal and illegal) of avocado plants and fruit from around the world will most likely act as a conduit for future introductions of both known and unknown exotic pests in California.

Intervention Strategies and Technologies for Managing New Avocado Pests

Intervention strategies include preventing establishment through exclusion, or eradication if exclusion measures fail, and regular natural enemy releases or pesticide treatment should avocado pests become established. From 1914 to 1997, the U.S. avocado industry was protected by a ban on the importation of avocados from

areas known to have pests of economic importance. In 1997, Mexico was able to gain partial access to the United States through the development of a set of mutually agreed upon SPS regulations known as the systems approach to pest management.

The systems approach is a set of safeguards and mitigation measures designed to individually and cumulatively reduce plant pest risk (but see Gray et al. 1998). Generally, two or more of the measures are independent, thereby creating the redundancy needed to protect the integrity of the system should one of the mitigation measures fail. The safeguards and mitigation measures can occur in the growing area, at the packinghouse, or during shipment and distribution of the commodity. The components of a systems approach may vary widely depending upon the commodity and the pests involved. The U.S. Department of Agriculture's (USDA's) Mexican avocado importation system is designed to reduce the risk posed by nine different species of quarantine pests, including four species of fruit flies, large and small weevils, and the avocado seed moth. The 1997 Hass Avocado Agreement has nine safeguards: (1) host resistance to fruit flies; (2) field surveys for stem and seed weevils and fruit flies; (3) trapping and field bait treatments for fruit flies; (4) field sanitation practices to decrease the chances of weevil or fruit fly establishment; (5) postharvest safeguards; (6) winter shipping only; (7) packinghouse inspection and fruit cutting to detect weevils or fruit flies; (8) port-of-arrival inspection to detect pests; and (9) U.S. distribution limited to 19 northeastern states from November through February (USDA 1995).

The fresh Hass avocados from Mexico that are shipped to the United States, according to the USDA, originate in an area of low pest prevalence for the nine known quarantine pests of concern. Four municipalities in the Mexican state of Michoacán qualify for this designation: Uruapan, Periban, Salvador Escalante, and Tancitaro. While there is still a heated and controversial debate about pest population levels in these areas and the susceptibility of Hass avocados to infestation by certain species of pests, there is little question that known and unknown pests are present in the avocado-growing regions of Michoacán. As long as this is the case, there is a certain level of risk associated with the importation of fresh Hass avocados from Mexico, and as volume increases, so does the risk of unintentionally importing new pests into the United States.

Mexico's first importation season in 1997 resulted in 13.3 million pounds of avocados being shipped to the United States. The fruit originated from 58 growers holding about 3,700 acres, and five packinghouses handled the Mexican fruit. In 1998-1999, 244 growers with 10,697 acres were certified for the export program, along with 14 packinghouses. USDA officials confirm that management of the importation system and monitoring for compliance were considerably more difficult in the second year, with oversight of field operations being the chief concern. The California avocado industry and other agricultural interests fear that there will be an economic incentive to illegally transship Mexican avocados from restricted northern markets to other parts of the United States. Transshipment is likely to occur if the profit margin realized by shipping to another market is greater than that which would be realized in the restricted area.

Hass avocados imported from Chile or the Dominican Republic do not pose the same risk because the fruit originates from pest-free areas. Nonetheless, an increase in commercial shipments from all sources may increase the likelihood of introduction of nonindigenous insects that are not considered quarantine pests. The persea mite and the avocado thrips, for example, are believed to have originated in Mexico and/or Central America, and it is likely that these pests may have first been introduced into California avocado groves through illegal commercial shipments of fruit smuggled into the country for personal use.

Despite the potential effectiveness of this USDA pest-prevention system, new pests, including avocado pests, may eventually be introduced and become established in the United States. Some of these pests may cause significant losses in avocados when conventional control tools are unavailable or unable to prevent the development of harmful pest population levels. Alternatively, biological control strategies may provide cost-effective and environmentally benign long-term control of pests, and may be integrated with other techniques to control additional avocado pests that become estab-

lished. This assumes that effective natural enemies can be located, introduced into the United States, and established, and that they are effective at reducing pest densities. With the establishment of such exotic avocado pests as avocado thrips and persea mite, a substantial financial burden is imposed on the California avocado industry because funds are diverted from plant breeding and management programs, which promote productivity, to investigating and developing strategies to control exotic arthropod pests. With the lack of effective indigenous biological control agents, Section 18 Emergency Registrations and Special Local Needs Permits have been sought and granted to provide growers legal access to pesticides not previously registered for use on avocados. These application processes are not inconsequential and are demanding of time and money. Increased reliance and use of pesticides increase the likelihood of resistance development, destruction of beneficial nontarget organisms, and environmental contamination.

Parties Potentially Affected by Pests and Their Subsequent Control

Rising pest-control costs from exotic pest introductions and changes in international SPS regulations affect growers, handlers, importers, exporters, and domestic consumers. The total farm-gate value of U.S. avocado production at the time of the Mexican Hass Avocado Agreement (1996/1997 season) was $273 million. The industry is composed of relatively low-value avocados, such as Fuerte, and high-value Hass avocados. The 1994-1997 average price growers received for Hass was $1,548 a ton, a little over twice the $758 a ton they received for other varieties. The southern coastal counties in California produced 87 percent of U.S. avocados, with Florida and Hawaii producing the remainder. However, as the only state producing the expensive Hass variety, the value of California production was 97 percent of the total U.S. value.

In California, avocados are primarily grown in the southern coastal counties and the inland counties of Riverside and Tulare. San Diego County produces 45 percent of the crop by value. The next largest producers are Ventura County with 22 percent, Riverside County with 13 percent, and Santa Barbara County with 12 percent. Hass avocados accounted for 80 percent of California avocado production from 1994-1997.

Successive waves of infestations by nonindigenous pests like the persea mite and avocado thrips have also contributed to rising input costs, because growers must treat these pests with insecticides. Costs associated with treatment often include one or more of the following: the purchase of bio-control agents or commercial pesticides, equipment and labor, and professional services from a pest control advisor or a flying service for the application of treatments by helicopter or fixed-wing aircraft. Aerial pesticide applications are frequently necessary because avocados are often grown on large parcels on steep hillsides.

The United States is a net importer of Hass avocados and a net exporter of other varieties of avocados. Prior to the Mexican Hass Avocado Agreement, Chile, and New Zealand were the major exporters of Hass avocados into the United States. Imports increased from 14,746 tons in 1990 to 60,831 tons in 1999. Chilean imports account for the biggest surges in imports, including a 300 percent increase in shipments to the United States in 1998, the year following the Mexican Hass Avocado Agreement. However, Chilean imports decreased by about a third in 1999, compared with the 1998 level. In 1994, U.S. exports were 12,940 tons. The 1995 to 1996 season average increased dramatically to 22,000 tons. Average exports have since decreased to about 11,350 tons a year.

Another trend emerged in 1998 with domestic producers shifting the time period in which shipments are made. Prior to the Mexican Hass Avocado Agreement, about 22 percent of annual domestic shipments were made from November through February. After Mexico was granted access to U.S. markets, the percentage of U.S. annual domestic shipments decreased to 18 percent for this same time period. Chile has also shifted when exports are made to the United States. After the Mexican Hass Avocado Agreement, Chile started exporting greater amounts to the United States from March through October, when California production is higher. Previously, it had exported avocados to

the United States almost exclusively during California's off season.

Per capita consumption of avocados was about 1.4 pounds during the early part of the 1990s. From 1995 to 1999, average per capita consumption was 1.66 pounds. This represents an increase of about 14 percent. As production costs increase because of exotic pests, the increase in costs may be passed along to consumers in the form of higher prices. The extent to which producers may pass on prices is influenced by how consumer demand changes in respond to higher prices and the availability of supplies from other countries. Research shows that in California, there is little change in demand for avocados for small changes in prices.

Description of the Economic Effects of Trade Liberalization and Exotic Pest Infestations

Traditionally, when trade barriers are removed, net social welfare increases. However, when the removal of trade barriers results in the introduction of an agricultural pest, the gains in welfare diminish, and may even become negative as production costs increase.

Graphically, domestic market supply, S, is equal to domestic quantity supplied, S_d, plus imports, S_m (Figure 12.5). If no pest enters, the removal of an SPS trade barrier will cause the import supply curve to shift out from S_m to S_m', causing the market supply curve to shift out from S to S' (Figure 12.5). Quantity supplied increases from Q to Q', price falls from P to P' and domestic production falls from Q_d to Q_d'. Increased market supply and lower prices leave consumers better off. However, lower domestic production and lower prices leave domestic growers worse off.

Should an exotic pest enter, however, the domestic supply curve shifts up from S_d to S_d' due to pest damage or treatment costs (Figure 12.6).

Depending upon the responsiveness of producers to price changes and the magnitude of pest damage or treatment costs, the domestic supply curve can shift up far enough such that the total market supply is less than the amount available before trade liberalization and pest infestations occurred. As producers become more responsive to price changes, the likelihood that market supplies will be lower and prices higher increases. The larger the magnitude of the domestic supply curve shift due to pest infestations, the more likely it is that consumer and producer welfare will decrease.

In both the trade liberalization and the trade liberalization and infestation scenarios, domestic producers are worse off. However, depending on the final market equilibrium, domestic consumers may be better or worse off. If final market supplies are greater and prices lower, consumers are better off. If final market supplies are lower and prices higher, consumers are worse off.

Policy Scenarios

Below, we estimate the welfare effects of the Mexican Hass Avocado Agreement and the establishment of avocado thrips to illustrate the effects of trade liberalization and exotic pest infestations. Since Mexico began exporting avocados to the United States, total avocado imports have increased by 60 percent and continue to rise. Therefore, the welfare effects of changes in trade regulations are simulated for a 60 percent increase in imports and a 100 percent increase in imports.

The first set of simulations estimates the effects of trade liberalization only. Because the major concern with removing trade bans is the possibility of exotic pests entering, these initial

simulations are compared to the case where both trade liberalization and exotic pest infestations occur. The short- and long-run welfare effects are estimated for each trade and pest scenario, for a total of eight simulations. In the short run, growers cannot easily move all (i.e., land, labor, etc.) resources into the production of other commodities. In the long run, all resources are moved to their most productive use.

Economic Effects on the U.S. and California Avocado Industry and Consumers

Methodology

The welfare effects of trade and pest shocks on the U.S. avocado industry are estimated using an equilibrium displacement model. In this model, a system of demand and supply conditions is laid out in log-linear form to determine how equilibrium quantities, prices, and other variables respond to shocks, e.g., partial removal of import restrictions and increases in production costs when an exotic pest becomes established. The model is parameterized with market and biological data.

The first set of equations characterizes the demand side of the market. Demand was separated into demand for Hass avocados, D_h, and other varieties, D_o. Quantity demanded depends on the prices of both Hass avocados and other varieties.

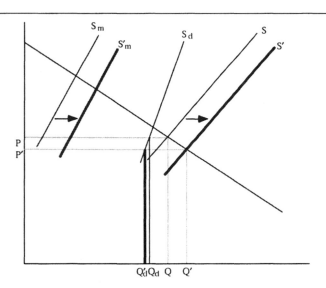

Figure 12.5 Market effects of trade liberalization.

$(12.1)\quad D_h = d_h(P_h, P_o)$

$(12.2)\quad D_o = d_o(P_h, P_o)$

The next set of equations characterizes the supply side of the market for Hass avocados. In the United States, Hass avocados are produced only in California or are imported. The total supply of Hass fruit is equal to California total Hass production, T_h, plus imports, M_h. California production depends on the price growers receive for their output, P_h, and the costs of production, C_h. Imports depend on the market price for Hass fruit and shifts in supply due to changes in trade regulations ϕ_n.

$(12.3)\quad S_h = T_{hc} + M_h$

$(12.4)\quad T_{hc} = t_{hc}(P_h, C_h)$

$(12.5)\quad M_h = m_h(P_h, \phi_h)$

The total supply of other varieties, S_o, is equal to total production from California, T_{oc}, and supply from the rest of the United States, S_{orus}. California supply depends on market price and costs of production, C_o. The rest of the United States exports other varieties, so supply by this region to the United States is equal to total production, T_{orus}, less exports, E_o. Total production and exports of other varieties depend on market prices of other varieties. Other varieties

of avocados are not imported into the United States.

$(12.6)\quad S_o = T_{oc} + S_{oc}$

$(12.7)\quad T_{oc} = t_{oc}(P_o, C_o)$

$(12.8)\quad S_{orus} = T_{orus} - E_o$

$(12.9)\quad T_{orus} = t_{orus}(P_o)$

$(12.10)\quad E_o = e_o(P_o)$

The final two equations are the market equilibrium conditions that state that demand of Hass must equal the supply of Hass and demand of other varieties must equal the supply.

$(12.11)\quad D_h = S_h$

$(12.12)\quad D_o = S_o$

Taking the log differential and converting into elasticities leaves the following set of equations:

$(12.13)\quad dnD_h - \eta_{hh}d\ln P_h - \eta_{ho}d\ln P_o = 0$

$(12.14)\quad dnD_o - \eta_{oh}d\ln P_h - \eta_{ho}d\ln P_o = 0$

$(12.15)\quad d\ln S_h - \lambda_{hc}d\ln T_{hc} - \lambda_{hm}\ln M_h = 0$

$(12.16)\quad d\ln T_{hc} - \varepsilon_{hc}d\ln P_h = -\varepsilon_{hc}d\ln C_h$

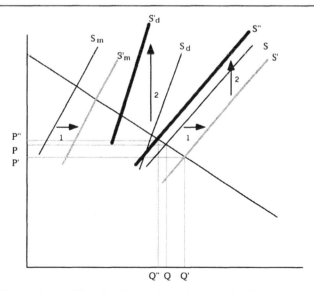

Figure 12.6 Market effects of trade liberalization and exotic pest infections.

(12.17) $d\ln M_h - \varepsilon_{hm}d\ln P_h = d\ln\phi_{hm}$

(12.18) $d\ln S_o - \lambda_{oc}d\ln T_{oc} - \lambda_{orus}d\ln S_{orus} = 0$

(12.19) $d\ln T_{oc} - \varepsilon_{oc}d\ln P_o = -\varepsilon_{oc}d\ln C_o$

(12.20) $d\ln T_{orus} - \gamma_{ous}d\ln S_{orus} - \gamma_{oe}d\ln E = 0$

(12.21) $d\ln T_{orus} - \varepsilon_{orus}d\ln P_o = 0$

(12.22) $d\ln E_o - \varepsilon_{oE}d\ln P_o = 0$

(12.23) $d\ln D_h = d\ln S_h$

(12.24) $d\ln D_o = d\ln S_o$

where η is the elasticity of demand, ε is the elasticity of supply, λ is the market supply share, and γ is the production share. For the demand equations, the elasticity of demand is negative for the own-price effects and positive for the cross-price effects. Therefore, if the price of Hass avocados increases, the demand for Hass avocados decreases and the demand for other varieties increases. The elasticities of supply for quantities destined for the U.S. market and output are positive. The elasticities of supply with respect to input costs and exports are negative. Therefore, if market prices rise, California supply increases, U.S. production increases, imports increase, and exports decrease. If costs increase, California quantity supplied decreases.

The proportional changes in all price and quantity variables were used to calculate the magnitude and direction of change in consumer and producer welfare for the United States as a whole, and for California only. The change in producer surplus (PS) is calculated as

(12.25) $\Delta PS = 0.5*((NP_j - OP_j) -$
$(OP_j * d\ln C_{ji}))*(NT_{ji} + OT_{ji})$

and the change in consumer surpluses (CS) is calculated as

(12.26) $\Delta CS = -0.5*(NP_j - OP_j)*(NS_{ji} + OS_{ji})$

where *NP* is the new price, *OP* the original price, *NT* the new production level, *OT* the original production level, *OS* the new market supply, *NS* the new market supply, j is equal to Hass or other varieties, and i is equal to California or the United States.

Changes in U.S. producer welfare were calculated by region (California or Florida and Hawaii) and variety, and then added together. The change in U.S. consumer welfare was calculated by variety using the change in the U.S. market supply and grower price. This estimate means that the consumer is the purchaser of avocados from the grower. For California, data on consumption of avocados do not exist, although there is a perception that more avocados are consumed per capita in California than elsewhere in the United States. Therefore, we calculated the changes in California consumer welfare as a percentage of the changes in U.S. consumer welfare. Two scenarios were used. The first assumes that California's consumption share is equal to its population share, 12 percent. The other assumes that Californians consume more avocados on average than elsewhere in the United States, or 25 percent of total U.S. consumption.

Model Parameters As stated previously, the assumed trade shocks to the system as a result of the Mexican Hass Avocado Agreement are a 60 percent and 100 percent increase in imports. The cost shocks come from increased pest control costs. With treatment, decreases in revenues due to scarring are avoided. The control costs reported in this study are those incurred for a typical California avocado grower in 2001. One treatment of abamectin is applied by helicopter per year. Total material and application costs for abamectin are estimated to be $180 per acre. This represents an increase in cost per ton of 4.4 percent for growers who treat with abamectin. Taking into account average pest densities across susceptible and nonsusceptible climatic zones, industry costs increase by 3.6 percent per ton.

Supply and demand elasticities were obtained from the literature (Carmen and Craft 1998). The short-run elasticity of supply was set at 0.15 for the United States and 1.0 for trade. The long-run elasticity of supply was set at 1.15 for the United States and 2 for trade. Demand elasticities in the literature were for Hass and other varieties combined (Carmen and Craft 1998). Techniques developed by Armington (1969) were used to obtain the own-price and cross-price elasticity of demand for Hass and other varieties. The own-price elasticity of demand for Hass was set at −1.2, and for other

Table 12.3 Price and quantity variables

	Price	Production	Imports	Exports
	$/ton		*Short tons*	
Hass - CA	1,548	147,900	20,500	0
Other - CA	758	21,000	0	0
Other – Rest of U.S.	758	20,800	0	16,800

Sources: Avocado Greensheet, USDA Fruit and Nut Report, FAO

varieties at −2.6. The elasticity of demand for Hass with respect to the changes in the price of other varieties was 0.4, and the elasticity of demand for other varieties with respect to the price of Hass was 1.8.

Supply shares were calculated based on a three-year average (1994–1997) of production (USDA 2000; California Avocado Commission 1994-2000), imports (Food and Agriculture Organization 2000), and exports (Food and Agriculture Organization 2000; California Avocado Commission 1994-2000) (Table 12.3). Production shares were also calculated using a three-year average (1994-1997) (USDA, 2000).

Results

Trade liberalization without exotic pest introductions results in the changes in market prices and quantities as described in figure 12.5. U.S. supply is greater due to the increased supply from Mexico. With greater market supplies the price for Hass avocados falls in the short run by 4.9 percent for a 60 percent trade shock (Table 12.4). California production decreases by 0.7 percent. In response to the lower U.S. prices, California production and imports from other countries decrease. In the long run, California Hass output declines by 4 percent. The decrease in domestic production and imports lowers market quantity supplied and raises prices from the short-run equilibrium. However, the long-run market equilibrium quantity increases by 3.1 percent and the price is lower by 2.7 percent compared with the preliberalization level due to the Mexican Hass avocado imports for a 60 percent trade shock. For a 100 percent trade shock, the direction of change in market supply, price and production is the same, but the magnitude of the change is greater.

The Mexican Hass Avocado Agreement also affects the market for other varieties. As the market price for Hass avocados falls, demand shifts from other varieties to Hass fruit, causing the market price of other varieties to fall also. Price falls by 0.5 percent in the short run for a trade shock of 60 percent (Table 12.4). As the price of other varieties falls, U.S. growers decrease their production. Production decreases by 0.1 percent in the short run and by 0.3 percent in the long run (Table 12.4). Florida and Hawaii producers also divert a portion of their production from the domestic to the international market. Therefore, market supply decreases by 0.7 percent in the short run.

Table 12.4 Simulation results: percentage change in price and quantity variables

Elasticity of supply		Shock		Price		Quantity							
						Hass			Other				
						Calif.		Total	Calif.	Rest of U.S.	Rest of U.S. to U.S.		Total
U.S.	Trade	Cost	Trade	Hass	Other	Output	Imports	U.S.	Output	Output	Supply	Exports	U.S.
								%					
0.15	1	0	60	−4.9	−0.5	−0.7	51.6	5.6	−0.1	−0.1	−3.9	0.9	−0.7
0.15	1	0	100	−8.1	−0.8	−1.2	86.0	9.4	−0.1	−0.1	−6.6	1.4	−1.2
1.5	2	0	60	−2.7	−0.2	−4.0	54.7	3.1	−0.3	−0.3	−2.6	0.3	−0.6
1.5	2	0	100	−4.5	−0.3	−6.7	91.1	5.2	−0.4	−0.4	−4.3	0.5	−1.0
0.15	1	3.6	60	−4.5	−0.3	−1.2	52.2	5.3	−0.6	−0.1	−2.7	0.6	−0.9
0.15	1	3.6	100	−7.8	−0.7	−1.7	86.6	9.1	−0.6	−0.1	−5.4	1.2	−1.4
1.5	2	3.6	60	−0.8	0.7	−6.7	58.3	1.3	−4.4	1.0	10.1	−1.2	−2.1
1.5	2	3.6	100	−2.6	0.6	−9.3	94.8	3.4	−4.6	0.8	8.4	−1.0	−2.5

Elasticities of demand: Hass own-price = −1.2; Other own-price = −2.6; Hass with respect to the price of Other = 0.4; Other with respect to the price of Hass = 1.8.

As was the case with Hass avocados, the long-run decrease in production causes prices of other varieties to partially recover from their short-run value. The long-run decrease is only 0.2 percent. With a smaller drop in prices, less fruit is diverted to international markets. Even though less fruit is produced, with a higher proportion of fruit going to the domestic market, the long-run decline in U.S. market supplies of other varieties is less than the short-run decline.

When trade liberalization occurs and an exotic pest enters, the direction of change in price and quantity variables cannot be predicted a priori. It depends on the time period and the magnitude of the trade and cost shocks. In the short run the market price for Hass avocados falls for both a 60 percent and a 100 percent trade shock, even with increased pest control costs. Prices do not fall as much as when no pest enters as the increased costs of production in California put upward pressure on prices for both Hass and other varieties. When a pest enters, the price fall for Hass avocados is only 4.5 percent for a 60 percent trade shock, compared with a fall of 4.9 percent when no pest enters and the trade shock is 60 percent (Table 12.4).

With a smaller change in prices, the change in the U.S. market supply is also smaller. However, the decline in California production of avocados is greater because growers have to cope both with lower market prices and higher costs of production. In the short run, and with a 60 percent trade shock, production of Hass avocados decreases from –0.7 percent when no pest enters to –1.2 percent when thrips establish. Florida and Hawaii growers do not face higher costs of production. With a smaller decline in short-run market prices as compared to the no pest-entering scenario, the decline in production of other varieties from these two states is also lower. For a 60 percent trade shock, the decline in production is –0.1 percent (Table 12.4).

In the long run when thrips become established the change in market prices is also negative for a trade shock of both 60 and 100 percent (Table 12.4). When the trade shock is 60 percent, the larger Hass import supply causes the market supply for Hass to increase by 1.3 percent and market prices to fall by 0.8 percent, even though California production decreases by 6.7 percent and pest control costs put upward pressure on prices. When the trade shock is 100 percent, market quantities increase by 3.4 per-

cent, market prices fall by 2.6 percent, and California production falls by 9.3 percent.

The long-run adjustments by California producers when a pest enters are sufficient to lower total market supplies of other varieties (Table 12.4). The increase in pest control costs causes California growers of other varieties to decrease production in the long run by 4.4 percent for a 60 percent trade shock. The decrease in California output causes the U.S. market supply to decline by 2.1 percent for a 60 percent trade shock. This decrease in production causes market prices for other varieties to increase by 0.7 percent. Higher market prices cause Florida and Hawaii growers to produce more and to move fruit away from international markets and into the United States.

When the trade shock increases to 100 percent, the changes in the price and quantity variables are very similar to the 60 percent levels, indicating that the cost shock has a more dominant effect on the market for other varieties than the trade shock (Table 12.4). The increase in market price is 0.6 percent for a 100 percent trade shock compared to 0.7 percent for a 60 percent shock (Table 12.4). The slight decrease in the market equilibrium price was due to the downward pressure put on market prices in response to the lower long-run market prices for Hass avocados with the 100 percent trade shock.

The differences in magnitude and direction of change in market price and quantity variables determine the net welfare effects on consumers and producers (Table 12.5). Under the no pest–enters scenario, Hass consumers gain from the lower prices and greater market supplies. Consumers gain $13 million annually in the short run and $7 million in the long run when the trade shock is 60 percent. California growers produce less and receive lower prices for their output. The annual decline in Hass producer welfare is $11.1 million in the short run and $6 million in the long run. The annual gain in consumer welfare and loss in producer welfare is lower in the long run due to the higher Hass avocado prices when compared with the short-run equilibrium values. The net change in welfare for the United States is positive because the gains to United States consumers are greater than the losses to producers (Table 12.5). However, given California's large share of the avocado market, gains to California consumers are less than the losses to California producers, for both the 12 percent and 25 percent consumption share, and the net change in welfare within California is negative.

For the Hass avocado market, consumer and producer welfare change significantly when a pest enters. For a 60 percent trade shock, total U.S. consumer welfare increases by $12.1 million in the short run. When the trade shock is 100 percent, the additional market quantities of Hass are sufficient to lower market prices, and consumers are better off by $21.1 million per year (Table 12.5). In the long run, the lower market quantities and higher market prices reduce the annual short-run consumer gains, but consumer welfare still increases. Within California, for the 60 percent trade shock, higher production costs and lower prices leave Hass growers in the long-run even worse off than when no pest enters. Producer losses increase by about 50 percent from $6 million when no pest enters to 9.8 million when thrips enter.

Even though California consumers are better off, the gain is not sufficient to overcome the much higher losses incurred by growers when a pest becomes established, and the change in net welfare for both the U.S. and California is negative. For a trade shock of 60 percent, the net change in U.S. welfare falls by 6.4 million in the short run and 7.6 million in the long run. When the trade shock increases to 100 percent, the decline in welfare fell to 4.6 million due to the increased benefits to consumers. Total welfare losses to California were greater than losses to the United States. The establishment of avocado thrips causes welfare for California to fall by $15.5 million in the short-run for a 60

percent trade shock and by $9.3 million in the long run. Contrary to the decline in welfare losses for the entire U.S., when the trade shock goes from 60 to 100 percent, welfare losses in California increase. In the long run, a 100 percent trade shock increases total losses from $9.3 million to $11.8 million (Table 12.5).

As was the case with Hass avocados, consumers of other varieties are better off with the Hass Avocado Agreement when no pest enters, and better off in the short run when thrips establish. However, the long-run increase in prices and decrease in market quantities leave consumers worse off (Table 12.6). Consumer welfare declines by $30,000 for a 60 percent trade shock. For other varieties of avocados, the change in welfare for producers is negative for both California and the United States as a whole. However, because market prices increase in the long run when thrips become established in California, growers in Florida and Hawaii are better off. Even though some groups benefit under certain scenarios, the net effect on United States and California welfare is negative for all pest scenarios and time periods. Total welfare losses for the United States are $130,000 in the long run for a 60 percent trade shock. For California the loss is $470 million.

The effect of aggregating the consumer and producer gains and losses for both Hass and other varieties is that the change for consumers is positive (Table 12.7). The net change in U.S. and California producer welfare is always negative when thrips enter, even though producers of other varieties in Florida and Hawaii are better off in the long run. For the U.S. as a whole,

Table 12.5 Simulation results—welfare changes for Hass avocados

Elasticity of supply		Shock		Producers	Consumers			Total		
U.S.	Trade	Cost	Trade	Calif.	U.S.	Calif. (12%)	Calif. (25%)	U.S.	Calif. (12%)	Calif. (25%)
		(%)					($ million)			
0.15	1	0	60	−11.1	13.0	1.6	3.3	1.9	−9.5	−7.8
0.15	1	0	100	−18.4	22.1	2.7	5.5	3.7	−15.8	−12.9
1.5	2	0	60	−6.0	7.1	0.8	1.8	1.1	−5.1	−4.2
1.5	2	0	100	−9.8	11.9	1.4	3.0	2.1	−8.4	−6.9
0.15	1	3.6	60	−18.5	12.1	1.5	3.0	−6.4	−17.0	−15.5
0.15	1	3.6	100	−25.8	21.1	2.5	5.3	−4.6	−23.3	−20.5
1.5	2	3.6	60	−9.8	2.2	0.3	0.5	−7.6	−9.5	−9.3
1.5	2	3.6	100	−13.6	6.9	0.8	1.7	−6.6	−12.7	−11.8

Elasticities of demand: Hass own price = −1.2; other own price = −2.6; Hass with respect to price of other = 0.4; other with respect to price of Hass = 1.8.

Table 12.6　Simulation results—welfare changes for other varieties of avocados

Elasticity of supply		Shock		Producers			Consumers			Total		
U.S.	Trade	Cost	Trade	U.S.	Calif.	Rest of U.S.	U.S.	Calif. (12%)	Calif. (25%)	U.S.	Calif. (12%)	Calif. (25%)
(%)							*($ million)*					
0.15	1	0	60	−0.15	−0.08	−0.08	0.02	0.01	−0.06	0.09	−0.07	−0.05
0.15	1	0	100	−0.26	−0.13	−0.13	0.04	0.02	−0.10	0.15	−0.11	−0.09
1.5	2	0	60	−0.05	−0.03	−0.03	0.01	0.00	−0.02	0.03	−0.02	−0.02
1.5	2	0	100	−0.09	−0.04	−0.04	0.01	0.01	−0.04	0.05	−0.04	−0.03
0.15	1	6.5	60	−0.68	−0.62	−0.05	0.02	0.01	−0.61	0.06	−0.62	−0.61
0.15	1	6.5	100	−0.78	−0.68	−0.10	0.03	0.01	−0.66	0.12	−0.66	−0.64
1.5	2	6.5	60	−0.35	−0.46	0.11	−0.03	−0.02	−0.48	−0.13	−0.47	−0.49
1.5	2	6.5	100	−0.39	−0.47	0.09	−0.03	−0.01	−0.49	−0.10	−0.49	−0.50

Elasticities of demand: Hass own-price = −1.2; other own-price = −2.6; Hass with respect to the price of other = 0.4; other with respect to the price of Hass = 1.8

the change in welfare is positive if no thrips establish and negative if they do. For all scenarios, any gains to California avocado consumers are less than the losses to California producers, and net welfare within California decreases.

General Discussion and Implications

The three most recent avocado pests to establish in California (avocado thrips, persea mite, and red-banded whitefly) were all species new to science at the time of their discovery in the United States. This fact highlights three important points:

1. There are probably additional potentially serious avocado pests in Central America that are unknown entities that may be able to establish in California and inflict severe damage to commercially grown avocados. Foreign exploration in Mexico for avocado thrips and its natural enemies has revealed one new species of *Frankliniella* (the western flower thrips group, which are serious disease vectors) and at least one other *Frankliniella* species, whose identity cannot be confirmed but could potentially be a new species as well. In addition, there is at least one species of *Scirtothrips* (same genus as the avocado thrips) whose taxonomic status is undetermined, which could be a new species (Hoddle, unpublished data) and a potential pest. Of the 578 slide-mounted thrips specimens collected from avocados in Mexico, the thrips fauna

was dominated by two species, the avocado thrips (*Scirtothrips perseae*) and *Neohydatothrips burangae* (Hood), neither of which was known from avocados in Mexico until foreign exploration work was undertaken in the period 1997–1999. Furthermore, just three species of thrips, *Frankliniella cephalica*, *Heliothrips haemorrhoidalis*, and *Pseudophilothrips perseae*, were listed as pest species by the USDA Animal and Plant Health Inspection Service (Firko 1995), and all three species have been collected during extensive exploration efforts for avocado thrips and its natural enemies (Hoddle et al., submitted). At least 38 species of thrips have been collected from avocados in Mexico, of which seven are confirmed pests. The insect fauna of Mexican grown avocados appears to be poorly documented and understood.

2. Border inspections intercepted both the persea mite and the avocado thrips on smuggled avocados from Mexico before either pest established in California. This strongly suggests that interception and exclusion policies are extremely valuable in preventing exotic avocado pests from entering from Central America and establishing in California. The biology of potentially serious pests, like thrips for example, makes detection very difficult. Thrips eggs are extremely small and are laid within the tissue of leaves, or the skin of fruit. The numbers of eggs laid within individual leaves and fruit in orchards infested with avocado thrips can exceed 40. Just one avocado fruit or leaf entering the United States with this number of viable eggs provides a good-sized cohort that could establish in a per-

Table 12.7 Simulation results—total welfare changes

Elasticity of supply		Shock		Producers		Consumers			Total		
U.S.	Trade	Cost	Trade	U.S.	Calif.	U.S.	Calif. (12%)	Calif. (25%)	U.S.	Calif. (12%)	Calif. (25%)
		(%)						($ million)			
0.15	1	0	60	−11.2	−11.2	13.1	1.6	3.3	1.9	−9.6	−7.9
0.15	1	0	100	−18.7	−18.6	22.3	2.7	5.6	3.6	−15.9	−13.0
1.5	2	0	60	−6.0	−6.0	7.1	0.9	1.8	1.1	−5.2	−4.2
1.5	2	0	100	−9.9	−9.9	12.0	1.4	3.0	2.0	−8.5	−6.9
0.15	1	6.5	60	−19.2	−19.1	12.2	1.5	3.0	−7.0	−17.6	−16.1
0.15	1	6.5	100	−26.6	−26.5	21.3	2.6	5.3	−5.3	−23.9	−21.1
1.5	2	6.5	60	−10.2	−10.3	2.1	0.2	0.5	−8.1	−10.0	−9.8
1.5	2	6.5	100	−13.9	−14.0	6.8	0.8	1.7	−7.1	−13.2	−12.3

Elasticities of demand: Hass own-price = −1.2; Other own-price = −2.6; Hass with respect to the price of other = 0.4; other with respect to the price of Hass = 1.8.

missive environment (i.e., abundant food, mild climate, lack of natural enemies).

3. The small numbers of pests intercepted on avocado plants and fruit that are moved into the United States from Central America suggest that founding populations of pests may often be very small. Work on thrips used for the biological control of weeds has demonstrated that 33 percent of releases of just 10 thrips into a permissive environment can result in establishment and proliferation (Memmott et al. 1998). The greater the frequency of small introductions, the higher the likelihood of establishment in comparison with few introductions of large numbers of thrips that can become extinct by chance (Memmott et al. 1998). This scenario from weed biological control may apply to the establishment of new avocado pests in California where frequent introductions (either through legal or illegal routes) of small numbers of pests may ultimately lead to their establishment.

As shown by this analysis, removal of trade restrictions can have a penurious effect, if not for the whole country, then for reasonable regional boundaries such as a state. For California, the net change in welfare is always negative. Should an exotic pest become established under the conditions described in the above scenarios, welfare for the United States can decline.

References

Aponte, O., and J.A. McMurtry. 1997. "Damage on Hass Avocado Leaves, Webbing and Nesting Behavior of *Oligonychus perseae* (Acari: Tetranychi-dae)." *Experimental and Applied Acarology.* 21:265–272.

Armington, P.S. 1969. "A Theory of Demand for Products Distinguished by Place of Production." IMF Staff Papers 16:159–178.

Bailey, J.B., M.P Hoffman, and K.N Olsen. 1988. "Black Light Monitoring of Two Avocado Insect Pests." *California Agriculture.* 42:26-27.

Bender, G.S. 1993. "A New Mite Problem in Avocados." *California Avocado Society Yearbook.* 77:73–77.

Bergh, B., and N. Ellstrand. 1986. "Taxonomy of the Avocado." *California Avocado Society Yearbook.* 70:135–145.

California Avocado Commission. 1997. "Blueprint for a Record Year." Annual Report of the California Avocado Commission. Santa Ana: California Avocado Commission. 15 pp.

California Avocado Commission. 1998. "Agricultural Advisory Policy Study Field Data Collection Report." Santa Ana: California Avocado Commission.

California Avocado Society and California Avocado Commission. 1994-2000. California Avocado Greensheets. Santa Ana: California Avocado Commission.

Carmen, Hoy F., and R. Kim Craft. 1998. "An Economic Evaluation of California Avocado Industry Marketing Programs, 1961-1995." Giannini Foundation Research Report 345. Oakland: University of California, Division of Agriculture & Natural Resources. 66 pp.

Condit, I.J. 1932. "Check List of Avocado Varieties." *California Avocado Society Yearbook.* 17:11–21.

Faber, B. 1997. The Persea Mite Story. *Citrograph.* 82:12–13.

Firko, M.J. 1995. "Transportation of Avocado Fruit (*Persea americana*) from Mexico: Supplemental Risk Assessment." www.aphis.usda.gov/ppq/avocados/PRAmemo.pdf.

Fleschner, C.A. 1953. "Biological Control of Avocado Pests." *California Avocado Society Yearbook.* 38:125–129.

Fleschner, C.A., J.C. Hall, and D.W. Ricker. 1955. "Natural Balance of Mite Pests in an Avocado Grove." *California Avocado Society Yearbook.* 39:155–162.

Food and Agriculture Organization. 2000. FAOSTAT Agricultural Data. http://apps.fao.org/page/form? collection=Trade.CropsLivestockProducts& Domain=Trade&servlet=1&language=EN& hostname=apps.fao.org&version=default.

Gray, G.M., J.C. Allen, D.E. Burmaster, S.H. Gage, J.K. Hammitt, S. Kaplan, R.L. Keeney, J.G. Morse, D.W. North, J.P. Nyrop, M. Small, A. Stahevitch, and R. Williams. 1998. "Principles for Conduct of Pest Risk Analyses: Report of an Expert Workshop." *Risk Analysis.* 18:773–780.

Hoddle, M.S., and J.G. Morse. 1997. "Avocado Thrips: A Serious New Pest of Avocados in California." *California Avocado Society Yearbook.* 81:81–90.

Hoddle, M.S., and J.G. Morse. 1998. "Avocado Thrips Update." Citrograph. 83:3–7.

Hoddle, M.S., S. Nakahara, and P.A. Phillips. 2002. "Foreign Exploration for *Scirtothrips perseae* Nakahara (Thysanoptera: Thripidae) and Associated Natural Enemies on Avocado (*Persea americana* Miller)." *Biological Control.* 24:251–265.

Kimball, M.H., and F.A. Brooks. 1959. "Plant Climates of California." *California Agriculture.* 13:7–12.

McMurtry, J.A. 1992. "The Role of Exotic Natural Enemies in the Biological Control of Insect and Mite Pests of Avocado in California." *Proceedings of the Second World Avocado Congress.* pp. 247–252.

McMurtry, J.A., H.G. Johnson, and S.J. Newberger. 1991. "Imported Parasite of Greenhouse Thrips Established on California Avocado." *California Agriculture.* 45:31–32.

Memmott, J., S.V. Fowler, and R.L. Hill. 1998. "The Effect of Release Size on the Probability of Establishment of Biological Control Agents: Gorse Thrips (*Sericothrips staphylinus*) Released Against Gorse (*Ulex europaeus*) in New Zealand." *Biocontrol Science and Technology.* 8:103–115.

Mound, L.A. 1997. "Biological Diversity." In T. Lewis, Ed., *Thrips as Crop Pests.* Wallingford, United Kingdom: CAB International. pp. 197–215.

Mound, L.A., and J.M. Palmer. 1981. "Identification, Distribution and Host-Plants of the Pest Species of *Scirtothrips* (Thysanoptera; Thripidae)." *Bulletin of Entomological Research.* 71:467–479.

Mound, L.A., and D.A. Teulon. 1995. "Thysanoptera as Phytophagous Opportunists." In B. L. Parker, M. Skinner, and T. Lewis, Eds., *Thrips Biology and Management.* New York: Plenum Press. pp. 3–19.

Nakahara, S. 1995. "Revision of the Genus *Tetraleurodes* in North America." *Insecta Mundi.* 9:105–150.

Nakahara, S. 1997. "*Scirtothrips perseae* (Thysanoptera: Thripidae), a New Species Infesting Avocado in Southern California." *Insecta Mundi.* 11:189–192.

Nakasone, H.Y., and R.E. Paull. 1998. *Tropical Fruits.* Wallingford, United Kingdom: CAB International.

Popenoe, F.O. 1915. "Varieties of the Avocado." *California Avocado Society Yearbook.* 1:44–69.

Popenoe, F.O. 1952. "Central American Fruit Culture." *Ceiba.* 1:269–311.

Rose, M., and J.B. Woolley. 1984a. "Previously Imported Parasite May Control Invading Whitefly." *California Agriculture.* 38:24–25.

Rose, M., and J.B. Woolley. 1984b. "Previously Imported Parasite May Control Invading Whitefly." *California Avocado Society Yearbook.* 68:127–131.

Storey, W.B., B. Bergh, and G.A. Zentmyer. 1986. "The Origin, Indigenous Range, and Dissemination of the Avocado." *California Avocado Society Yearbook.* 70:127–133.

USDA. 1995. "Importation of Avocado Fruit (*Persea americana*) from Mexico." Washington, D.C.: U.S. Dept. of Agriculture, Supplemental Pest Risk Assessment.

USDA. 2000. *Fruit and Nut Situation and Outlook Yearbook.* Washington, D.C.: U.S. Dept. of Agriculture, Market and Trade Economics Division, Economic Research Service. FTS-290.

13

Ash Whitefly and Biological Control in the Urban Environment

Timothy D. Paine, Karen M. Jetter, Karen M. Klonsky, Larry G. Bezark, and Thomas S. Bellows

Introduction

The ash whitefly was introduced into urban neighborhoods in Southern California in 1988. A combination of broad host range, high reproductive rate, and short generation time enabled the ash whitefly to produce dramatic population increases in the absence of natural enemies. Within three years the insect was distributed throughout the state. Its primary effect was defoliation of ornamental trees. In addition, sticky whitefly-produced honeydew covered sidewalks, lawns, automobiles, patio furniture, carpeting, draperies, and windows, and significantly reduced the quality of life in the affected urban areas (Bellows et al. 1992a). In some regions ash whitefly numbers were large enough that people became concerned over the potential harmful health effects of inhaling them.

How best to manage the infestation was a challenge. Traditionally, once an exotic pest has become established in urban environments, it is the responsibility of local governments, households, and private firms to manage the pest. However, the quickness with which the whitefly multiplies means that treatments are needed every three to four weeks to keep it under control. The costs of this type of extensive chemical control program are beyond the available resources of those groups usually responsible for control. Therefore, the severity of the ash whitefly infestation resulted in a large demand for some type of public pest control program.

An alternative to chemical controls is to locate, import, and distribute a natural enemy that attacks only the exotic pest. The goal of an introductory biological control program is to release a natural enemy that will multiply and spread throughout the environment wherever the exotic pest is located. As the natural enemy attacks the pest, pest populations fall. Accordingly, the numbers of natural enemies will also fall. Eventually, the pest and natural enemy will exist in equilibrium at very low levels. At the low levels, no noticeable damage occurs.

In the case of the ash whitefly, previous research had identified several potential biological control agents. However, few incentives exist for private pest control firms to complete an introductory biological control program. With no way to limit the spread of the natural enemy and no need for further releases once it was established in the environment, commercial marketing of the natural enemy would have been difficult.

Furthermore, the nonexclusionary nature of urban aesthetic beauty makes control of exotic ornamental pests a public good. Therefore, the California Department of Food and Agriculture and the University of California, Riverside, implemented a collaborative project to import, rear, and distribute two biological control agents, a parasitic wasp and a predatory beetle.

The wasp, *Encarsia partenopia*, proved to be particularly effective against the ash whitefly. The first releases of the wasp were in 1989. By 1992 almost all California counties (91%) had reported that the infestation was under control (Jetter and Klonsky 1994).

Biology and Ecology

Ash whitefly, *Siphoninus phillyreae* (Haliday), is native to Europe, northern Africa, western Asia, and India. Eggs are usually laid on the lower surface of leaves. Following egg hatch, the first-stage nymphs crawl a short distance from the egg, insert their tubular mouthparts into the leaf tissue, and begin to feed on plant fluids. With a net reproductive rate of 49 and a mean generation time of about 28 days at 25°C (Leddy et al. 1995), the whitefly was capable of dramatic increases in number, and all life stages could be found throughout the year. Direct injury to susceptible hosts includes chlorosis, premature leaf abscission, and reduced growth (Gould et al. 1992a).

The ash whitefly feeds on many plant species, and its host range includes deciduous and evergreen woody shrubs and ornamental trees and agricultural crops (Bellows et al. 1990; Leddy et al. 1993). The ash whitefly's most favored urban hosts are ash (*Fraxinus* spp.) and ornamental pear (*Pyrus* spp) trees. Other ornamental hosts include crape myrtle, lilacs, and hawthorn. The agricultural crop most susceptible to ash whitefly is pomegranate (*Punica granetum*). Pear (*Pyrus* spp.) and apple (*Malus* spp.) are also hosts. Citrus has been identified as an overwintering host. However, the primary concern is the ash whitefly's potential as a pest in the urban landscape (Bellows et al. 1992a).

Introduction and Spread

How the ash whitefly entered California is unknown. It did not spread naturally from adjacent areas, but was somehow imported into the state either on legally or illegally imported host material. Therefore, it is not known whether unidentified pathways exist for exotic landscape pests in general or if greater enforcement of current border regulations could prevent their introduction.

Los Angeles, Orange, and San Bernardino counties were the first areas to detect infestations of ash whitefly in 1988. The following year the infestation had spread south to the Mexican border, north into the southern end of the San Joaquin Valley, and up along California's Central Coast. 1990 saw the greatest increase in the number of counties reporting discoveries of ash whitefly as it continued to spread throughout the San Joaquin Valley and up the coast. By 1991, the ash whitefly had spread to northern California. Since 1991, no further counties in California have detected ash whitefly. In all, 46 of California's 58 counties had ash whitefly infestations. Only the mountainous and coolest regions of California did not become infested (Pickett et al. 1996).

Intervention Strategies

When ash whitefly was discovered in California, it was quickly determined that eradication was not a feasible control option. The whitefly was already too widely distributed, was dispersing rapidly, and had a broad host range that prevented targeted chemical treatments. California also had experienced a notable lack of success in trying to eradicate other whiteflies, such as the citrus whitefly and woolly whitefly, and Florida was not able to eliminate the citrus blackfly with pesticides. Because species of whiteflies had been successfully controlled with parasites, experts recommended a biological control program.

Researchers began a major effort to describe the biology of the whitefly and to introduce natural enemies. European literature on the geographic distribution, damage potential, and natural enemies of *Siphoninus phillyreae*, as well as personal contacts with European biological control scientists, helped determine where to search for parasitoids and predators that might be suitable for release in California. An accessible literature base, museum records, personal experience of scientists in California, and local expertise in Europe proved invaluable in mounting a rapid response to the pest invasion.

Two species of natural enemies, the hymenopteran parasitoid *Encarsia inaron* (Walker) and the coleopteran predator *Clitostethus arcuatus* (Rossi), were obtained in September 1989 through a collaborative biological control program between the University of California, Riverside, and the California Department of Food and Agriculture. The project included the development of mass production techniques for whitefly hosts and natural enemies, and a release program was initiated within eight months.

Each natural enemy controls the ash whitefly differently. The parasitoid female wasps search for leaves infested with whitefly nymphs. Once located, the wasp injects an egg into the body of the nymph. The larval wasp consumes the host from the inside. When development is complete, the adult wasp cuts its way out of the body of the host. The predatory beetle adults and larvae consume developing nymphs.

A cooperative project developed by the California Department of Food and Agriculture (CDFA) and the University of California, Riverside (UCR) included a research field release component and a statewide distribution component. During the initial stages of the infestation, ash whitefly was found only in southern California. Consequently, the parasitoid- and predator-rearing effort was initially located at UCR and used both UCR and CDFA facilities. A protocol was developed between the two teams to provide parasites for research needs and statewide field distribution. The first parasites released were part of the laboratory and field research programs conducted by UCR scientists to evaluate the impact of *E. inaron* on ash whitefly populations (Gould et al. 1992a, 1992b).

CDFA staff, with guidance from the California Agricultural Commissioners and Sealers Association Ash Whitefly Subcommittee, developed additional protocols for statewide distribution of parasitoids and worked with county personnel to select field release locations. Releases of 250 parasites each were first made as part of this effort during 1990–1991 at 26 locations on *Fraxinus* spp. or *Pyrus* spp. in 18 communities ranging from coastal areas to interior valleys to foothills.

Nine locations were sampled one to three months later to verify reproductive success. An attempt was made to measure dispersal at four sites by collecting samples from host plants at increasing distances along the four cardinal directions from the release point. From some of those locations, *E. inaron* had spread at least a mile eight months after release. Subsequent releases were made in all southern California counties. Parasites were subsequently recovered at all sites and had dispersed three to four miles at several locations a year later.

As part of the process of facilitating wide distribution following parasitoid establishment, project release sites in each county were designated as parasitoid nursery sites. Bio-control agents could be redistributed from these sites after the populations had increased in size. Staff members from county Departments of Agriculture were assigned responsibility for subsequent parasite movement in their specific jurisdictions. In addition to the parasitoids distributed by CDFA and county staffs, parasites were made available to private individuals, institutions, and organizations throughout southern California from UCR Cooperative Extension for the cost of production.

By 1990 populations of the whitefly in the release areas in Riverside were reduced 10,000-fold (Bellows et al. 1992a). This dramatic reduction in whitefly populations can be attributed, in part, to a parasitoid net reproductive rate of 69.3 and a mean generation time of about 19 days in the preferred life stage of the host at 25°C (Gould et al. 1995). By 1991 populations of the ash whitefly and the parasitoid, *E. inaron,* were at very low equilibrium densities in Riverside (Bellows et al. 1992a; Gould et al. 1992a, 1992b). In central and northern California, parasites were released into four widely spaced ornamental trees in each of 36 counties. Ash whitefly density and parasitism were monitored at each of these trees from the day of release up to three years afterward. Summer infestation density of the ash whitefly before release of *E. inaron* averaged 8 to 21 individuals/cm² leaf. Within two years of *E. inaron* releases, the infestation density of the ash whitefly averaged 0.32 to 2.18 individuals/cm² leaf.

The coleopteran predator was released at different sites and quickly became established at those locations. The beetle had proven to be highly effective in the laboratory (Bellows et al. 1992b). However, the impact of the predator on whitefly populations in the field was overshadowed by the effects of the parasitoid.

Potentially Affected Parties

The ash whitefly primarily attacks urban woody shrubs and trees. Ash and ornamental pear trees are the most susceptible species and are widely planted in urban environments, accounting for 17 percent of the street trees based on city tree databases from 14 cities in the affected areas (Jetter and Klonsky 1994). The main effects of the ash whitefly are chlorosis of leaves, premature defoliation, and sooty black mold. The chlorosis, defoliation, and mold reduce the aesthetic beauty of trees in the urban landscape.

The extent of urban tree damage caused by ash whitefly infestations varied geographically. Regions in California with relatively hot summers had greater densities of ash whiteflies and, therefore, greater damage than regions with relatively cool summers. Counties with high damage are in the Sacramento Valley, the San

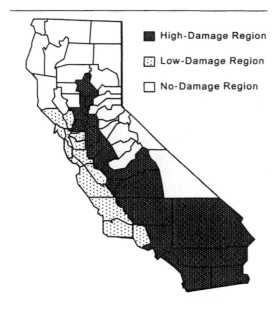

Figure 13.1 Extent of ash whitefly damage by region.

Joaquin Valley, and in Southern California (Figure 13.1). Defoliation of ash and ornamental pear trees in the high-damage region reached 70 to 90 percent during peak infestations in late summer and early fall.

The counties with low damage are along California's central coast (Figure 13.1). The ash whitefly caused 40 to 50 percent defoliation of susceptible trees in the low-damage region during peak infestations. The remaining counties in California have climates too cold to support the ash whitefly and consequently suffered no damage.

In addition to the degradation of urban landscapes, honeydew produced by the ash whitefly became a nuisance for people near host trees. Copious honeydew production damaged car finishes, drapes, and carpets. To decrease the effects of the honeydew, cars were washed more frequently or not parked under shady ash trees. The city of Modesto developed a voucher program allowing residents to take their cars to car washes to remove honeydew. People also removed shoes to prevent tracking in honeydew and ruining carpets and closed windows to keep honeydew off interior drapes. Before enjoying outdoor patios, furniture and decking needed to be hosed off, often on a weekly basis.

Concerns were also expressed about the potential harmful health effects of the ash white-

fly. The main concern was that in some regions the number of whiteflies was large enough that people inhaled them as they passed by host trees. Some people were also concerned about the health effects of the honeydew.

The severity with which urban residents were affected by the ash whitefly invasion is apparent in the number of calls made to the California County Agricultural Offices (Jetter and Klonsky 1994). During the peak of the infestation, over 19,000 phone calls were received. In Los Angeles County a separate phone line was installed solely for the purpose of responding to questions about the ash whitefly. The main question asked was how to control the whiteflies (46.9%). This was followed by questions on how to control the honeydew (13.7%) and concerns over the possibility of trees dying (12.6%). In addition, some health concerns were raised with respect to the ash whitefly (7.8%) and the honeydew (4.9%). UC Cooperative Extension and city public works departments also fielded calls on ash whitefly related problems.

Residents were advised to hose down trees to remove whiteflies and clean off the honeydew. Other recommendations were to give extra water and fertilizer to trees and to spray with Safer soap or not spray at all to maintain a healthy habitat for the *Encarsia* wasps.

Local government agencies were also potentially affected by ash whitefly infestations. Had the biological control program not been successful, city governments would have had the extra costs of hiring personnel to handle public complaints; spray, prune, cut or replant trees; clean streets; and develop alternative pest control strategies. In addition, the aesthetic value of any removed trees also would be lost.

While the ash whitefly is primarily an urban pest, several agricultural crops are also at risk. The ash whitefly is especially detrimental to pomegranates. Damage from the nymphs and honeydew reduces fruit size and yields. One case of a farmer who did not harvest his pomegranate crop due to the reduced size and yield of the fruit is documented, and a reduction in fruit size and crop yields for pomegranates was observed in other cases (Jetter and Klonsky 1994).

Apple and pear trees are secondary hosts of the ash whitefly, and citrus is an overwintering host. There are no documented yield decreases due to ash whiteflies on these crops. In the case of citrus, many varieties mature during the winter months when ash whitefly populations are low, and the ash whitefly does not use citrus trees during the summer. Also, citrus crops are sprayed regularly for other pests, and the pesticides would also help keep the ash whitefly populations below damaging thresholds.

Policy Scenarios

Eradication was not a viable policy option. The only way to eradicate the ash whitefly is to remove all primary, secondary, and overwintering hosts. This would have meant removing 30 percent of the street trees and trees from parks, private residences, and golf courses, as well as destroying California's pear, apple, citrus, and pomegranate industries. The combined farm receipts of the host crops were over $2.65 billion in 1990.

Treatment by city governments with chemical pesticides was also not an option. Repeated chemical sprayings would have quickly exhausted local budgets and prevented the routine maintenance and care of urban forests.

Given the extent of the public reaction to the ash whitefly invasion and knowledge about effective natural enemies, the biological control program was the only viable option. Consequently, the costs and benefits of only the collaborative biological control project between the CDFA and the UCR to import and distribute a natural enemy are estimated in this case study.

Economic Effects

Biological control preserves the aesthetic beauty of urban landscapes and does not increase the level of pesticides in the environment. The benefits are quantified only for preserving the aesthetic value of ash and ornamental pear trees. Potential additional benefits from using biological controls will also be discussed.

Methodology

Value of Preserving the Aesthetic Beauty of Street Trees   The benefits of the ash whitefly biological control program are calculated as the difference between the appraised value of primary host trees when the ash whitefly is controlled and no defoliation occurs, and the appraised value when ash whiteflies are present and defoliation results. Only ash and ornamental pear trees, the two trees most severely damaged by the ash whitefly, are included in the analysis. The change in the appraised value is calculated as a one-time increase in aesthetic beauty from when the ash whitefly populations are at their greatest to when no visible damage is present after *E. inaron* establishes. The total benefits are equal to the change in the appraised value per host tree in the high- and low-damage regions times the number of host trees in each region. Only street trees in California are included in this analysis. Not included in this analysis are the additional benefits to the change in aesthetic values for trees located on private property, parks and other public areas, golf courses, and trees in adjacent states that would benefit from the California releases and spread of *E. inaron*.

A widely used landscape tree appraisal technique, the trunk formula method developed by the Council of Landscape and Tree Appraisers, was used to estimate the appraised values for each primary host (ISA, 1992).

The formula is

(13.1)   The Appraised Value = Basic Value × Location Factor × Condition Factor

where

(13.2)   Basic Value = Replacement Cost + [(Trunk Area$_a$ – Trunk Area$_r$) × Basic Price × Species Factor]

where Replacement Cost = cost (retail or wholesale) to buy and install the largest normally available transplantable tree in the region

Trunk Area$_a$ = trunk area in square inches of a cross section of the tree being appraised at a height of 4.5 feet

Trunk Area$_r$ = Trunk Area in square inches of a cross section of the largest normally available transplantable tree at a height of 6 to 12 inches

Basic Price = Cost (retail or wholesale) per unit trunk area of a replacement tree measured at the height prescribed by the American Nursery standards. In this analysis the replacement tree is the largest normally available transplantable tree

Species Factor = Percentage adjustment

based on the type of tree being appraised

Location Factor = Percentage adjustment based on where (street, yard, park, highway, etc.) the tree is planted in the urban landscape

Condition Factor = percentage adjustment based on the plant's health and any structural defects.

The appraised value formula (Equation 13.1) determines a basic value for a landscape tree based on its size and species, and then adjusts that value according to where the tree is located and its overall condition. The basic value formula (Equation 13.2) is made up of two parts. The first part is the replacement cost. The replacement costs are equal to the market price to purchase and install the largest available nursery tree. The second part of the basic value formula calculates the additional value of a mature landscape tree because market prices do not exist for them.

The value of a mature landscape tree is first determined by taking the total area of a cross section of the tree, subtracting the area of a cross section of the replacement tree, and multiplying the difference by the per unit basic price. This figure determines the value of the landscape tree based on its size. Subtracting the area of the replacement tree prevents double counting of the replacement tree's value.

Replacement costs at both retail and wholesale prices are for a 15-gallon container-grown plant with a 1.5-inch diameter at 1 foot, the largest normally available transplantable tree (Table 13.1). Replacement costs are calculated at both wholesale and retail prices because

cities could pay the wholesale price or the retail price depending on the number of trees purchased and the source of the trees. The wholesale costs represent a lower bound to the estimated benefits and the retail costs an upper bound. The average trunk area of the trees being appraised is calculated from measurements of the circumference of over 100 ash and ornamental pear trees from several different locations in Davis, California, at a height of 4.5 ft (Table 13.1).

The value of the tree based on its size is then adjusted by the species factor, which allows a tree's value to vary by species because different species have different characteristics. A species factor adjustment of 50 percent for ash trees and 70 percent for ornamental pear trees was obtained from the *Species Classification and Group Assignment* handbook published by the Western Chapter of the International Society of Arboriculture (Table 13.1).

Once the basic value of the landscape tree is calculated, it is adjusted according to the tree's location in the landscape. The location factor allows trees planted along streets to be evaluated differently from trees in parks or backyards. The location factor adjustment is 60 percent for street trees (Table 13.1) (Nowak 1993).

Finally, the tree's value is adjusted by its condition factor. The condition factor reflects the tree's structural integrity, pest damage, and overall state of health. The condition factor is determined from the condition rating, which is calculated as the sum of the rating scores in five categories: roots, trunk, scaffold branches, smaller branches/twigs, and foliage (ISA 1992). Each

Table 13.1　Data for calculating appraised value of primary host trees

	Ash		Ornamental pear	
	Wholesale	Retail	Wholesale	Retail
Replacement costs	$70	$88	$70	$94
Trunk area$_a$	169 in²	169 in²	87 in²	87 in²
Trunk area$_r$	2 in²	2 in²	2 in²	2 in²
Trunk Area$_a$ – Trunk area$_r$	167 in²	167 in²	85 in²	85 in²
Basic price per square inch	$35	$44	$35	$47
Species factor	50%	50%	70%	70%
Location factor	60%	60%	60%	60%
Condition factor with no AWF[a] damage	71%	71%	71%	71%
Condition factor with AWF damage:				
High-damage region	56.5%	56.5%	56.5%	56.5%
Low-damage region	64%	64%	64%	64%

[a]AWF, ash whitefly.

category is ranked on a scale of 0 to 5 with a total of 25 points possible. The sum of the number of points given in each category determines the final value assigned to the condition factor.

The average condition rating for landscape trees in California is 19 points (Nowak 1993). The 19 points were allocated as approximately four points in each of the five categories. A rating of 4 indicates that there are no apparent problems (ISA, 1992). A condition rating of 19 corresponds to a condition factor of 71 percent (Nowak 1993). The 71 percent condition factor is used to estimate the appraised value of a host tree when ash whiteflies are controlled.

Ash whitefly damage affects the condition rating in the foliage category, because the principal damage is defoliation. In high-damage areas, where defoliation was 70 to 90 percent, the rating for foliage ranges from 4 for healthy trees to 0.5 for extremely defoliated trees. As a result, the total condition rating decreased from 19 points to 15.5. The corresponding condition factor decreased from 71 percent to 56.5 percent. The new condition factor number is extrapolated based on the 71 percent starting value. In low-damage regions, where defoliation was 40 to 50 percent, the rating for foliage decreases from 4 to 2. The 2 indicates that major problems exist in the appearance of the foliage. The total condition rating decreased from 19 points to 17, and the corresponding condition factor from 71 percent to 64 percent (Table 13.1).

The change in aesthetic value (CAV) is

(13.3) $CAV_{ijp} =$ Appraised Value without defoliation $-$ Appraised Value with defoliation

where i is equal to the geographical region, j is equal to ash or ornamental pear tree and p is equal to the wholesale or retail price of the replacement tree.

The total benefits (B) of preserving the aesthetic qualities of ash and pear trees is

(13.4) $B_p = \Sigma_i \Sigma_j CAV_{ijp} \times T_{ij}$

where T is equal to the total populations of ash or ornamental street trees in each region. The total estimated benefits are the sum of the aesthetic value change for each species in each region at each price, multiplied by the number of street trees for each species in each region.

Street tree populations are extrapolated from street tree inventories of 14 cities throughout the affected areas of California (Jetter and Klonsky 1994). The inventories included data only on street tree populations planted and maintained by a public agency and did not include trees in other public areas (e.g., parks, golf courses, and freeways) or trees on private property. The inventories are separated into the regions in which they are located.

First, the average tree density per square kilometer for each species in each region is calculated as the total number of ash or ornamental pear trees listed on the inventories for each region, divided by the land area of the cities that furnished the inventories. Then the total number of ash and pear trees throughout California is estimated by multiplying the average street tree density per square kilometer by the total land area of all urbanized centers in each affected region. The urban land areas of the affected regions are available from the *United States 1990 Census Data on Urbanized Areas.*

The costs of the ash whitefly biological control program are provided by the CDFA and the UCR. Costs include salaries of employees hired for the ash whitefly project, the time that permanent employees of CDFA and UCR spent working on the project, their travel expenses to collect and import the parasitic wasp, materials to rear the wasp in greenhouses, and travel expenses to release the wasp at selected sites and subsequent trips to monitor its spread. These costs do not include any overhead expenses for administration or depreciation of greenhouses and buildings. Furthermore, the long-term research expenses previously incurred by European scientists to identify the parasite are not included.

Not included in this analysis are the additional benefits to the change in aesthetic values for trees located on private property, parks and other public areas, golf courses, and trees located in adjacent states. Also, the losses to growers and consumers of California pomegranates cannot be estimated due to a lack of data.

Additional Costs and Benefits of Using Biological Pest Controls   The potential costs and benefits of using controls that do not increase the level of pesticides in the environment are discussed using the results of a survey of California County agricultural commissioners. The survey asked questions regarding the ash white-fly infestation and the public response to the use of biological controls. All county agricultural commissioners responded to the survey.

Results

Value of Preserving the Aesthetic Beauty of Street Trees The appraised value of an ash tree with no ash whitefly damage is between $1,279 dollars at wholesale prices, $1,607 at retail prices, and between $922 and $1,238 for an ornamental pear tree (Table 13.2). Even though ornamental pear trees have a larger species factor adjustment, ash trees are appraised at about $360 more per tree due to their larger size.

In the high damage region, the appraised value of an ash tree decreased by $261 at wholesale prices and $328 at retail prices due to ash whitefly defoliation (Table 13.2). The loss in appraised value of ash trees was about $75 more than for ornamental pear trees due to the lower base value of the pear trees (Table 13.2). Even though the absolute value of the decrease varied by species and price used, the decline is consistently 20 percent of the initial value.

As expected, in the low-damage region the decrease in the appraised value of the suscepti-ble hosts was much lower than in the high-dam-age region. The appraised wholesale value of ash trees decreased by $126, and retail value decreased by $158. The appraised value of ash trees is about $35 more than for ornamental trees. The percentage decrease in value is equal to 10 percent for each species at each price, half the percentage decrease than in the high-dam-age region.

As stated earlier, ash and ornamental pear trees represent a significant part of the urban landscape, comprising 17 percent of all street trees. Ash trees are more prevalent in the high-damage region than the low-damage region and make up 15 percent of all trees for the high-damage region and 3.3 percent for the low-dam-age region. The high damage region was the area with hotter summers. Ash trees have a wide crown and are often planted to provide shade. There are fewer ornamental pear trees in both the high-damage (4.1 percent of all trees) and low-damage (2.6 percent of all trees) regions.

The average street tree densities were 86 ash trees per square kilometer for the high-damage region and 20 ash trees for the low-damage re-gion, 23 ornamental pear trees per square kilo-meter for the high-damage region and 16 orna-mental pear trees for the low-damage region. The total square kilometers of urban centers in the high-damage region (11,364 km^2) were twice the total square kilometers for the low-damage region (5,065 km^2). In all, there were an estimated 974,848 ash trees and 262,894 pear trees in the high-damage region and 101,914 ash trees and 79,987 ornamental pear trees in the low-damage region. The estimated total number of primary host street trees equaled 1,419,643 (Table 13.3).

The change in appraised value per tree per region is multiplied by the number of trees in each region to estimate the total benefits of the ash whitefly biological control program. The total benefits from the biological control pro-gram to preserve the aesthetic value of street trees were between $255 million at wholesale and $320 million at retail prices for ash trees and between $50 million and $66 million for

Table 13.2 Calculation of change in appraised value (CAV) per tree

	Ash		Ornamental pear	
	Wholesale	Retail	Wholesale	Retail
		($)		
Appraised value using condition factor for no AWF damage	1,279	1,607	922	1,238
Appraised value using condition factors for AWF damage				
High-damage Region	1,017	1,279	734	985
Low-damage Region	1,152	1,449	831	1,116
Change in appraised value				
High-damage region	261	328	188	253
Low-damage region	126	158	91	122

Table 13.3 Aesthetic benefits

Tree species	Number of trees	Average CAV per tree		Total benefits[a]	
		Wholesale	Retail	Wholesale	Retail
High-damage region					
Ash trees	974,848	$261	$328	$254,541,345	$319,994,833
Pear trees	262,894	$188	$253	$49,511,617	$66,487,029
Total trees:	1,237,742	$246	$312	$304,052,962	$386,481,862
Low-damage region					
Ash trees	101,914	$126	$158	$12,846,573	$16,149,978
Pear trees	79,987	$91	$122	$7,272,353	$9,765,732
Total trees:	181,901	$111	$142	$20,118,926	$25,915,709
Total regions					
Ash trees	1,076,762	$248	$312	$267,387,918	$336,144,811
Pear trees	342,881	$166	$222	$56,783,971	$76,252,760
Total trees:	1,419,643	$228	$290	$324,171,888	$412,397,571

[a]Benefits may not equal due to rounding of CAV.

ornamental pear trees in the high-damage region (Table 13.3). In the low-damage region the total benefits are substantially lower and range from $13 million to $16 million for ash trees and $7 million to $10 million for ornamental pear trees.

Total estimated benefits from the biological control program range between $324 million at wholesale and $412 million at retail replacement costs (Table 13.3). Over three-quarters of the economic benefits are from preserving the scenic beauty of ash trees in the high-damage region, and ash trees in both regions combined account for over 80 percent of total benefits. Ornamental pear trees in the high-damage region account for 16.6 percent of the total economic benefits, whereas in the low-damage region they account for only an additional 2.4 percent.

As stated earlier, these benefits represent a one-time change in the aesthetic beauty of the host trees that is achieved when the ash whitefly populations are at their highest in early fall and defoliation is greatest. This was the situation when the parasitic wasp was released. Had a viable biological control agent not been found, stress on trees from the feeding and defoliation would have led to tree death over time, and benefits would be greater.

The direct costs of the ash whitefly biological control program totaled $1,224,342 (Table 13.4). The net benefits (total benefits less total costs) are between $323 million at wholesale values and $411 million at retail values. The

rate of return for each dollar spent to import, rear, release, and monitor the parasitic wasp is between $265 and $337. If the overhead costs of the biological control program and the long-term research costs are also included, total costs would be higher and the rate of return would be lower. Had the additional trees been located in parks, golf courses, public areas, and residences been included, the benefits would increase and the rate of return would be higher.

Additional Costs and Benefits of Using Biological Pest Controls The survey of California agricultural commissioners included questions on the public response to the introductory biological control program (Jetter and Klonsky 1994). In general, people were highly supportive of the program. What they were most satisfied about depended upon whether they lived in a high- or low-damage region and how long the ash whitefly had been present before being controlled. People who lived in the hotter region and who had experienced the damage the longest (mostly residents in Southern Califor-

Table 13.4 Ash whitefly biological control program costs

Item	Costs ($)
Salary	772,492
Collection and importation of parasite	4,000
Rearing and monitoring	447,850
Total costs	1,224,342

nia) expressed more satisfaction over the ash whitefly being controlled than how it was controlled. Experiencing the effects of large infestations over an extended period left many residents in that area impatient for control. Many purchased parasitic wasps for release in their own trees in lieu of waiting for them to spread on their own. While supportive of the use of biological controls, they would also have been satisfied if control had been achieved with chemical pesticides.

People who live in a coastal county expressed greater satisfaction from the use of a biological alternative than in just achieving control. In this region the negative effects of the ash whitefly were milder than in Southern California. Also, the time between the establishment of the ash whitefly and the establishment of the parasitoid was much shorter, so people endured the ash whitefly for less time. Finally, as the wasp spread from the south, scientists and county agricultural commissioners were able to relay more information to the public on its effects in California.

Some people expressed disapproval over the importation and distribution of a parasitic wasp to control ash whitefly. In general, their disapproval centered on perceived rather than actual possible negative effects. The main concerns were fear of wasp stings and what would happen to the parasitic wasp after the ash whiteflies were gone. Once informed that the wasp was stingless and so tiny that it was barely visible to the naked eye, most people had no problem with releases of the parasitoid.

The concern over what would happen when ash whiteflies were gone was addressed by explaining that the ash whitefly is never entirely gone. It continues to exist in equilibrium with the natural enemy at levels that do not cause visible damage. It was also explained that the parasitic wasp is adapted only to the biology of the ash whitefly and cannot survive on any other insects. While a handful of people continued to remain skeptical, most concerns were answered satisfactorily, and the ash whitefly biological control program received widespread public support.

General Discussion

The cost of arthropod pest introductions is difficult to calculate, but may include loss of plant value (aesthetic quality, value-added to real property, loss of appeal/value as a nursery product, etc.), plant removal costs, plant replacement costs, and less-suitable alternative plant choices for specific uses. The ash whitefly program represents an excellent example of how a permanent, cost-effective, and environmentally acceptable solution can be implemented when an exotic pest insect has been introduced into an urban environment.

The successful biological control program is the result of effective collaboration between universities, government, agricultural industries, and homeowners. However, the case of ash whitefly also demonstrates a failure to identify the pathways through which exotic pest introductions of urban landscape plants occur. In the intervening 10 years, there have been introductions of more than a dozen important insect pests of woody ornamental plants and landscape trees into California.

One problem is that the collaboration between diverse groups that existed during the ash whitefly program is not always assured. The ornamental nursery industry in California is made up of producers of extreme diversity, thus they are difficult to organize into a collective voice to address issues of regulations and pathways. To initiate effective interception programs or institute appropriate preventative efforts to limit introductions of exotic urban pests, it is important to develop risk assessments for potential pests of critical landscape plants.

Conflicts can arise when the most effective control is not feasible. For example, the penalty for airline passengers violating quarantine regulations in Australia is AU$100,000. The high penalty provides a more effective deterrent than the $2,500 penalty for violating U.S. quarantines. Stiffer U.S. penalties would not increase program costs and may increase the effectiveness of quarantines, yet a $100,000 penalty would be considered excessive by most in the United States.

Effective monitoring and interception efforts have been implemented for some high-profile insects (e.g., gypsy moth). However, the vast majority of insects are not particular problems within their native range because their populations are regulated by biotic and abiotic factors. Thus, it is difficult to predict potential pest status if introduced without natural controls and, consequently, virtually impossible to develop focused monitoring programs aimed at millions of individual species. There have been limited efforts to develop a priority list of potential pests with a high risk of introduction; they have focused primarily on key plants. This model of first identifying the plant species at risk and then determining potential pest introduction has been implemented by forestry agencies and appears to be working well for assessing risk of introductions of exotic timber pests. With increased movement of plants, plant parts, and people in California's expanding global trade networks, a plant-based (or phytocentric) pre-evaluation of risk may provide new insights for limiting the introduction of exotic insect and disease pests.

References

Bellows, Jr., T.S., T.D. Paine, K.Y. Arakawa, C. Meisenbacher, P. Leddy, and J. Kabashima. 1990. "Biological Control Sought for Ash Whitefly." *California Agriculture*. 44:4–6.

Bellows, Jr., T.S., T.D. Paine, J.R. Gould, L.G. Bezark, J.C. Ball, W. Bentley, R. Coviello, A.J. Downer, P. Elam, D. Flaherty, et al. 1992a. "Biological Control of Ash Whitefly: A Success in Progress." *California Agriculture*. 46:24-28.

Bellows, Jr., T.S., T.D. Paine, and D. Gerling. 1992b. "Development, Survival, Longevity, and Fecundity of *Clitostethus arcuatus* (Coleoptera: Coccinellidae) on *Siphoninus phillyreae* (Homoptera: Aleyrodidae) in the Laboratory." *Environmental Entomology*. 21:659-663.

Gould, J.R., T.S. Bellows, Jr., and T.D. Paine. 1992a. "Population Dynamics of *Siphoninus phillyreae* (Haliday) in California in the Presence and Absence of a Parasitoid, *Encarsia partenopea*." *Ecological Entomology*. 17:127–134.

Gould, J.R., T.S. Bellows, Jr., and T.D. Paine. 1992b. "Evaluation of Biological Control of *Siphoninus phillyreae* (Haliday) by the Parasitoid, *Encarsia partenopea* (Walker), Using Life-Table Analysis." *Biological Control*. 2:257–265.

Gould, J.R., T.S. Bellows, Jr., and T.D. Paine. 1995. "Preimaginal Development and Adult Longevity and Fecundity of *Encarsia inaron* (Walker) (=*Encarsia partenopea* Masi) (Hymenoptera: Aphelinidae) Parasitizing *Siphoninus phillyreae* (Haliday) (Homoptera: Aleyrodidae)." *Entomophaga*. 40:55-68.

ISA (International Society of Arboriculture). 1992. *Guide for Plant Appraisal*. Urbana, IL: International Society of Arboriculture. 150 pp.

Jetter, K., and K. Klonsky. 1994. "Economic Assessment of the ash whitefly (*Siphoninus phillyreae*) biological control Program." Final Report to California Department of Food and Agriculture. California Department of Food and Agriculture, Sacramento, CA.

Leddy, P.M., T.D. Paine, and T.S. Bellows. 1993. "Ovipositional Preference of *Siphoninus phillyreae* and Its Fitness on Seven Host Plant Species." *Entomologia Experimentalis et Applicata*. 68:43–50.

Leddy, P.M., T.D. Paine, and T.S. Bellows. 1995. "Biology of *Siphoninus phillyreae* (Haliday) (Homoptera: Aleyrodidae) and Its Relationship to Temperature." *Environmental Entomology*. 24:380–386.

Nowak, David J. 1993. "Compensatory Value of an Urban Forest: An Application of the Tree-Value Formula." *Journal of Arboriculture*. 19(3):173–177.

Pickett, C.H., J.C. Ball, K.C. Casanave, K.M. Klonsky, K.M. Jetter, L.G. Bezark, and S.E. Schoenig. 1996. "Establishment of the Ash Whitefly Parasitoid *Encarsia inaron* (Walker) and Its Economic Benefit to Ornamental Street Trees in California." *Biological Control*. 6:260–272.

14

Economic Consequences of a New Exotic Pest: The Introduction of Rice Blast Disease in California

Jung-Sup Choi, Daniel A. Sumner, Robert K. Webster, and Christopher A. Greer

Introduction

After many years in which it was thought to be unsuited to California conditions, rice blast disease was detected for the first time in California in 1996 (Greer et al. 1997). It spread to a larger area in 1997. The severity of the disease was limited in 1998 and 1999, but it continues to have the potential to become the most serious disease affecting California rice.

This chapter measures the economic effects of introducing rice blast disease in California. We assess the economic impact of rice blast on the price and quantity of rice production and related economic variables. We also analyze the economic benefits and costs of integrated blast control measures. Our analysis provides a case study of assessing economic consequences of the introduction of an exotic pest. We isolate private cost issues, industry-wide public goods, and the impact on producers and consumers. The results of this analysis may provide a basis for a better understanding of the economic prospects and for public policy or industry-wide actions that could mitigate some of the negative consequences of exotic diseases.

Industry Context

Approximately 500,000 acres are planted annually to rice in California. The total annual market revenue generated by California's rice sales has varied between $180 and $338 million during the last 10 years, with government payments adding $88 million to $161 million more (Sumner and Lee 1998). California's rice production is concentrated in the Sacramento Val-

ley, which accounted for 92 percent of the state's production in 1998 (California Agricultural Statistical Service 1999). California is one of the few regions in the world that has the capability of producing and exporting high-quality Japonica rice (Sumner and Lee 2000).

In addition to standard economic concerns, the rice industry is affected by a number of challenges, including environmental issues related to water and air quality (Lee, Sumner and Howitt 1999; Carter et al. 1992). The use of pesticides and herbicides in the production system is increasingly restricted. Disposal of crop residue is also highly regulated. Traditionally, rice straw was burned after harvest, but the already restricted rice straw burning is being phased down and is scheduled for elimination (except to protect against disease) by 2003 (Carey et al. 2000).

Rice Blast Disease in California

Blast was first detected in California in the fall of 1996 in about 12,000 acres in Colusa and Glenn counties (Webster 1997). More extensive observations in 1997 showed that the infested areas encompassed more than 55,000 acres, with about 30,000 acres in Colusa County, 24,000 acres in Glenn County, and 1,200 acres in Sutter County. In 1998 the disease was less severe than in 1996 or 1997, but the infested area expanded further east and south (Webster 1999). In 1999, blast frequency and severity were about the same as in 1998 (Webster 1999), but the disease continued to spread and was found for the first time in Butte County.

Blast may affect several parts of the plant (Ou 1985). Yield losses may reach 50 percent, usually with reductions in quality of the grain that is produced. The blast pathogen infects all cultivars presently grown in California, with the relatively early maturing M-201 being the most susceptible. M-201 was planted to 9.2 percent of the California rice acreage in 1996 (Agricultural Experiment Station 1997), but as a response to this susceptibility, its share fell to 2.6 percent by 1999 (California Rice Commission 2000).

The causal organism, *Pyricularia grisea,* is a highly variable fungus with more than 200 documented physiological races, presenting difficult challenges to plant-breeding efforts to produce resistant cultivars. Thus far only one race is known to be established in California (Webster 1999). As a result, there is an industry-wide interest in ways to avoid the introduction of additional races.

Economic Analysis

Model

We now consider the effects of blast and blast control on prices, revenues, and welfare aggregates. The biological and other industry information indicates that supply-side effects of rice

blast appear as declines in yield per acre and lower milling yield per unit of paddy rice. Blast control improves yield per acre and milling yield toward the no-disease levels.

In Figure 14.1, introduction of blast causes the rice supply curve, SS, to shift left all the way to S'S' (arrow a). As a result, the price rises from P to P', and quantity falls from Q to Q'. The magnitudes of these impacts depend on the degree of shifts, of supply elasticity, the elasticity of demand curve, DD. Successful blast control measures move the supply curve back to S"S" (arrow b). This shift results in increased quantity from Q' to Q" and lower price from P' to P".

We developed an equilibrium-displacement simulation model to project changes in the economic impacts of blast on the rice industry based on ranges of variables for the biological variables and economic parameters. We focused on the intermediate-run impacts without examining the path of adjustment. The model provides a reasonable, yet simple, way to build in plausible ranges of estimates. The model requires specification of the underlying equations that represent supply, demand, and market equilibrium.

Consider the following equilibrium displacement model system specified in log-linear differential form:

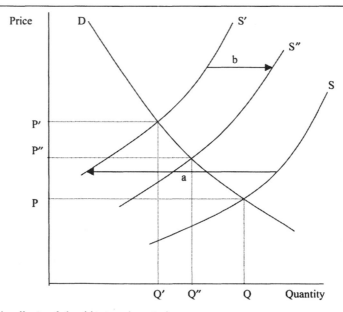

Figure 14.1 Supply effects of rice blast and control.

(14.1) dlnY = δ (change in milled rice per acre)

(14.2) εdlnL = εdlnP − εdlnC + εdlnY (acreage shifts)

(14.3) dlnS = εdlnP − εdlnC + (1 + εδ) (total quantity supplied)

(14.4) dlnD = − ηdlnP (quantity demanded)

(14.5) dlnS = dlnD (market clearing condition)

Change in yield per acre, Y, is represented by Equation 14.1, where δ denotes the percentage of change in the milled rice-adjusted paddy yield per acre. (Blast reduces head rice, and this effect is converted to a quantity effect by adjusting for the milling rate.) For our analysis, we assume that rice yield is not significantly responsive to price change over the ranges considered (Choi and Helmberger 1993). Equation 14.2 denotes planted area, L, as a function of price, P, and production costs, C, where ε is price elasticity of area planted. The elasticity of area with respect to marginal cost per acre is the negative value of ε under constant returns to scale. In this model, we allow interaction between acreage and yield, since rice farmers would adjust acreage when there is expectation of yield change. Market supply, S, is given by Equation 14.3, as the sum of percentage of change in acreage and percentage of change in yield from Equations 14.1 and 14.2. Equation 14.4 represents the market demand, D, as a function of price, where η is the absolute value of price elasticity of demand. Finally, Equation 14.5 is the market clearing condition.

The government program for rice provides substantial payments to growers, but these payments are not affected directly by market price or quantity of rice produced in California. This is especially true in the case of the California industry, because it is isolated from the rest of the industry and tends to face different market price patterns (Sumner and Lee 2000). Thus, government farm program payments are unaffected by the blast disease.

The simulation begins by noting that blast has an exogenous shock shown in Equation 14.1. The model is solved to get equilibrium values. In a second simulation, we note that

blast control measures undertaken by growers affect production costs and mitigate some of the yield impact. We then solve the equation system with new shocks to get the equilibrium values after incorporating control measures.

Solution of the equation system gives formulas for changes in price, acreage, equilibrium quantity, and revenue.

dlnP = {1/(ε + η)} {εdlnC − (1 + ε)δ}

dlnL = {ε/(ε + η)} {δ(η − 1) − ηdlnC}

dlnQ (=dlnS = dlnD) = {− η/(ε + η)} {εdlnC − (1 + εδ}, and

dln(PQ) = dlnP + dlnQ = {(1 − η/(ε + η)} {εdlnC − (1 + εδ}

Approximate changes in producer surplus and consumer surplus are:

ΔPS = (1 + dlnP)(1 + dlnQ)(K − Z)(1 + 0.5Zη) and

ΔCS = (1 + dlnP)(1 + dlnQ)Z(1 + 0.5Zη),

where Z = K{ε/(ε + η)} and K = − {εdlnC − (1 + εδ}

Note that the changes in producer surplus and consumer surplus are calculated as the ratio to the industry revenue (Alston and Larson 1993). Producer surplus loss represents the loss in net returns to growers and farm input suppliers. A consumer surplus loss represents the loss in net benefits to processing and marketing firms as well as to those who ultimately eat the rice.

To set the plausible range of parameters, we used biological data from experiments and observations in California. The following section explains the method used to acquire statewide data required for the estimation of bio-economic model.

Biological Data

Statewide Damage Caused by Rice Blast Disease Occurrence and severity of blast have been sporadic in the areas infested since its introduction into California. Some fields have lost 35 to 40 percent of yield whereas others have only had a trace of disease. For the first-order economic analysis, we need aggregate, industry-wide effects of blast on rice yield and quality. Yield loss depends on the disease severity in individual infested fields and the portion of infested plants and acreage. Estimates published by the California Rice Industry Association in 1997 indicate that, in California, the average reduction in yield on an infested field attributable to blast ranges from 15 to 30 percent. The geographic spread depends heavily on the ratio of acreage planted to cultivars that differ in susceptibility. Agricultural specialists believe that the total infested acreage may have reached 100,000 acres, or 20 percent of the total acreage. It is probable that the disease will continue to spread throughout the rice-producing area of the Sacramento Valley. Considering the yield reduction of infested fields and the total infested acreage, the statewide yield reduction from blast may reach 3 to 6 percent (i.e., 20 percent of 15 to 30 percent) (California Rice Industry Association 1997).

Rice blast also causes quality deterioration by uneven kernel moisture at harvest and smaller grain on infested plants. For the economic analysis, it is convenient to convert the quality deterioration, measured by the milling percentage of head rice and total rice, into the equivalent yield loss. Monitoring results of the impact of blast on rice yield and quality can be seen in Table 14.1 where blast reduced the head/total milling yield from 48/68 to as low as 29/59 (Webster 1997).

Losses in total and head-milling rate reduce rice value. Considering that the normal milling ratio for short/medium grain in California is 58/68, we find that blast reduces the value of rice produced in an infested field by 10 to 20

percent. Applying again the potential ratio of infection (20 percent), the average statewide additional yield loss attributable to the milling quality loss may reach between 2 to 4 percent (10 to 20 percent of 20 percent). Adding this to the statewide paddy yield per acre reduction from blast, i.e., 3 to 6 percent, the industry-wide, head-rice–adjusted yield loss ranges from 5 to 10 percent. We apply these figures as the potential direct effects of blast with no effort to control the disease.

Impacts of Blast Control Methods Blast control methods have both costs and benefits to the rice industry. Management of the blast disease requires an integrated approach, including the development of cultivars with improved resistance, manipulation of cultural practices, and the judicious use of fungicides. Rice blast control methods include:

1. destruction of infested crop residue, which limits initial inoculum but will not protect fields from airborne conidia from other fields;
2. planting pathogen-free seed, which aids in preventing spread of the pathogen into uninfested areas but will not protect fields from other inoculum sources;
3. water seeding, which reduces disease transmission from seed to seedlings;
4. continuous flooding from planting to maturity, which minimizes blast;
5. assuring that nitrogen fertilizer levels do not exceed levels needed for maximum yield;
6. fungicide applications to minimize reductions in yield and quality in years when conditions favor blast;
7. resistant cultivars (it will take several

Table 14.1 Impact of rice blast disease on yield, grain moisture, and quality: results from blast infested fields

Field	Cultivar	Yield (cwt, dry)	Moisture at harvest(%)	Milling quality (% head/% total)
1	M-202	71.18	18.6	45/66
2	M-201	59.08	13.6	29/59
3	M-201	44.03	13.9	29/60
4	M-201	50.38	14.7	45/65
5	M-201	63.59	15.4	48/68
6	SP-211	61.40	15.7	41/65
7	M-201	62.05	12.7	41/64

Source: R.K. Webster (1997).

years to incorporate resistance into California cultivars); and

8. monitoring seed imports to reduce the chance that additional races of the blast pathogen enter California.

Among the control methods, the application of fungicide and the development of resistant cultivars are most costly. Other control methods may incur additional costs in the form of changes in cultural practices. The currently available fungicide in California (Quadris) costs about $40 per acre, including application costs (California Rice Industry Association 1997). The efficacy is not known with certainty, but the experimental results show that Quadris lowers the yield loss by 80 to 90 percent with two applications (about $80 per acre) per year. (Refer to Table 14.2 where we present 1998 data from Webster.)

Another costly control method is the breeding of resistant cultivars. Costs required to develop and maintain blast-resistant cultivars are estimated at $508,000 annually (Brandon 1998). Spread over 500,000 acres, this amounts to about $1 per acre. Considering that per acre sample cost to produce rice in 1998 was about $842 (Williams et al. 1998), blast control costs by individual growers may reach between 5 to 10 percent of production costs per acre, depending on the intensity of control methods. Those costs include the chemical and application costs and breeding costs. Experiments show that these control methods may recover the quality-adjusted yield up to 80 percent (with one application of chemical) and 90 percent (with two applications of chemical) of the original yield.

In the simulation, quality-adjusted yield losses of 5, 10, and 15 percent (the parameter δ in Equation 14.1) are considered as plausible cases. The increase in per acre production costs from the control for blast (parameter dln in Equation 14.2) is postulated as 5 and 10 percent of the production costs. The benefit of blast control is the reduction in yield loss. We examine the cases for which yield loss is decreased by 80 percent, with a 5 percent per acre cost increase, and by 90 percent, with a 10 percent per acre cost increase.

Economic Parameters

For the simulation, we also need the supply and demand elasticities. We use the acreage supply elasticity of 0.5 and 1.0. McDonald and Sumner (1998) reviewed this literature (Chen and Ito 1992; Cramer et al. 1990; and Salassi 1995) and showed that estimates using data over the period for which farm programs restricted acreage allocations are biased downward for use during the period since 1996 when rice farmers have enjoyed more planting flexibility. They suggest a supply elasticity of 1.0 for California, with no farm program restrictions. The current program does keep some limits on what can be grown on rice base and maintains the marketing loan program, so we use 0.5 as a lower figure for the rice supply elasticity. The final parameter needed is demand elasticity facing California rice producers. We use values of -2.0, -4.0 and -6.0. It is generally accepted that the overall demand for rice is inelastic (say, -0.2) in almost all markets (Song and Carter 1996). But, for this discussion we use the demand facing California rice producers. The higher range of demand elasticities reflects the small share of California in the domestic market (less than 20 percent), the small share of California in international markets, and the extreme inelastic nature of demand in Japan under World Trade Or-

Table 14.2 Efficacy of fungicide on rice blast disease in California: Comparisons of treated and untreated fields

Field	Cultivar	Fungicide	Yield cwt/acre	% Grain moisture	Milling quality (% Head/% Total)
10	M-202	Untreated	88.3	15.3	43/69
	M-202	Quadris	100.8	19.9	64/70
11	M-202	Untreated	88.8	16.8	62/68
	M-202	Quadris	97.0	21.0	66/70

Source: R.K. Webster (1998).

ganization–imposed import quotas (Sumner and Lee 2000).

Simulation Results

This section reports on several simulations using the data and estimates discussed above. We provide ranges for most estimates to reflect uncertainties about parameters.

Impacts of Introduction of Blast with No Effort to Control Table 14.3 provides the simulated effect of rice blast on the rice industry under alternative parameters. For example, consider the case under which blast reduces yield by 10 percent when supply elasticity is 1.0 and demand elasticity is –2.0 (the 10th row of Table 14.3). In this case, the price of rice increases by 6.7 percent and the equilibrium quantity falls by 13.3 percent. As a result, the industry gross revenue falls by 6.7 percent. With a higher acreage-supply elasticity, price increases more and quantity falls more. With higher demand elasticity, price falls less and quantity falls more. Depending on the scenario, the industry revenue falls by between $8.6 million and $61.6 million (3.0 percent and 21.4 percent of the total annual market revenue of

about $288 million). The ratio of the change in the producer surplus to the initial revenue is 11.5 percent in the base-case scenario (10th row). The ratio of the change in the consumer surplus to the initial revenue is 5.8 percent (Table 14.3, row 10). Table 14.3 presents a wide range of results reflecting the realistic uncertainty about the precise costs of the disease and the supply and demand elasticities.

Impacts of the Recommended Control Strategy Table 14.4 shows simulated impacts of the disease when growers attempt control using conventional measures. Again, consider the case where yield loss is 10 percent, supply elasticity is 1.0 and demand elasticity is –2.0, and blast control measures add 5 percent to production cost and have 80 percent efficacy. In this case industry total revenue falls by 3.0 percent as the result of 3.0 percent price increase and 6.0 percent equilibrium quantity fall. The change in producer surplus as ratio of the initial industry revenue is –5.6. But note this is a smaller producer surplus loss than the loss of 11.5 percent when there is no attempt at control. Consumer surplus loss is now 2.8 percent of total revenue, so there is a welfare gain for both consumers and producers relative to no at-

Table 14.3 Simulated result: measuring costs of rice blast disease

| Yield loss | Acreage elasticity | Demand elasticity | Percentage of change in | | | | ΔPS[a] | ΔCS[a] |
			Acreage	Price	Quantity	Revenue		
–5	0.5	–2.0	–1.0	3.0	–6.0	–3.0	–5.71.3	–1.4
		–4.0	–1.7	1.7	–6.7	–5.0	–6.2	–0.8
		–6.0	–1.9	1.2	–6.9	–5.8	–6.4	–0.5
	1.0	–2.0	–1.7	3.3	–6.7	–3.3	–6.2	–3.1
		–4.0	–3.0	2.0	–8.0	–6.0	–7.2	–1.8
		–6.0	–3.6	1.4	–8.6	–7.1	–7.6	–1.3
–10	0.5	–2.0	–2.0	6.0	–12.0	–6.0	–10.9	–2.7
		–4.0	–3.3	3.3	–13.3	–10.0	–11.5	–1.4
		–6.0	–3.8	2.3	–13.8	–11.5	–11.8	–1.0
	1.0	–2.0	–3.3	6.7	–13.3	–6.7	–11.5	–5.8
		–4.0	–6.0	4.0	–16.0	–12.0	–12.9	–3.2
		–6.0	–7.1	2.9	–17.1	–14.3	–13.4	–2.2
–15	0.5	–2.0	–3.0	9.0	–18.0	–9.0	–15.4	–3.8
		–4.0	–5.0	5.0	–20.0	–15.0	–16.0	–2.0
		–6.0	–5.8	3.5	–20.8	–17.3	–16.1	–1.3
	1.0	–2.0	–5.0	10.0	–20.0	–10.0	–15.8	–7.9
		–4.0	–9.0	6.0	–24.0	–18.0	–17.0	–4.3
		–6.0	–10.7	4.3	–25.7	–21.4	–17.4	–2.9

[a]Change in producer surplus (ΔPS) or change in consumer surplus (ΔCS) is calculated as the ratio to the total industry revenue.

tempt at control. Using blast control measures that increase production costs by 10 percent and have 90 percent efficacy implies a reduction of industry revenue by 4.0 percent. In this case, the impact of blast and the control measures together reduce producer surplus by 7.3 percent of industry revenue. Again, note control results in a higher producer surplus relative to the loss of producer surplus of 11.5 percent when there is no attempt to control. However, this level of control reduces both producer and consumer surplus relative to the more moderate control measures that cost only 5 percent of cultural costs. Table 14.4 shows that the higher control costs (10 percent of cultural costs) do not pay for themselves in terms of producer surplus gains. Also, with higher disease severity (higher yield losses), there are larger producer surplus losses. The changes in producer and consumer surplus are in the same direction across these alternatives.

Overall, with appropriate control measures in place, the introduction of rice blast disease has cost the industry annual producer surplus losses of from $12.7 million to $19.6 million (4.4 percent to 6.8 percent of base total revenue). We consider the annual loss of $16.1 million in producer surplus as the base scenario. Consumer surplus losses range from $6.3 million to $9.8 million, with an annual loss of $8.1 million as the base scenario.

Public Policy for Eradication

Since blast disease was only recently introduced to California, it is natural to consider the possibility of eradication. However, a brief review indicates that the full economic costs of eradication may exceed the potential benefits, even if eradication could be achieved. Blast is now distributed over at least 100,000 acres and occurs sporadically and at differing severities, depending on environmental conditions and cultural practices. The only potential method of eradication would be to eliminate rice production in the whole region where blast has occurred, create a buffer zone around the region, and burn all levees and associated areas where straw residue might occur. Such a ban on production would need to last for three seasons. The total area affected would be approximately 40 percent of the whole rice-growing area in California.

To consider the costs of this radical eradication scenario, we consider the effects of cutting rice production to 300,000 acres each year for three years. Using a demand elasticity of −2.0, the price of California rice would rise by 20 percent with a 40 percent cut in quantity supplied. Producers would gain an additional $34.6 million in revenue from the higher price. Against this benefit producers would forego producer surplus of between $23 million and $46.1 million on the 200,000 acres that was left

Table 14.4 Changes in economic indicators by the rice blast control effectiveness

| Yield | Control costs | Yield with control | Percentage of change in | | | | ΔPS | ΔCS |
			Acreage	Price	Quantity	Revenue		
−5	0	n.a.	−1.7	3.3	−6.7	−3.3	−6.2	−3.1
	5	−1.0	−3.7	2.3	−4.7	−2.3	−4.4	−2.2
	10	−0.5	−6.8	3.7	−7.3	−3.7	−6.8	−3.4
−10	0	n.a.	−3.3	6.7	−13.3	−6.7	−11.5	−5.8
	5	−2.0	−4.0	3.0	−6.0	−3.0	−5.6	−2.8
	10	−1.0	−7.0	4.0	−8.0	−4.0	−7.3	−3.7
−15	0	n.a.	−5.0	10.0	−20.0	−10.0	−15.8	−7.9
	5	−3.0	−4.3	3.7	−7.3	−3.7	−6.8	−3.4
	10	−1.5	−7.2	4.3	−8.7	−4.3	−7.9	−4.0

1) n.a. is not applicable.

2) Values in the table are for a supply elasticity of 1.0 and demand elasticity of −2.0.

3) Control costs of 5 percent and 10 percent of per acre cultural costs are assumed to reduce yield loss by 80 percent and 90 percent, respectively.

4) Change in producer surplus or change in consumer surplus is calculated as the ratio of the total industry revenue.

out of production, depending on the cost of production associated with the land removed. In the case with smaller losses, industry-wide producer surplus would actually rise by $11.5 million per year because the higher price on the remaining production would more than offset the loss of producer surplus on the foregone output. In the case of higher losses on foregone production, producers in aggregate would lose a net $11.5 million under eradication. Of course, the consumer surplus would fall both because of lost benefit on the foregone output and the higher price. Consumer surplus loss is $46.1 million per year. The net loss of producer and consumer surplus together ranges from $34.6 million per year to $57.7 million for each of three years. Using the central figure of 46.1, this loss is approximately $138.3 million. To this we should add costs for administration of the program and for burning the area to eliminate blast in volunteer rice and weeds. Burning costs no more than $5 per acre, so these costs are likely to be only about $3 million for three years. Administrative costs are also likely to be no more than a few million dollars per year. The total direct and indirect costs are in the range of $150 million.

Benefits of eradication are reversing the losses of living with blast disease that are outlined above. Under the base case outlined above, we estimate those costs as a producer surplus loss of $16.1 million per year and a consumer surplus loss of $8.1 million per year. Remember, these losses include the direct cost of control, the effects associated with the remaining yield loss, and the equilibrium price and quantity impacts. Again, using a 5 percent interest rate, but now using an infinite horizon to reflect the fact that the disease is permanent, we get a capital value of producer surplus loss of $322 million and a consumer surplus loss of $162 million for a total welfare loss of $482 million.

This calculation suggests that if the radical approach on 200,000 acres would permanently eliminate rice blast from California, the economic impact would be strongly positive as a capital investment at a 5 percent interest rate. Indeed, the rate of return to investing in blast eradication is about 20 percent.

Despite these calculations there has been no serious discussion of eradication as a response to this exotic pest infestation. Why? A main reason that eradication has not been attempted is likely to be the uncertainty about success of the eradication program or about the success in keeping the disease out if it were successfully eradicated. However, given the calculations above, even if the probability of success were only 50 percent, the investment would still pay a relatively high expected rate of return of about 10 percent.

A second reason that eradication has not been attempted is likely that many growers and others in the industry may not expect the significant price increase and, therefore, the producer surplus gain from the lower output that would follow from eliminating rice production on 200,000 acres. Third, a very significant part of rice farm income derives from government payments. The industry would need to carefully assess how an eradication program would affect eligibility for program payments. The calculations above assume that government program payments would be unaffected by the introduction of blast disease or by eradication. That may be true, but farmers and landlords would need assurance that their current and future eligibility for payments was not jeopardized by taking land out of production to comply with eradication rules. This may require a designation of land withdrawn as officially prevented from being planted as defined in the applicable farm program. Fourth, there may be environmental objections to the field burning required for successful eradication. However, given the reduction in pesticide use that would follow from successful eradication, field burning would likely be a manageable issue.

An additional concern that makes an eradication program difficult is the complicated distribution of costs and benefits. Growers and landowners who had land removed from production would need to be compensated. Growers and landlords who had land remaining in production would gain substantially from the higher market price. Given uncertainty inherent in commodity markets, it would be complex to decide the amount of transfer between the two groups and to develop a mechanism to accomplish the transfer. Furthermore, with eradication and idled acreage in specific locations, there would be losses among processors that would not be distributed equally. Therefore, even though eradication would cost consumer surplus less in present-value terms than allowing blast to continue, significant distribution issues arise here as well. The result is that blast is now treated like any other established pest or disease that is dealt with by individual growers.

Nonetheless, at least three major public policies related to rice blast disease are important. The first relates to regulations on pesticides used to control the disease. The only chemical presently approved for use in California is effective for control, but it is also relatively expensive. Regulators are faced with balancing environmental concerns with economic health of the industry. Careful but expeditious regulatory review of chemicals and other control methods is particularly appropriate in this case. Second, research on blast is also important and may demand public or industry-wide support. This may include development of new methods of chemical control and, especially, resistant cultivars. Efforts to develop resistant cultivars are considered the most important and profitable industry-wide response to blast. The industry has already begun funding such research. Although blast seems to be established in California, thus far only one pathogenic strain of the fungus is known to be present. This should facilitate the development of resistant cultivars and further reductions in losses. New races of blast pathogen are likely to evolve, making breeding resistance an ongoing project.

Research to find resistant cultivars emphasizes the need for measures to ensure that additional pathogenic strains of the blast fungus are not introduced into California. This means that it is necessary to continue monitoring and limiting seed imports or other possible means of introduction. Thus, while blast is established, government restrictions remain to protect from further infestations with new strains.

References

Agricultural Experiment Station, UC Davis. 1997. "Agronomy Progress Report." No. 257.

Alston, J.M., and D.M. Larson. 1993. "Hicksian vs. Marshallian Welfare Measures: Why Do We Do What We Do?" *American Journal of Agricultural Economics.* 75(August):764–769.

Brandon, M. 1998. California Rice Experiment Station. Personal communication, October, 1998.

California Agricultural Statistics Service. 1999. *Agricultural Commissioners' Data.* (August):60-61.

California Rice Commission. 2000. "California Rice Statistics and Related National and International Data." Statistical Report.

California Rice Industry Association. 1997. "Emergency Exemption Request from California for the Use of Azoxystrobin (Quadris) on Rice to Control Rice Blast." Sacramento: California Rice Industry Association.

Carey, M., D. Sumner, and R. Howitt. 2000. "An Economic Analysis of Tradable Rice Straw Burn Credits." *Issues Brief.* No. 12. University of California Agricultural Issues Center (May).

Carter, H.O., et al. 1992. *Maintaining the Competitive Edge in California's Rice Industry.* Davis: University of California, Agricultural Issues Center.

Chen, D., and S. Ito. 1992. "Modeling Supply Response with Implicit Revenue Functions: A Policy-Switching Procedure for Rice." *American Journal of Agricultural Economics.* 74:186-196.

Choi, J., and P.G. Helmberger. 1993. "How Sensitive Are Crop Yields to Price Changes and Farm Programs?" *J. Agr. and Applied Econ.* 25(July):237-244.

Cramer, G.L., E.J. Wailes, B. Gardner, and W. Lin. 1990. "Regulation in the U.S. Rice Industry, 1965-1989." *American Journal of Agricultural Economics.* 72:1056-1065.

Greer, C.A., R.K. Webster, and S.C. Scardaci. 1997. "Rice Blast Disease Caused by *Pyracularia grisea* in California." *Plant Disease.* 81:1049.

Lee, Hyunok, D.A. Sumner, and R. Howitt. 1999. *Economic Impacts of Irrigation Water Cuts in the Sacramento Valley.* Davis: University of California, Agricultural Issues Center.

McDonald, J.D. and D.A. Sumner. 2003. In Press. "The Influence of Commodity Programs on Acreage Response to Market Price: With an Illustration Concerning Rice in the United States." *American Journal of Agricultural Economics.*

Ou, S.H. 1985. *Rice Diseases,* 2nd ed. Kew, United Kingdom: Commonwealth Mycological Institute.

Salassi, Michael E. 1995. "The Responsiveness of U.S. Rice Acreage to Price and Production Costs." *Journal of Agricultural and Applied Economics.* 27:386–399.

Song, J., and C.A. Carter. 1996. "Rice Trade Liberalization and Implications for U.S. Policy." *American Journal of Agricultural Economics.* 78(November):891–905.

Sumner, D.A., and H. Lee. 1998. *Economic Prospects of the California Rice Industry Approaching the 21st Century.* Sacramento: California Rice Promotion Board.

Sumner, D.A., and H. Lee. 2000. "Assessing the Effects of the WTO Agreement on Rice Markets: What Can We Learn from the First Five Years?" *American Journal of Agricultural Economics.* 82 (August):709–717.

Webster, R.K. 1997. "Cause and Control of Rice Diseases." *Annual Report of Comprehensive Rice Research*. California Rice Research Board Project No. RP-2. Davis: University of California and USDA.

Webster, R.K. 1998. "Investigations on Rice Blast Disease in California." *Annual Report of Comprehensive Rice Research*. California Rice Research Board Project No. RP-8. Davis: University of California and USDA.

Webster, R.K. 1999. "Investigations on Rice Blast Disease in California." *Annual Report of Comprehensive Rice Research*. California Rice Research Board Project No. RP-9. Davis: University of California and USDA.

Williams, J. et al. 1998 "Sample Costs to Produce Rice-Sacramento Valley, Rice Only Rotation." University of California Cooperative Extension. http://agronomy.ucdavis.edu/uccerice/bpric981. htm.

15

Biological Control of Yellow Starthistle

Karen M. Jetter, Joseph M. DiTomaso, Daniel J. Drake, Karen M. Klonsky,
Michael J. Pitcairn, and Daniel A. Sumner

Introduction

Yellow starthistle has been a problem in California for more than 70 years and is one of the most significant weed species in the state. Yellow starthistle is a winter annual widely distributed in the Central Valley and adjacent foothills of California. It is spreading in mountains below 7,000 feet and in the Coast Range but is less common in desert, high mountains, and moist coastal sites. It is typically found in full sunlight and deep, well-drained soils where annual rainfall is between 10 and 60 inches.

Yellow starthistle interferes with grazing and lowers yield and forage quality of rangeland, increasing the cost of managing livestock and lowering land values. Ranchers can treat yellow starthistle infestations; however, in many cases these degraded rangelands are unlikely to return to their preinfestation quality unless ranchers incur additional costs to restore the land. Restoration improves rangeland forage quality, increases annual rental rates and land values, but requires large capital outlays.

Yellow starthistle control and land restoration by private landowners also provides public benefits by preventing the spread of other invasive weeds and increasing water availability in watersheds. However, when the private costs of control are greater than the private benefits, adoption of private control techniques may be insufficient to realize the public benefits. Therefore, when private restoration results in large public benefits, public subsidies of private control may be beneficial for society if the public benefits are greater than the public subsidies.

Biology and Ecology

Reproduction

Yellow starthistle reproduces only by seed. Yellow starthistle seedheads have two types of seed. The dark outer ring lacks a pappus (the small hairs found on thistles and dandelions). Most seed, however, is brown with a bristly pappus ring at the apex. These seeds account for approximately 67 to 75 percent of the total seed production.

Plants typically begin flowering in early June and continue through October and even later in some areas. Almost all plants are self-incompatible and require pollen from a genetically compatible plant to produce seed. European honeybees are an important pollinator and in some populations are responsible for over 50 percent of the seed set. Bumblebees are the second most important visitors to starthistle flowers, but several other insects also contribute to fertilization. Cross-fertilization ensures a high degree of genetic variability within populations. Large plants can produce nearly 75,000 seeds. Seed production in heavily infested areas varies from 50 million to 200 million seeds per acre.

The time from flower initiation to the development of mature viable seed is only eight days. Thus, any late season control strategy such as hand weeding, mowing, herbicides, burning, or tillage must be performed earlier than eight days after flowering initiation to prevent seed production.

Germination and Seedbanks

Both seed types germinate over an extended time, beginning with the first fall rains and ending after the last seasonal rainfall in late spring or early summer. Provided that adequate moisture is available, germination can occur with 24 hours of imbibition (uptake of water by seed). Yellow starthistle seeds can germinate over a wide range of temperatures, but appear to depend on light. Germination responses are greatly reduced in dark environments and by exposure to light enriched in the far-red portion of the spectrum. Most seeds germinate with the first fall rains, usually in November. In exposed areas, high germination can result in extremely dense seedling populations. In central California, even in high-density situations, seedling survival usually exceeds 30 percent, provided adequate moisture is available.

The average longevity of yellow starthistle seeds in the soil is reported to be more than six years in Idaho. However, California studies show that the yellow starthistle seedbank is depleted by about 75 percent after a single year, 94 percent after two years, and between 96 and 99 percent after three years. Thus, in California it seems possible to deplete the seedbank of yellow starthistle within a few years, as long as new seed recruitment does not occur from neighboring populations or plants escaping control.

Growth

Roots Following germination, the growth strategy of yellow starthistle is to allocate resources initially to root growth, secondarily to leaf expansion, and finally to stem development and flower production. Root growth during the winter is rapid and can extend well beyond 4 feet in depth. During this same time period, rosettes (basal leaves) expand slowly. Rapid germination and deep root growth in yellow starthistle allow plants to avoid late season competition with other annual species and survive into late summer, long after seasonal rainfall has ended and annual grasses have senesced (i.e., produced viable seed and begun to die). By extending the period of resource availabili-

ty, competition is reduced at the reproductive stage. This can greatly benefit the plant by ensuring ample seed production.

Since the root systems of most annual species are comparatively shallow, there is little competition for moisture between yellow starthistle and annual grasses during late spring and early summer. In contrast, the use of soil moisture by yellow starthistle is similar to that of perennial grasses. Like yellow starthistle, perennial grasses have an extended growing season. Thus, they compete more for water with yellow starthistle than annual species.

Shoots Seeds typically germinate in late fall or early winter and overwinter as basal rosettes (leaves clustered near the soil surface and without an elongated stem, e.g., dandelion). Rosettes continue to develop throughout the early spring. In grasslands where rosettes are exposed to low light (for example due to shading by other vegetation), the leaves are larger and more erect, whereas developing leaves are flatter and more compacted in full sunlight. Dense starthistle seedling cover can significantly suppress the establishment of annual grasses and flowering plants. However, yellow starthistle rosettes are also very susceptible to light suppression and will produce short roots, larger leaves, more erect rosettes, and fewer flowers than plants in full sunlight. Consequently, survivorship and reproduction are significantly reduced in shaded areas, and yellow starthistle is less competitive in areas dominated by shrubs, trees, taller perennial flowering plants and grasses, or late season annuals. For this reason, infestations are nearly always restricted to disturbed sites or open grasslands dominated by annuals. Even in areas dominated by yellow starthistle, the level of competition for light can be so intense that seedlings will vigorously compete with each other, accounting for the low rate of seedling survival.

In the Central Valley and foothills of California, bolting (production of an elongated stem) typically occurs in April, and by May spines appear on developing seedheads. At the more mature stages of development, the soft down and waxy grayish coating on the foliage of yellow starthistle reflect a considerable amount of light. This reduces the heat load and the transpiration demand during the hot and dry summer months. The winged stems add surface area and also act to dissipate heat like a radiator. These characteristics, as well as a deep root system, allow yellow starthistle to thrive under full sunlight in hot and dry conditions. Vigorous shoot growth coincides with increased light availability due to senescence and desiccation of neighboring annual species. Moreover, the presence of spines on the bracts surrounding the seedhead protects against feeding by grasshoppers, seed predation by birds, and grazing by large animals (e.g., deer and cattle). This is particularly important during the vulnerable flowering and seed development stages.

Environmental Effects

In deep Central Valley soils of California, starthistle can significantly reduce soil moisture reserves to depths greater than 6 feet, and in 3-foot-deep foothill soils it can extract soil moisture from fissures in the bedrock (Gerlach et al. 1998). The ability of yellow starthistle roots to draw moisture from soils below competing vegetation has allowed this plant to readily invade California grasslands. Because the root systems of most annual species are comparatively shallow and actively grow during winter and early spring, there is little competition for moisture between yellow starthistle and annual grasses during late spring and summer.

Seasonal moisture can also influence the competition advantage between yellow starthistle and annual grasses. Under dry spring conditions, early maturing annual grasses have an advantage over late season annuals, because they use the available moisture and complete their life cycle before later-maturing species, such as starthistle (Larson and Sheley 1994). In contrast, under moderate or wet spring conditions, starthistle has an advantage by continuing its growth later into the summer and fall, and producing more seed.

Thus, in grassland systems, the greatest advantage for yellow starthistle occurs in areas with deep soil dominated by annual grasses and in years with moderate to wet spring rainfall (Sheley and Larson 1992). Under these conditions, yellow starthistle matures later, has increased seed production and has little competition for deep soil moisture. In annual grasslands, the least competitive situation for yellow starthistle is in areas with shallow soils and in years with low spring rainfall.

The characteristics that enable yellow starthistle to invade grasslands can threaten native species and ecosystem processes. Native species such as blue oak (*Quercus douglasii*) and purple needlegrass (*Nassella pulchra*) depend on summer soil moisture reserves for growth and survival (Gerlach et al. 1998). Yellow starthistle, however, uses deep soil moisture reserves earlier than blue oak or purple needlegrass. Thus, when starthistle infestations are high, native species can experience drought conditions, even in years with normal rainfall (Gerlach et al. 1998)

Introduction and Spread

Origin and Worldwide Distribution

Yellow starthistle is native to southeastern Europe and was first collected near Oakland, California, in 1869. It was most likely introduced after 1824 as a contaminant of alfalfa seed. Introductions prior to 1909 were most likely from Chile, whereas introductions from 1909 to 1927 appear to have been from Argentina, France, Italy, Spain, and Turkestan (Gerlach 1997).

By 1917 yellow starthistle had become a serious weed in the Sacramento Valley and was spreading rapidly along roads, trails, streams, ditches, overflow lands, and railroad right-of-ways. Yellow starthistle had spread to more than 1 million acres in California by the late 1950s and nearly 2 million acres by 1965. In 1985 it was estimated to cover almost 8 million acres in California (Maddox and Mayfield 1985). Today it is estimated to have spread to more than 12 million acres in California and can be found in 56 of the 58 counties in the state. Potentially, yellow starthistle expansion without significant changes in control efforts may extend to 40 million acres of California grasslands. Yellow starthistle is also a serious problem in eastern Oregon, eastern Washington, and Idaho, but not at levels of infestation common in California.

Modes of Dispersal

The pappus-bearing seeds are usually dispersed soon after the flowers senesce and drop their petals. However, nonpappus-bearing seeds can be retained in the seedhead into the winter. Wind does not contribute to long-distance dispersal of pappus-bearing yellow starthistle seed. Over 90 percent of the seeds fall within 2 feet of the parent plant, with very little seed dispersing beyond 30 feet. By comparison, birds such as pheasants, quail, house finches, and goldfinches feed heavily on yellow starthistle seeds and can disperse seeds long distances.

Human activities are the primary mechanisms for the long-distance movement of yellow starthistle seed. Seed is transported in large amounts by road maintenance equipment and on the undercarriage of vehicles. The movement of contaminated hay and uncertified seed are also important long-distance transportation mechanisms. Hay used as mulch along roadsides or disturbed areas can be a source of yellow starthistle introduction. The Bureau of Land Management, U.S. Forest Service, and the National Park Service have combined forces to institute hay certification programs in a number of states. This weed-free forage program is designed to reduce the spread of noxious weeds, particularly yellow starthistle.

Once at a new location, seed is transported in lesser amounts and over short to medium distances by animals and humans. The short, stiff, pappus bristles are covered with microscopic, stiff, appressed, hair-like barbs that readily adhere to clothing, hair, and fur. The pappus is not an effective long-distance wind dispersal mechanism, because wind dispersal moves seeds only a few feet [maximum wind dispersal 16 ft (<5 m) over bare ground with wind gusts of 25 miles/hour (40 km/hour)] (Roché 1992).

Intervention Strategies and Technologies

Quarantine, exclusion, and eradication are not possible with yellow starthistle. Populations are so widespread in California that the primary goal for public regulatory agencies, private industries, and people in most locations is to prevent new large-scale infestations and to manage existing populations. It is important to prevent large-scale infestations by controlling new invasions. Spot eradication is the least expensive and most effective method of preventing establishment of yellow starthistle in new areas; however, its effectiveness depends on rigorous widespread monitoring of grasslands and wilderness areas, as well as human-disturbed areas such as roads and housing developments.

Several yellow starthistle control methods have been developed in California, including tillage and mowing; animal management such as timed grazing by sheep, goats, and cattle; competitive planting of grasses and clovers to prevent seedling recruitment; large-acre burns; pre- and postemergent herbicides; and introduction of biological control agents. Development of these methods has been by private industries and public regulatory agencies, except for the biological control program. The activity of the released biological control agents is self-sustaining and not site specific; therefore, benefits spread regionally. As such, a successful biological control program is a public good. Because developing a biological control program is a public good, as well as a risk to nontarget species, it is undertaken only by public agencies. In established stands, any successful control strategy will require (1) dramatic reduction or, preferably, elimination of new seed production, (2) multiple years of management, and (3) follow-up treatment or a restoration program to prevent rapid reestablishment. Effective control using any of the available techniques depends on proper timing.

Mechanical Removal

Although tillage can control yellow starthistle, it will expose the soil for rapid reinfestation from newly germinating seeds if subsequent rainfall occurs. Under these conditions, repeated cultivation is necessary. During dry summer months, tillage practices designed to detach roots from shoots prior to seed production are very effective. For this reason, the weed is rarely a problem in agricultural crops. Weedeaters or mowing can also be used effectively. However, mowing too early, during the bolting or spiny stage, will allow increased light penetration and more vigorous plant growth and high seed production. Mowing is best when conducted at a stage where 2 to 5 percent of the total population of seedheads are flowering. Mowing beyond this period will not prevent seed production, because many flowerheads will already have produced viable seed. In addition, mowing is only successful when the lowest branches of plants are above the height of the mower blades. Under this condition, recovery is minimized.

Cultural Control

Yellow starthistle can be a good forage species when grazed at the bolting stage. Intensive grazing by sheep, goats, or cattle prior to the production of seedhead spines (spiny stage) but after bolting can also reduce seed production. To be effective, grazing needs to mimic mowing and accomplish the objectives described for mechanical removal (described above). Often, high densities of animals must be used for short durations. Grazing is best in May and June, but depends on the location. Revegetation with competitive perennial grasses and annual legumes can provide effective control of yellow starthistle in pastures and grasslands. The choice of legume or perennial grass species depends on the conditions at a particular site.

Prescribed Burning

Under certain conditions, burning can provide effective control and increase the survival and cover of native flowering plants and perennial grasses. Yellow starthistle control can be achieved most effectively by burning after native or desirable species have dispersed their seeds but before starthistle produces viable seed (June–July). The dried vegetation of senesced plants serves as fuel for the burn. Three consecutive annual burns have been shown to reduce the seedbank of yellow starthistle by 99.5 percent and provide 98 percent control of the weed, while increasing both native plant diversity and perennial grass cover.

Chemical Removal

Although several nonselective preemergence herbicides will control yellow starthistle, few of these can be used in rangeland or natural ecosystems. The exception, however, is chlorsulfuron, which provides very good control in winter when combined with a broadleaf selective postemergence compound. Although it is registered in most noncrop areas, it is not registered for use in pastures or rangeland. The primary options for control in noncrop areas are postemergence herbicides; 2,4-D, triclopyr, dicamba, and glyphosate. With the exception of glyphosate, these other compounds are selective on broadleaf species and are best applied in late winter or early spring to control starthistle seedlings. Once plants have reached the bolting stage, the most effective control can be achieved with glyphosate.

For early season control of yellow starthistle, the most effective compound is clopyralid (Transline). This herbicide has excellent pre-emergence and postemergence activity on yellow starthistle. In contrast to other broadleaf selective herbicides, clopyralid has a much narrower range of selectivity. It is primarily effective on members of the Asteraceae (sunflower family) and herbaceous members of the Fabaceae (pea family), but provides little control of most other broadleaf weeds. Excellent control can be achieved with applications from December through April. However, winter treatments may lead to significant increases in the quantity of other species, particularly grasses. When the grasses are good forage species, this may be desirable. On the other hand, in some cases undesirable noxious grasses may increase.

The continuous use of a single herbicide may lead to other problems. For example, legume species are important components of rangelands, pastures, and wildlands, and repeated clopyralid use over multiple years may have a long-term detrimental effect on their populations. Another possible drawback to the continuous use of clopyralid is the potential to select for other undesirable species, particularly annual grasses such as medusahead (*Taeniatherum caput-medusae*), ripgut brome (*Bromus diandrus*), or barb goatgrass (*Aegilops triuncialis*). Furthermore, a Washington population of yellow starthistle developed resistance to repeated use of picloram, and this population was also cross-resistant to clopyralid, which has a similar mode of action. Thus, the potential exists for the development of resistance to clopyralid if the herbicide is used year after year.

Biological Control

The U.S. Department of Agriculture's Agricultural Research Service (USDA-ARS) and the California Department of Food and Agriculture (CDFA) are pursuing biological control of yellow starthistle in a cooperative effort. Six exotic insects have been approved and released in California. Five of these have become established, and three have become widespread: *Bangasternus orientalis*, *Urophora sirunaseva*, and *Eustenopus villosus*. A fourth insect, the false peacock fly, *Chaetorellia succinea*, was accidentally introduced in 1991 in southern Oregon. It has a strong affinity to yellow starthistle and has dispersed throughout California. It is now found wherever yellow starthistle occurs. All of these exotic insects attack the seedheads of yellow starthistle and destroy developing seeds.

The gall fly (*Urophora sirunaseva*) and the bud weevil (*Bangasternus orientalis*) were the earliest insects released for the control of yellow starthistle and have been widely distributed in California by CDFA. Unfortunately, these insects have failed to build up populations to densities needed to damage a significant number of seedheads and alone will not control yellow starthistle. Because of their widespread distribution and limited impact on starthistle populations, distribution of these insects by CDFA has been discontinued.

The hairy weevil (*Eustenopus villosus*) was introduced into California in 1990 and has shown excellent potential. Its distribution is expanding rapidly through the efforts of CDFA. It is relatively sedentary and does not migrate long distances. Unlike the gall fly and the bud weevil, it has built up high populations that attack 50 to 70 percent of the seedheads. In addition, adult weevils feed on and damage a high percentage of developing flowerheads. However, attack by the hairy weevil is limited to June through August, whereas yellow starthistle flowers from June well into October in central California. While the hairy weevil activity coincides with peak flowering, the long flowering period allows plants to compensate for some of the seed loss by producing flowers in late summer, limiting the effectiveness of this insect.

The peacock fly (*Chaetorellia australis*) and the flower weevil (*Larinus curtus*) are recent introductions. The peacock fly emerges before yellow starthistle head buds are available and generally requires an early flowering secondary host (bachelor's button, *Centaurea cyanus*), to survive. Because bachelor's button has naturalized only in the extreme northern portion of California, this insect has not become widespread. The flower weevil has established in several locations throughout California, but its populations have failed to build to significant numbers (usually less than 10 percent of the seedheads are attacked). The false peacock fly was accidentally introduced with the peacock fly, but, unlike the peacock fly, the false peacock fly has a strong affinity for yellow starthistle. Because it is multivoltine (more than one generation per year) and adults oviposit in seedheads from May to October, it will potentially complement the hairy weevil by attacking late-season flowerheads. In addition, it is a very mobile insect that distributes quickly over large areas.

Unfortunately, even in areas where the hairy weevil and the false peacock fly have built up high population levels, no reduction in yellow starthistle abundance has been observed. It is estimated that as much as 50 to 60 percent of total seed production can be destroyed by these insects. For successful control, however, a reduction of 80 to 90 percent may be necessary. These exotic insects that attack the seedhead are the first group of natural enemies being used to develop a successful biological control. Research is under way to obtain new natural enemies that attack the roots and stems. The most promising candidates include *Ceratapion bassicorne*, a rosette weevil that attacks rosettes in early spring; *Puccinea jaceae* var. *solstitialis*, a rust disease that attacks the leaves and stems of young flowering plants; *Psilloides* nr. *chalcomera*, a flea beetle that attacks rosettes in early spring; and *Aceria* sp., a blister mite that galls the growing tips. These new natural enemies are undergoing host specificity studies to determine their safety for use as biological control agents. The rust disease is the furthest along, because petition for its release has been submitted and is pending approval by USDA-APHIS. The com-

Table 15.1 Exotic biological control agents released in California for control of yellow starthistle

Species	Common name	Status
Bangasternus orientalis	Bud weevil	Established—widespread
Urophora jaculata	Gall fly	Failed to establish
Urophora sirunaseva	Gall fly	Established—widespread
Eustenopus villosus	Hairy weevil	Established—widespread
Larinus curtus	Flower weevil	Established—limited
Chaetorellia australis	Peacock fly	Established—limited
Chaetorellia succinea	False peacock fly	Established—widespread (accidental release)

bined attack of seedhead insects and stem and root feeders will likely be necessary to achieve successful biological control of yellow starthistle in California.

Combinations

Integrating or rotating control methods into a management strategy can minimize the probability of selecting for other noxious weed species, developing herbicide resistance, or suppressing legume populations. For example, prescribed burning followed by spot application of postemergence herbicides to surviving plants can prevent the rapid reinfestation of the treated area. Similarly, combinations of mowing and grazing, revegetation and mowing, or herbicides and biological control may provide better control than any single method. In many cases, the most effective integrated approach includes a first-year prescribed burn followed by a second-year clopyralid treatment. Effective combinations may depend on the particular location or the objectives and restrictions imposed on land managers.

Potentially Affected Parties

Agriculture, land managers, recreationalists, homeowners, horse owners, ecosystems, and taxpayers are all potentially affected by yellow starthistle invasions. How each group is affected depends on how it uses, or its interests in, the land that becomes infested with yellow starthistle. Land in California is heterogeneous and used differently according to the characteristics of a particular plot of land. Some land is more fragile and susceptible to invasion by yellow starthistle and other invasive species. Land is also used for different economic activities such as ranching or recreation. Consequently, some land or activities are at higher risk of, or are more likely to have large economic costs associated with, yellow starthistle infestations than other areas.

The agricultural industry most seriously affected by yellow starthistle is ranching. Perennial grass sites will have a reduced carrying capacity from a shorter green forage season, reduced forage quality, and, perhaps, reduced forage quantity. These effects are exacerbated with increasing levels of infestation, at lower-quality sites characterized by shallow soils, and

in low rainfall areas or during period of drought. In addition, yellow starthistle thorns and often-impenetrable growth modify grazing behavior by restricting cattle access to other forage. Land management practices may have to change following yellow starthistle invasions by supplanting the area with higher-quality vegetation if cattle are to continue grazing the land.

Rangeland values are determined in part by the carrying capacity of the land. A reduction in carrying capacity causes the rental rate for land to decrease. As the rental rate decreases, land values decrease. Ranchers can treat the yellow starthistle infestations using a combination of methods; however, once the land is disturbed, it is unlikely that it will return to its predisturbance quality unless ranchers incur additional costs to restore the land.

Public land managers, homeowners, recreationalists, horse owners, and even farmers must also contend with the noxious effects of yellow starthistle. Heavy infestations prevent the movement of people and animals. Access to hiking trails is blocked, and animals have greater difficulty finding food and water sources. Consumption of yellow starthistle by horses leads to the development of brain lesions and, over time, death.

Control of yellow starthistle is generally expensive and can pose some risk to wildlife, homeowners, and land managers. For example, prescribed burning can decrease air quality, compromise establishment of biocontrol agents and the health of wildlife, and lead to catastrophic wildfires should a prescribed burn escape containment. Herbicides can contaminate water or lead to unwanted population shifts to other invasive species that also threaten grasslands.

When yellow starthistle invades new areas, it disrupts the native flora as well as the native insect species that depend on those plants or environments. Endangered species are particularly susceptible, and risk of their extinction may increase. While there are no reliable measurements of the value of changing the risk of extinction for endangered species, a significant value exists, because people are willing to donate time and money for the protection of endangered species.

Yellow starthistle may indirectly affect ecosystems due to its high water demands compared to native plants and annual grasses. An important source of water for urban, agricultural, environmental, and recreational purposes in California is runoff from mountain watersheds. Water draining from the mountains is captured in dams for use during California's long dry season. The depletion of soil moisture by yellow starthistle on invaded sites is equivalent to a loss of 15 to 25 percent of mean annual precipitation. Consequently, yellow starthistle infestations can create drought conditions even in years with normal rainfall (Gerlach et al. 1998). The consequences of a reduced water supply depend on the amount of precipitation received during the rainy season. Under drought conditions, water deliveries to agriculture are reduced, and secondary markets in water sales become active, raising the costs of agricultural production. Water levels in streams and lakes decrease, reducing the demand for recreational activities. Decreased stream flows may also reduce or delay spawning of anadromous fish and degrade fisheries water quality through effects of reduced flow on water temperature.

Finally, taxpayers incur costs for both regional and statewide control of yellow starthistle by public agencies on public lands, including costs of chemicals, prescribed burnings, and mowing. Taxpayers also fund the biological control program for statewide management of this noxious weed.

Substantial costs may be imposed by any of the potentially affected parties onto another when yellow starthistle is not controlled. If a rancher, public land manager, or homeowner does not control yellow starthistle, it will potentially spread onto the surrounding land, whether that is rangeland, a farm, roadside, or wilderness area. When high private control costs deter the control of yellow starthistle on private land, public subsidies may be worthwhile when private restoration results in large public benefits. In addition, if the public biological control program is successful, it will reduce private costs to restore rangeland and may also reduce the level of government subsidies needed to adopt rangeland restoration methods. With improved rangelands, the land may be less susceptible to invasion by other exotic species.

Policy Scenarios

The policy scenarios in this chapter focus on both public and private management of yellow

starthistle and interactions between the two groups. The first scenario examines the costs and benefits of a publicly supported biological control program. The second examines the effects of a successful biological control program on the private costs to remove yellow starthistle and restore rangeland in California's intermountain region. A statewide program to control yellow starthistle through a combination of prescribed burning and chemical controls is not viable in California. With a potential 12 million acres that would need to be treated, achieving statewide control is highly unrealistic. In addition, the ecological effects of widespread pesticide use and risks associated with burning wilderness areas would hinder public support for this type of control program. Finally, given the public benefits of improved watersheds, this analysis estimates the level of public subsidies needed to promote the adoption of private rangeland restoration programs in California's intermountain regions, with and without a biological control program.

Biological Control Program

Federal and state agencies have been involved in the research and dissemination of biological control agents for a number of years and have released several that have led to reductions in annual seed production by yellow starthistle (Pitcairn et al. 2002). While no one agent has shown potential to effectively control yellow starthistle in California, it is anticipated that introduction of the complete suite of necessary biological control agents will result in successful control of this weed. This study estimates the costs and benefits of a more intensive biological control program specifically designed for yellow starthistle.

A biological control program introducing host-specific biological agents may be a long-term solution to the problem of yellow starthistle infestations. Researchers are exploring new habitats for natural enemies in their areas of origin and determining the host specificity of newly identified natural enemies. However, research takes time and is expensive. Exploration of yellow starthistle habitats is complicated because the most successful natural enemies result in a scarcity of the weed in its native environment, making it difficult to find plants and control organisms. Once potential natural enemies are identified, it takes time to test them and determine what, if any, negative consequences for native plants or agricultural commodities would occur if the biocontrol agent was released.

The objective of a successful biological control program is to establish self-perpetuating populations of the biological control agents that will eventually spread throughout the infested region. Thus, all land infested with yellow starthistle is available to the natural enemies, so everyone in California benefits. Therefore, the costs will be compared to the benefits of removing yellow starthistle from rangelands, wilderness, roadsides, and other areas, and of preventing its spread onto uninfested but susceptible land. The total amount of infested and susceptible land is estimated to be 40 million acres.

Even with concerted efforts to identify viable biological control agents, at the end of the biological control program there is still some probability that the imported agents will only be able to diminish the severity of the California infestations instead of achieving complete control. Because some probability exists that the biological control program will not be successful, the expected benefits of the biological control program will be compared to the costs.

The Effects of a Biological Control Program on Private Rangeland Restoration Costs

Ranchers will adopt a rangeland restoration project only if their cost is less than their benefits from increased forage availability, higher forage quality, and greater accessibility. This analysis examines how a biological control program influences the private costs for rangeland improvements by comparing the change in rangeland values when yellow starthistle undergoes private control with chemical pesticides to the change in rangeland values when a biological control program is successful. A successful biological control program lowers restoration costs by removing the need to control yellow starthistle with an herbicide or other conventional control methods.

Land restoration is an investment decision in which the benefits (increases in annual rental rates) are capitalized into land values. Private investment in rangeland will take place if the

increase in land value is greater than the capital costs of improving land. As a result, changes in land values reflect both the costs and benefits of restoration (Plantinga and Miller 2001; Clark et al. 1993).

Public Policy To Subsidize Private Rangeland Control

When private costs to restore rangelands are greater than the private benefits, public agencies may choose to subsidize private landowners to increase the restoration adoption rate sufficiently to provide a public benefit such as improving watersheds and protecting native species. Land subsidy values are calculated as the amount that will leave a rancher indifferent between restoring land and doing nothing. It is equal to the loss in land values when the capital costs of improvement are greater than the benefits. Because the publicly supported biological control program influences the private costs, we examine the level of public subsidy needed with and without a successful biological control program.

Not included in this analysis are the potential benefits from the use of biological alternatives to chemical pesticides. Research has shown that willingness to pay for biological pest controls is significantly higher than chemical controls for the same pest (Jetter 1998). However, estimation of that value is beyond the scope of this project.

Methodology To Estimate the Yellow Starthistle Control and Rangeland Restoration Policies

Introductory Biological Control Program

The economic analysis of the biological control program consists of estimating the present value of the costs and benefits and comparing them to determine if the benefits of controlling yellow starthistle are greater than the costs.

Costs of a Statewide Introductory Biological Control Program The costs of the biological control program are for travel abroad to search for natural enemies, initial host testing, and importation from overseas. Within California, costs are for quarantine, host testing, release, and monitoring in the environment. This program is determined by scientists familiar with previous work on biological control of yellow starthistle as having a high probability of successfully achieving statewide control in 10 years. Because this is a multiyear project, future costs (*FC*) need to be discounted into current dollars to determine their present value (*PV*). The present value of future costs is calculated as

$$PV = \sum_{t} \frac{FC_t}{(1+r)^t}$$

where *r* is the discount rate and FC_t are future costs at time *t*, with *t* starting at 1 for the current year and continuing through year 10. An annual discount rate of 7 percent is used in the analysis.

Benefits of the Biological Control Program
Total benefits are calculated as the 40 million acres of land, either infested or susceptible to infestation, times the average benefit per acre. The average benefits per acre of managing yellow starthistle are calculated as only those accruing from eliminating this invasive species with no restoration of land. However, a problem arises because the benefits from controlling yellow starthistle where it is already established may not be equal to the benefits of preventing its spread into new areas. This potential asymmetry in benefit values exists because the removal of yellow starthistle does not necessarily result in the recovery of land to its preinfestation quality. Consequently, an average benefit per acre across both infested and susceptible land is used in the analysis. In addition, high and low average benefit values reflecting the wide variations in land quality in California are calculated.

The average benefit per acre is calculated as the difference in land value with and without yellow starthistle. This land valuation approach primarily captures benefits from changes in rangeland productivity, improved land access, changes in weed roadside management, and enhanced aesthetics. It does not capture benefits such as ecosystem preservation or improved watersheds and increased water availability to recreation, agriculture, etc. One study has estimated economic benefits of $16 to $75 million per year in the Sacramento River watershed alone should yellow starthistle be controlled in

annual grasslands (Gerlach et al. 1998). These additional benefits may be significant; however, insufficient data exist to estimate them for all of California.

Data on land values with and without yellow starthistle were obtained from interviews with agricultural land appraisers in counties with relatively high starthistle infestations, and with land purchasers for The Nature Conservancy, an organization that purchases land for wilderness restoration. In all cases the interviews revealed that the presence or absence of yellow starthistle may have some effect, but is not a significant factor in determining land values. This is due in part to the general degradation of range and other land throughout California.

The estimate of the lower-bound average benefit level per acre is largely for land already infested with yellow starthistle or for lower quality land. Based on the interviews with land appraisers, the difference in land values between degraded land and land infested with yellow starthistle is not easily measurable; however, if two parcels have identical prices and characteristics (terrain, water availability, etc.), except for the presence of yellow starthistle, the land without the starthistle would sell faster than the infested parcel. Therefore, the benefit of yellow starthistle removal is equal to the lost interest for the extra time it takes to sell the property with yellow starthistle. The minimum lost interest (*LI*) for a one-month delay in selling an infested parcel is calculated as

$$LI = \frac{r}{12} * \text{average land value}$$

where *r* is the annual interest rate. The one-month lost interest is equal to the lower-bound average benefit per acre.

The interest rate is equal to 7 percent. Total lower-bound benefits are calculated by multiplying all acres infested or susceptible to infestation by yellow starthistle by the lower-bound average benefit per acre.

The estimate of the upper-bound average benefit level per acre is based on the change in land values as yellow starthistle spreads and infests relatively higher-quality rangeland. The change in land values was provided by land appraisers and is based on appraisals completed before and after ranches became infested. Total upper-bound benefits are calculated by multiplying the 12 million infested acres by the lower average benefit per acre and the 28 million susceptible acres by the upper-bound average benefit per acre and adding the two sums together.

Cost/Benefit Analysis of the Public Biological Control Program

Even a well-funded biological control program may or may not result in a successful introduction of biological controls. Because the program may not be successful, the probability of success needs to be incorporated into the analysis.

If the probability of a successful program was known for the characteristics of that program, the expected benefits could be estimated directly by multiplying the total benefits by the probability of success. However, the exact probability of success for a yellow starthistle biological control program is not known. Given that the probability is unknown, a break-even probability at which the expected benefits will be equal to the costs is calculated by dividing the total costs by the total benefits at the upper and lower benefit level.

The break-even probability level is then compared to a qualitative assessment of the likely probability of success to determine if the project is economically feasible. Even though an exact value, or range of values, of the probability of success is unknown, scientists can provide a qualitative assessment as to whether a project has a high, medium, or low probability of success. In some instances, the estimated probability at which the expected benefits equal the costs is significantly different from the qualitative assessment and can therefore assist in decision making. In other cases, they will be close, and additional variables will need to be included to determine whether the expected benefits of a biological control program are likely to be greater than the costs. For the yellow starthistle biological control program, the funding level of $1.2 million per year for 10 years is estimated to result in a high probability of success.

Rangeland Restoration

The intermountain rangeland restoration scenario examines the effect of the biological control program on the private costs of controlling yellow starthistle and restoring the land for cattle grazing. Rangeland restoration is a four-year program involving herbicide treatments, reseeding, and changes in grazing patterns. In the first year, ranchers treat yellow starthistle infestations with clopyralid (Transline), other weed infestations with glyphosate (Round-up), and reseed with wheatgrass. Cattle are not grazed during this year. Treatments are completed the second year as needed. Based on weed infestations in the intermountain region, between 50 and 75 percent of seeded land will need to be retreated with herbicides the second year. Light grazing may be acceptable. During years three and four, the wheatgrass is becoming fully established. No further herbicide treatments are likely to be necessary. Grazing at about 75 percent of restored capacity may be done during this time. From year five onward the land may be fully grazed. Rangeland needs to be monitored once the perennial grasses become established to prevent overgrazing, which may lead to reinfestation by yellow starthistle and other invasive plants. However, with proper management, further yellow starthistle treatments should not be necessary. This method of restoration will work only in the intermountain region where there is some (3-4 inches) summer rainfall.

Rangeland Valuation Model

The price received for a parcel of land reflects the discounted stream of income that will be received over time on that land, less discounted costs incurred to obtain that income. The income received during each time period is equal to the rental rate received. A rangeland restoration program improves the quality of the land, mainly through lengthening the forage season, and increases the rental rate. Therefore, the stream of income received over time can be broken down into two periods. The first period is the time it takes to treat the rangeland, including the time the land cannot be grazed or may be only lightly grazed. The second period is when the land is fully restored and is grazed at sustainable levels. Total income from a plot of land is equal to the present value of income received in each year. Therefore, the income stream is capitalized into the land value as

$$\text{Total income} = \sum_{t=1}^{t^*} \frac{R_0}{(1+r)^t} + \sum_{t=t^*+1}^{\infty} \frac{R_1}{(1+r)^t}$$

where R_0 is the rental rate for degraded land, R_1 is the rental rate for the improved, restored land, r is the discount rate and t is the time period.

For this analysis $t*$ is equal to two years because cattle are grazed at 75 percent of restored rangeland capacity in years three and four.

Costs must be incurred to restore land. Total costs during each year of the restoration program are equal to weed treatment costs plus the opportunity cost of lost grazing. Because cattle cannot be grazed on land while it is being treated, landowners lose the income they would have received for renting out that land for grazing (this holds even when the landowners rent to themselves and graze their own cattle). The total costs of the restoration program are equal to the present value of yearly costs and capitalized into land values as

$$\text{Total rangeland restoration costs} = \sum_{t=1}^{T} \frac{C_t}{(1+r)^t}$$

where C_t is the annual cost of the rangeland restoration program and T is the total number of years it takes for the land to become fully restored. In this analysis T is equal to four years.

Land values are estimated as the total discounted income stream less the total discounted costs or

$$LV = \sum_{t=1}^{t*} \frac{R_0}{(1+r)^t} + \sum_{t=t*+1}^{\infty} \frac{R_1}{(1+r)^t} - \sum_{t=1}^{T} \frac{C_t}{(1+r)^t}$$

Solving land values may be calculated as

$$LV = R_0 \frac{1-1/(1+r)^{t*}}{r} + R_1 \frac{1/(1+r)^{t*+1}}{r} - \sum_{t=1}^{T} \frac{C_t}{(1+r)^t}$$

The increase in rental rate from R_0 to R_1 reflects the benefits to rangeland restoration. By normalizing R_0 to equal 0, R_1 is now only the increase in rental rates due to increases in carrying capacity, and the equation estimates the net change in land values associated with land restoration. For clarification, the increase in rental rates will be denoted as ΔR. The equation estimated in the analysis is

$$\text{Change in land values} = \Delta R \frac{1/(1+r)^{t*+1}}{r} - \sum_{t=1}^{T} \frac{C_t}{(1+r)^t}$$

If this value is negative, then the costs of restoring land are greater than the benefits, and ranchers would not adopt this program.

Estimating ΔR, the Benefits to Rangeland Restoration

The benefits to restoring rangeland are estimated as the increase in rental rates due to greater forage availability. The main increase in forage availability is from an extension of the time during which cattle may be grazed. A typical 100-acre pasture of yellow starthistle–infested rangeland in California's intermountain region supports 20 animal units (cow and calf pairs) for 2 months of grazing. Conversion of the site to perennial grasses would allow the same 20 pairs to graze for 3.5 months. The site improves from a carrying capacity of 30 acres per animal unit to 17 acres per animal unit.

Landowners are paid a grazing fee per animal unit (AU) per month. This fee is a random variable that depends on beef prices, forage, and such land characteristics as the terrain and availability of water. Consequently, grazing fees of $10, $15, and $20 are used in this analysis. These fees are typical for California's intermountain region for yellow starthistle–infested land.

Grazing fees must be converted into rental rates per acre. Rental rates are calculated as the total fees received by the 100-acre ranch for a year of grazing 20 animal units, divided by the 100 acres. Total fees are equal to 20 AU times the number of months the AU may be grazed times the fee per month. Subtracting the rental rate for infested land from the rental rate for restored land is the increase in rental rate per acre.

Additional benefits accrue to improved animal health from improved diet quality and reduced confinement feeding, but were not quantified. These are difficult to quantify. Furthermore, improved accessibility to land for management purposes and improved aesthetics due to reduced yellow starthistle is of benefit but also not quantified. In some cases, yellow starthistle infestations preclude horseback riding and other management options.

Costs for Rangeland Improvements

Costs are calculated for the herbicide applications, reseeding, and the opportunity cost of restricted grazing during the restoration program. Treatment costs were obtained from ranchers who participated in the restoration trials and from pest control companies. In the first year, material and application costs per acre to restore rangeland are $15.75 for yellow starthistle using 4 ounces of Transline, $8 for the removal of other weeds using 1 pint of Round-up, and $46 for reseeding with 14 pounds of wheatgrass (Table 15.2).

In addition to the direct costs of restoration, ranchers can no longer graze their animals or rent out land for grazing by others. This is the opportunity cost of rangeland restoration and varies according to the rental rate value.

Lost rental income is the opportunity cost of lost grazing. The first two years the opportunity cost of lost grazing is equal to the rental rate for degraded rangeland, R_0. For grazing fees of $10, $15, and $20 per pair per month for two months, the cost of lost grazing is $4, $6, and $8 per acre per year (Table 15.2). For the second two years, the opportunity cost is equal to the percentage of reduction in grazing while the perennial grasses are establishing times the rental rate of restored land. For the intermountain region, cattle can be grazed at 75 percent of restored capacity in years three and four. The opportunity cost during this time, therefore, is 0.25 times R_1. For grazing fees of $10, $15, and $20 per pair, the cost of lost grazing is $2.45, $2.63, and $2.70 per acre per year (Table 15.2). The annual costs were discounted at a rate of 7 percent per year.

Public Subsidies

If the rangeland restoration program causes a decline in land values, then a landowner will not restore the land. The decline in land values is the amount of the capital loss a landowner would incur if land is restored. The landowner would need to be subsidized at least this amount before a rangeland restoration program is adopted. Therefore, the subsidy level is estimated to be equal to the capital loss incurred by a landowner if they were to restore the land. Total taxpayer costs of the subsidies and biological control program will then be estimated for different levels of adoption in California's intermountain region.

Results

Statewide Biological Control Program

Costs of the Biological Control Program
The total costs of the biological control program with characteristics that would have a high probability of success are $1.2 million a year for 10 years. The current annual cost for the biological control program against yellow starthistle is $670,000. This consists of $500,000 to support USDA and CDFA domestic implementation and quarantine research and $170,000 for overseas research at the USDA European Biological Control Laboratory. An increase of $530,000, to $1.2 million a year, should result in completion of the program in 10 years. In the revised program, funding for foreign exploration is increased to $500,000, while domestic programs are increased slightly to $700,000. Using a discount rate of 7 percent, the present value of the biological control program costs is estimated to be $8.4 million.

Benefits of the Biological Control Program
The lower-bound benefits level is based on the lost interest for a one-month delay in selling property due to the presence of yellow starthistle. Average values for degraded rangeland are $200 to $300 an acre. At an annual interest rate of 7 percent, lost interest would be approximately $1.17 to $1.75 a month. For the analysis the benefit is set at the lower value of $1 an acre.

The upper-bound benefit level is based on the change in land values after yellow starthistle invades new sites. Using land appraisals completed before and after ranches became infested, an infestation of yellow starthistle on higher-quality rangeland causes land values to decline by $50 per acre.

Table 15.2 Summary of annual costs per acre

Fee per animal pair	Year 1				Year 2 (75%/50% retreated)			Year 3 & 4
	Trans-line	Round-Up	Wheat grass	Lost grazing	Trans-line	Round-Up	Lost grazing	Lost grazing
				($)				
10	15.75	8	46	4	11.8/7.88	6/4	4	2.45
15	15.75	8	46	6	11.8/7.88	6/4	6	2.63
20	15.75	8	46	8	11.8/7.88	6/4	8	2.70

Total statewide benefits vary in proportion to the value per acre. For a benefit level of $1 per acre for all acres infested or susceptible to infestation, total benefits to controlling yellow starthistle in California are $40 million. For a benefit level of $1 per acre for infested sites and $50 per acre for land susceptible to yellow starthistle, total benefits are $1.412 billion. Because there are large variations in land quality and measurement of the economic values of controlling yellow starthistle is difficult, these two benefit levels provide a reasonable range in which to assess the benefits of the biological control program.

Cost/Benefit Analysis of the Biological Control Program The cost/benefit analysis estimates the break-even probability of success and compares it to a qualitative assessment of the actual probability to determine if the expected benefits of the biological control program will be greater than the costs. This break-even probability of success depends on the level of benefits per acre (Table 15.3). When the benefit level is $1 for all acres, the biological control program needs to have a 21 percent probability of success for the expected benefits to equal the costs.

Experts familiar with biological control of yellow starthistle estimate that once completed, the biological control program has a high probability of success. If that probability is higher than 21 percent, then the expected benefits would be greater than the costs. If it is uncertain whether a high probability of success is greater than 21 percent, then it would be useful to gather more detailed information on specific regions infested with yellow starthistle to more precisely estimate the benefits of eradicating this invasive plant.

As the benefits per acre increase, the probability at which the expected benefits equal costs

Table 15.3 Cost/benefit analysis for the public biological control program[a]

Benefit per acre	Total benefits	Break-even probability
($)	($ million)	(%)
1 for all	40	21.0
1 for infested	1,412	0.60
50 for susceptible		

[a]For a present value of program costs of $8.4 million.

decreases. When the benefits per acre reach the upper-bound value of $50 per susceptible acre, the break-even probability is 0.60 percent (Table 15.3). At this probability of success, it would be unnecessary to gather more specific information of the regions infested with yellow starthistle. The expected benefits of the biological control program would be greater than the costs.

Rangeland Restoration in the Intermountain Region

The benefits of rangeland restoration are reflected in annual increases in the rental rate per acre. For yellow starthistle land that rents for $10 per pair per month, restoration would increase rental rates by $3 per acre annually (Table 15.4). Land that rents for $15 per pair per month would have an increase in the rental rate of $4.50 an acre, and when the fee is $20 per pair, the increase in rental rates is $6 per acre.

Costs are aggregated over the four-year project period. When yellow starthistle is controlled with chemical herbicides and 50 percent of the land is retreated, the total discounted cost per acre to restore land is $86 when the increase in rental rates is $3 per acre (Table 15.5). Restoration costs are $91 when the rate is $4.50 per acre and $96 for a rate of $6 per acre. When 75 percent of the land is retreated, total discounted costs increase by $5 at each rental rate. The increase in costs as the rental rate increases is due to the opportunity costs. As the rental rate increases, opportunity costs increase.

A successful biological control program lowers the costs to restore yellow starthistle–infested land. When 50 percent of the land is retreated, the total discounted costs will fall by $22 an acre. For a rental rate of $3 an acre, costs fall from $86 an acre for land treated with chemicals to $64 an acre with biological controls. Costs fall from $91 to $69 an acre when the rental rate is $4.50 and from $96 dollars to $74 dollars when the rate is $6 per acre. When 75 percent of the land is retreated, total discounted costs will fall by $25 an acre. On average, restoration costs when the biological control program is successful are 25 percent lower than when chemical controls are used to treat yellow starthistle.

When the private costs and benefits are compared, under every scenario the costs of restora-

Table 15.4 Benefits of rangeland restoration

Acres	Pairs	Months grazing	Cost per month per pair	Total cost per year	Rental rate per acre	Increase in rental rates following restoration
					($)	
Infested land						
100	20	2.0	10	400	4.00	
100	20	2.0	15	600	6.00	
100	20	2.0	20	800	8.00	
Restored land						
100	20	3.5	10	700	7.00	3.00
100	20	3.5	15	1,050	10.50	4.50
100	20	3.5	20	1,400	14.00	6.00

tion are greater than the benefits, and the change in land values is negative (Table 15.5). The decline in land values is greatest when chemical controls are used to eradicate yellow starthistle and for a rental rate of $3. When the benefits of restoration are higher (as reflected in higher rental rates), the decline in land values is not as great; however, the change in land values is still negative. While a successful biological control program lowers the costs to restore land, it is not sufficient to result in a net increase in land values (Table 15.5). Consequently, landowners do not have sufficient private incentives to restore rangeland in California's intermountain region.

Government Subsidies

Public subsidies would be needed to provide landowners incentives to restore rangeland to obtain the public benefits of increased water availability and land that is more resistant to invasion by weedy plants. The amount of subsidy per acre is equal to the decline in land values. This subsidy level varies, depending on the rental rate and type of yellow starthistle control. As rental rates increase, the subsidy level decreases. For example, when chemical controls are used, 50 percent of the land is retreated and the rental rate is $3 an acre, the subsidy level is $51 an acre. When the rental rate increases to $6 an acre, the subsidy level is only $26 an acre, almost half. If the biological control program is successful, when 50 percent of the land is retreated and the rental rate is $3 an acre, the subsidy level is only $29 an acre. The biological control program decreased the level of subsidies needed to encourage ranchers to adopt the restoration program by 42 percent, a significant decline in taxpayer costs.

Of interest is how a successful biological control program influences the total costs to

Table 15.5 Change in land values for a restoration program[a]

Yellow starthistle control method	Increase in rental rate	Percent of land retreated	Total treatment costs	Change in land values
	($)			*($)*
Chemical	3.00	50	86	−51
	4.50	50	91	−38
	6.00	50	96	−26
	3.00	75	91	−56
	4.50	75	96	−44
	6.00	75	101	−31
Biological	3.00	50	64	−29
	4.50	50	69	−16
	6.00	50	74	−4
	3.00	75	66	−31
	4.50	75	71	−18
	6.00	75	76	−6

[a]Assumes a 7 percent interest rate.

taxpayers to eradicate yellow starthistle and re-store rangeland. How much the biological control program reduces costs depends on the number of acres restored. The more acreage that is restored, the greater the cost savings. For simplicity, the analysis will focus on the case where 50 percent of land is retreated and the increase in rental rate is $4.50 an acre.

When chemical herbicides are used to eradicate yellow starthistle, subsidy levels are $38 per acre and $16 per acre when the biological control program is successful. If 10,000 acres are restored, total taxpayer costs are $380,000 with chemical controls and $11.16 million with biocontrol (Table 15.6). Total taxpayer costs to subsidize rangeland restoration are less when chemical controls are used to manage yellow starthistle. When the number of acres restored increases to 200,000 acres, total taxpayer costs with chemical methods are $7.6 million, and with biological controls they are $14.2 million. Subsidies for rangeland improvement are again less when chemical controls are used in place of biological controls to manage yellow starthistle (Table 15.6).

When the number of acres restored is 500,000, the savings in subsidies paid due to a successful biological control program make the total costs for the program and subsidies just equal to the subsidies paid when chemical methods are used.

For restoration of land in excess of 500,000 acres, the taxpayer costs when the biological control is successful are less than the costs to subsidize restoration using chemicals to remove the yellow starthistle. As the number of acres increases, the cost savings increase. When 1 million acres are restored, the cost savings due to decreases in subsidies needed by landowners

to restore rangeland are sufficient to pay for the biological control program. The 1 million acres is just under the total number of acres estimated to be infested with yellow starthistle in California's intermountain region.

The remaining question then is what are the total public benefits to California of removing yellow starthistle and restoring intermountain watersheds. While no value has been determined for improved watersheds, intermountain watersheds are a primary source of water supplies and hydroelectric power in California. Therefore, increases in water availability due to the absence of yellow starthistle may have significant benefits.

General Discussion and Implications

In the past 10 years a number of effective control strategies have been developed for yellow starthistle management. In the future, biological control strategies may continue to improve the management outlook, especially with the introduction of new, more successful, agents. Currently, however, the decision on whether to manage yellow starthistle or other invasive weed species and the choice of control strategy will depend on the economic feasibility of implementing a long-term management approach. This decision will consider the cost of control methods, value of the land, and potential recovery in forage production or land use. A successful management program will not only require the control of yellow starthistle, but also that the desired land-use goals and objective be achieved. In some cases, restoration of severely degraded grasslands may necessitate revegeta-

Table 15.6 Taxpayer costs for subsidies and the biological control program

Acres restored	Subsidies		Total costs		Difference between biological and chemical
	Chemical ($38/acre)	Biological ($16/acre)	Chemical[a]	Biological[b]	
(1,000s)				*($1,000s)*	
10	380	160	380	11,160	−10,780
200	7,600	3,200	7,600	14,200	−6,600
500	19,000	8,000	19,000	19,000	0
1,000	38,000	16,000	38,000	27,000	11,000

[a]Subsidy costs.
[b]Subsidy costs plus the cost of the public biological control program.

tion with desirable species. These efforts can greatly increase costs. To be cost effective, a land manager's decision to manage yellow starthistle and restore degraded grasslands may require financial support beyond the capability of private landowners or agencies.

References

Clark, J. Stephen, K.K. Klein, and Shelley J. Thompson. 1993. "Are Subsidies Capitalized into Land Values? Some Time Series Evidence from Saskatchewan." *Canadian Journal of Agricultural Economics.* 40:155–168.

Gerlach, Jr., J.D. 1997. "How the West Was Lost: Reconstructing the Invasion Dynamics of Yellow Starthistle and Other Plant Invaders of Western Rangelands and Natural Areas." *Proceeding, California Exotic Pest Plant Council Symposium.* 3:67-72.

Gerlach, J.D., A. Dyer, and K.J. Rice. 1998. "Grassland and Foothill Woodland Ecosystems of the Central Valley." *Fremontia.* 26:39–43.

Jetter, Karen. 1998. "Estimating Household Willingness to Pay for Urban Environmental Amenities From a Combined Contingent Valuation/Contingent Ranking Survey." Unpublished Ph.D. Dissertation, Department of Agricultural and Resource Economics, University of California, Davis. 175 pgs.

Larson, L.L., and R.L. Sheley. 1994. "Ecological Relationships Between Yellow Starthistle and Cheatgrass." *Ecology and Management of Annual Rangeland.* pp. 92–94.

Maddox, D.M., and A. Mayfield. 1985. "Yellow starthistle infestations are on the increase." *California Agriculture* 39(11/12):10–12.

Pitcairn, M.J., D.M. Woods, D.B. Joley, and V. Popescu. 2002. "Seven-year population buildup and combined impact of biological control insects on yellow starthistle." In D.M. Woods, Ed., *Biological Control Program Annual Summary, 2001.* Sacramento: California Department of Food and Agriculture, Plant Health and Pest Prevention Services. pp. 57–59.

Plantinga, Andrew J., and Douglas J. Miller. 2001. "Agricultural Land Values and the Value of Rights to Future Land Development." *Land Economics.* 77:56-67.

Roché, Jr., B.F. 1992. "Achene DISPERSal in Yellow Starthistle (*Centaurea solstitialis* L.)." *Northwest Science.* 66:62–65.

Sheley, R., and L. Larson. 1992. "Is Yellow Starthistle Replacing Cheatgrass?" *Knapweed.* 6(4):3.

Glossary of Terms and Acronyms

Agreement on Agriculture: Part of the Uruguay Round Agreement covering three major areas related to agriculture: market access, export subsidies, and internal support.

AHEPSP: California Department of Food and Agriculture and USDA, jointly. Animal Health Emergency Preparedness Strategic Planning process initiated in 1996.

Animal Health Protection Act: Passed as part of 2002 Farm Bill, Title X, Subtitles E and F; signed into law on 13 May 2002 by President Bush.

APHIS: Animal and Plant Health Inspection Service of U.S. Department of Agriculture (USDA), formed in 1972.

ARS: Agricultural Research Service of USDA.

Biological control: Same as biocontrol. Pest control strategy making use of living natural enemies, antagonists, or competitors, and other self-replicating biotic entities (FAO Code and PPA). *Classical biological control* refers to the intentional introduction of an exotic biocontrol agent for long-term pest control.

Biological pesticide (biopesticide): A term generally applied to a pathogen or biologically derived material, formulated and applied in a manner similar to a chemical pesticide for short-term pest control.

Black list: Also known as the *dirty list*. Listed organisms (plant and animal) may not be imported because they have been proven harmful. In contrast, a white list indicates those species that are allowed because they have been proven harmless.

BLM: Bureau of Land Management of the U.S. Department of the Interior.

BSE: Bovine spongiform encephalopathy, also knows as mad cow disease. A progressive, degenerative, fatal brain disease of cattle thought to be associated with prions.

CAHFS: California Animal Health and Food Safety Laboratory System. Laboratories are located in Davis, Turlock, Fresno, Tulare, and San Bernardino, California.

Cairns group: An alliance of nations advocating agricultural trade liberalization (members include Argentina, Australia, Brazil, Canada, Chile, Colombia, Fiji, Hungary, Indonesia, Malaysia, New Zealand, Paraguay, the Philippines, Thailand).

Cal-EPA: California Environmental Protection Agency.

CDC: U.S. Centers for Disease Control and Prevention. Human health is its major emphasis (including diseases vectored or transferred from animals).

CDFA: California Department of Food and Agriculture.

CEFTA: Central European Free Trade Agreement.

CFR: Code of Federal Regulations (CFR). The codified rules of U.S. executive departments and agencies that have been published in the *Federal Register*.

CITES: World Convention on International Trade in Endangered Species of Wild Flora and Fauna.

Codex: United Nations FAO/WHO Codex Alimentarius Commission. Deals with food safety and residue standards.

Consumer welfare: Net benefits to consumers from the purchase and consumption or use of a good or service. Commonly measured by the difference between the willingness to pay and the price actually paid. Consumer welfare often includes welfare of intermediate handlers of a good or service. Changes in consumer welfare are a part of the net welfare effect of a policy change.

CVDLS: California Veterinary Diagnostic Laboratory System, recently renamed California Animal Health and Food Safety Laboratory (CAHFS).

Depopulate: Purposeful destruction of infected or disease-exposed animals.

Developing countries: Countries with relatively low incomes. There is no strict criterion and listing may vary by source or purpose.

DFG: California Department of Fish and Game, Resources Agency.

DOC: U.S. Department of Commerce.

DOI: U.S. Department of Interior.

DPR: California Department of Pesticide Regulation, a department of the California Environmental Protection Agency.

DSB: Dispute Settlement Body. The General Council of the WTO, composed of representatives of all member countries, convenes as the Dispute Settlement Body to administer rules and procedures agreed to in various agreements.

EFTA: European Free Trade Association.

Elasticity (e.g., of demand or supply): Used in economics to designate a ratio of percentage changes. For example the own-price elasticity of demand denotes the percentage increase in the quantity purchased of a good relative to the percentage change in the price of that good.

Embargo: Prohibition on departure or entrance of shipments into or out of a particular country or region.

Endemic: Describes a disease that is constantly present within a population, usually at low incidence.

Enquiry point: Sanitary and Phytosanitary Agreement of the 1994 Uruguay Round Agreements requires that each country designate an enquiry point office to receive and respond to any requests for information regarding that country's sanitary and phytosanitary measures.

EPA: U.S. Environmental Protection Agency.

Epidemiology: The study of epidemics. Involves the study of diseases and pests within a population.

Equivalency: The principle that recognizes that an equivalent level of protection for public, animal and plant health can be provided through the different (alternative), but scientifically justifiable requirements of the different exporting nations.

ERS: Economic Research Service of USDA.

EU: The European Union created in 1993 with the implementation of the Maastricht Treaty on European Union (1991). In 2002 there are 15 member countries: Belgium, Germany, France, Italy, Luxembourg, the Netherlands, Denmark, Ireland, the United Kingdom, Greece, Spain, Portugal, Austria, Finland, and Sweden.

EURO: Common currency of 12 of 15 members of the European Union. Launched in 1999.

Exotic pest or disease: The same as non-indigenous, introduced, non-native, alien, invasive, foreign, immigrant, transboundary pest or disease. Intentionally or inadvertently introduced by humans, other species or natural forces.

Externality: Typically used to refer to a cost or benefit of an action that is not borne by the actor. For example, an external cost of production may include costs of water pollution that occurs as a result of production but that is not borne by the producer. Such externalities often arise when property rights are not fully developed for some resource such as groundwater quality or when transaction costs of imposing full costs on actors are particularly high. Positive externalities refer to external benefits. Negative externalities refer to external costs.

FAD: Literally, a foreign animal disease. An exotic or foreign transmissible animal disease believed to be absent from the United States or its territories that has potential significant health or economic impact.

FADD: Foreign animal disease diagnostician.

FADDL: Foreign animal disease diagnostic laboratory.

FAO: Food and Agricultural Organization of the United Nations.

FAS: Foreign Agricultural Service, USDA.

Fast track authority: With reference to international trade agreements, Congress agrees to vote yes or no on agreements negotiated by the president, rather than make amendments after negotiations are concluded, now known as Trade Promotion Authority (TPA).

FDA: Food and Drug Administration, U.S. Department of Health and Human Services.

Federal Register: (FR) Published daily, amends Code of Federal Regulations (CFR) which contains rules promulgated by executive departments and agencies of the U.S. government.

Fomites: Objects that mechanically transfer infectious organisms from one individual to another, for example undisinfected shoes.

FSIS: Food Safety and Inspection Service, USDA.

FWS: Fish and Wildlife Service, U.S. Department of the Interior.

GATT: General Agreement on Tariffs and Trade, established in 1947 to increase international trade by reducing tariffs and other trade barriers, revised in 1994. Beginning 1 January 1995, the GATT is administered by the World Trade Organization (WTO).

GATT 1994: The Uruguay Round of Multilateral Trade Negotiations signed in Marrakech on 15 April 1994 which established the (WTO). GATT 1994 constitutes an integral part of the GATT. GATT 1994 includes an agreement on agriculture and a sanitary and phytosanitary agreement among many others.

Genotype: The genetic makeup of a particular organism, particular alleles.

GMO: Genetically modified organism; same as living modified organism. Genetic material has been modified through modern biotechnologies.

HACCP: Hazard analysis critical control points. Used in food processing to improve food safety.

Harmonization: "The establishment, recognition and application of common sanitary and phytosanitary measures by different Members" (Annex A. Definitions. The WTO Agreement on the Application of Sanitary and Phytosanitary Measures; effective January 1995). The SPS Agreement provides incentives to governments to harmonize their measures for food safety based on those of the FAO/WHO Codex Alimentarius Commission; for animal health, on the Office International des Épizooties; and for plant health on the FAO International Plant Protection Convention.

HHS: U.S. Department of Health and Human Services; includes CDC, OPHS, and FDA.

ICPM: IPPC Interim Commission on Phytosanitary Measures. (The New Revised Text of the International Plant Protection Convention approved by FAO in November 1997 provides for a Commission on Phytosanitary Measures [Article XI] to serve as new governing body.)

IEA: Investigative and Enforcement Services, USDA-APHIS.

IICA: Inter-American Institute for Cooperation on Agriculture.

Import barriers: Regulations imposed by governments to restrict the quantity or value of a good that may enter that country. Tariffs, embargoes, import quotas, and sanitary and phytosanitary restrictions are examples of such barriers.

International trade barriers: See *import barriers*.

Invasive species: Nonindigenous species that are especially successful at establishing viable populations in their new location.

IPPC: International Plant Protection Convention, a treaty deposited with UN-FAO to protect plant health in the provision of trade. As of April 2002, there are 117 contracting governments.

IS: International Services, USDA-APHIS.

ISPMs: International Standards for Phytosanitary Measures for the protection of plant health.

ITA: International Trade Administration, USDOC.

MERCOSUR: The Southern Common Market, the customs union between Argentina, Brazil, Paraguay, and Uruguay.

NAFTA: North American Free Trade Agreement between Canada, the United States and Mexico.

NAHEMS: National Animal Health Emergency Management System. A state-federal-industry effort to prevent and respond to emergencies.

NAHRS: The voluntary National Animal Health Reporting System. A joint effort of the U.S. Animal Health Associations (USAH), the American Association for Veterinary Laboratory Diagnosticians (AHVLD), and USDA Animal and Plant Health Inspection Service (APHIS). NAHRS replaced the Veterinary Diagnostic Reporting System (VDLRS).

NANPOA: Nonindigenous Aquatic Nuisance Prevention and Control Act of 1990.

NAPPO: North American Plant Protection Organization, a regional plant protection organization constituted under Article VIII of the International Plant Protection Convention (IPPC),

Food and Agricultural Organization of the United Nations. NAPPO is comprised of the national plant protection organizations of Canada, the United States, and Mexico. It was created in 1976 to prevent introduction and spread of quarantine pests of plants in North America.

NASDA: National Association of State Departments of Agriculture.

NASS: National Agricultural Statistics Service of USDA.

NCIE: National Center for Import and Export, USDA-APHIS Veterinary Service. Applications to import animals or animal products may be obtained from NCIE.

NIPP: National invasive plant pest.

NIS: Nonindigenous species (exotic, alien, etc.).

NISA: National Invasive Species Act of 1996; amended the Nonindigenous Aquatic Nuisance Prevention and Control Act. Purpose is to prevent unintentional introduction and dispersal of nonindigenous species into waters of the United States through ballast water management and to coordinate removal of zebra mussel.

Non-tariff trade barriers: Government measures other than tariffs that restrict trade flows. Examples of non-tariff barriers include quantitative restrictions, import licensing, variable levies, import quotas and technical barriers to trade.

Noxious weed: Any plant or plant product that can directly or indirectly injure or cause damage to crops (including nursery stock or plant products), livestock, poultry, or other interests of agriculture, irrigation, navigation, the natural resources of the United States, the public health, or the environment.

NPB: National Plant Board. An organization of all plant pest regulatory agencies of each U.S. state plus Puerto Rico.

NSHA: National Seed Health System of USDA-APHIS. NSHS established by APHIS (18 July 2001 rule). Accredits nongovernmental inspection entities to report preharvest phytosanitary inspection and seed health testing results to APHIS to issue phytosanitary certificates required for seed export.

NTA: New Transatlantic Agenda of the European Union and United States to promote cooperation, partnership, and joint action in areas ranging from trade liberalization to security.

nvCJD: New variant Creutzfeldt Jakob disease, a human spongiform encephalopathy that appears linked to BSE.

NVSL: National Veterinary Services Laboratory, USDA-APHIS.

OES: Governor's Office of Emergency Services, California.

OHA: Outdoor household articles. An acronym used by APHIS.

OIE: Office International des Épizooties (International Office of Epizootics), also known as the World Animal Health Organization.

OPHS: Office of Public Health and Science in U.S. Department of Health and Human Services. May provide surveillance data, scientific analysis, and advice to USDA-APHIS.

OTA: Former U.S. Office of Technology Assessment, U.S. Congress.

Pathogen: Disease-producing microorganism; causative agent of a disease, including a pathogenic virus, mycoplasm, bacterium or fungus.

Pathotype: In plant pathology, members of a pathogenic species that differ in their genetically determined virulence. Frequently, synonymous term for *physiological race*.

Pest: Any organism, plant (e.g. weeds), animal (e.g. insects, nematodes, birds, rodents) or microorganism (e.g. fungus, virus, or prion) that causes damage or economic loss or transmits or produces disease. Classification of a specific organism depends on the context.

Pesticide: A product that is either (1) intended to prevent, destroy, repel, or mitigate a pest; (2)

a plant growth regulator to alter rates of growth or maturation; (3) a plant desiccant; and/or (4) a defoliant.

Phenotype: The physical characteristics, i.e., appearance of an organism as determined by the genotype and environment.

Physiological race or biotype: A term used in microbiology and botany for a race differing from other races in its physiological, biochemical, pathological or cultural properties, rather than in morphology.

Phytosanitary measure: Legislation, regulation, or procedure intended to prevent the introduction and/or spread of plant pests and diseases.

PPA: Federal Plant Protection Act of 2000, which amended or repealed 10 laws pertaining to plant protection, including biological control.

PPO: Plant protection organization.

PPQ: Plant Protection and Quarantine within USDA-APHIS is responsible for inspecting ships, planes and their cargo, passengers, and luggage arriving from foreign countries. Also responsible for domestic programs to manage or eradicate infestations.

Precautionary principle: Used by the European Union and others in the context of regulation of risks. In the context of exotic pests and SPS rules, it refers to the idea that when scientific evidence remains uncertain and difficult to quantify, and when costs of relaxing restrictions may be irreversible, regulations should be restrictive, even though the net benefits cannot be demonstrated by balancing expected costs and benefits using the best evidence available at the time.

Preclearance programs: Preclearance phytosanitary inspections, treatments and/or other mitigation measures conducted in the country of origin, performed under the direct supervision of qualified APHIS personnel.

Producer welfare: The net benefit to producers from the production or sale of a good or service. Typically measured by the difference between market revenue and total cost. The change in producer welfare from a policy or market change is part of the net change in total welfare.

Program diseases: Animal diseases that are the object of cooperative federal-state eradication programs.

Public good: Used in economics to refer to goods or services for which it is particularly costly to exclude consumption benefits from those who have not paid for the good (referred to as *excludability*) and for which consumption by one user does not hamper consumption by another user (referred to as *nonrivalry*). For such goods, such as certain exotic pest exclusion services, provision of the good or service to an additional beneficiary entails no additional costs, and once the good is available, benefits by one party do not reduce the benefit by another party.

Quarantine: Restraints because of health and safety concerns; refers either to the (1) prohibition of movement from a particularly defined region; (2) period of time during which plants or animals or products are held, observed, or tested prior to treatment, destruction, return, or release; or (3) law or regulation prohibiting entrance or allowing entrance contingent upon meeting requirements. Federal quarantines are either foreign, territorial, or domestic; state quarantines are exterior or interior.

Quarantine 37: 7 CFR 319.37. United States quarantine regulation pertaining to entrance of nursery stock and other plant propagative materials from foreign countries.

Quarantine 56: 7 CFR 319.56 et seq. United States quarantine regulation pertaining to entrance from various countries of fruits and vegetables for food.

REDEO: Regional Emergency Animal Disease Eradication Organization infrastructure of USDA-VS.

Regional trade agreements: Agreements among nations made generally to reduce trade barriers among the members. The North American Free Trade Agreement (NAFTA) is an example.

Regionalization: A WTO principle that, when scientifically justified, regions within nations or states be recognized as pest free or disease free for purposes of trade.

Regulated nonquarantine pest: In reference to plant pests, as defined by NAPPO in 1998, "A nonquarantine pest whose presence in plants for planting affects the intended use of those plants with an economically unacceptable impact and which is therefore regulated within the territory of the importing contracting party."

RPL: Regulated pest list posted on the Internet by USDA-APHIS. Largely derived from pests identified in Title 7, Code of Federal Regulations, Parts 300–399. Because of the dynamic nature of pest status, does not include all pests for which APHIS might take action.

RPPOs: Regional Plant Protection Organizations of IPPC: Asian and Pacific Plant Protection Association (APPPC, established 1956); Caribbean Plant Protection Commission (CPPC, established 1967); Comite Regional de Sanidad Vegetal Para el Cono Sur (COSAVE, established 1980); and North American Plant Protection Association (NAPPO, established 1976).

Serotype or serovar: Terms used to identify subspecies of a pathogen.

Smuggling: Clandestine importation.

SPS Agreement: Agreement on the Application of Sanitary and Phytosanitary Measures. An international agreement of the GATT concerning application of food safety and animal and plant health regulations. Sanitary pertains to human and animal health; phytosanitary pertains to plant health. Contained in the Final Act of the Uruguay Round of Multilateral Trade Negotiations signed in Marrakech on 15 April 1994.

Stamping-out policy: A policy of eradication.

Subsidy: Typically used to refer to government programs that add to the benefit of producing or consuming some good or service. A negative subsidy is a tax.

Surveillance: For pests and diseases, monitoring or inspection.

Systems approach: As defined in the federal Plant Protection Act in 2000, "a defined set of phytosanitary procedures, at least two of which have an independent effect on mitigating pest risk associated with movement of commodities."

T&E permit: Transportation and exportation permit, wherein routing and other conditions are specified by APHIS-PPQ.

TBT: Technical barriers to trade. TBTs include technical regulations and voluntary standards and procedures except those whose purpose is sanitary and phytosanitary (SPS). In contrast to SPS measures, which must be scientifically-based, countries may base their TBTs on other considerations.

TPA: Trade Promotion Authority. See Fast track authority.

Transparency: The rule-making or decision-making requirement that the process and decision points are made available for review and scrutiny. A fundamental concept of the WTO Sanitary and Phytosanitary Agreement.

TSEs: Transmissible spongiform encephalopathies. Characterized by progressive, degencrative, fatal brain disease. Scrapie in sheep and bovine spongiform encephalopathy (BSE) in cattle are TSEs.

Uruguay Round: See *GATT 1994*. The latest concluded round of multilateral trade negotiations to facilitate international trade.

USC or U.S.C.: The United States Code. The statutes enacted by Congress are codified in the U.S.C.

USDA: United States Department of Agriculture.

USDA actionable plant pests: Those pests for which the USDA has authority to require some type of quarantine action: refuse entry, treat, or destroy.

Vector: In biology, an agent capable of transmitting the causative agent of disease, e.g., mosquitoes and fleas.

Virulence: A measure of the ability of a pathogen to cause disease, whereby the more virulent the pathogen is, the more serious the disease it can cause.

VS: Veterinary Service of USDA-APHIS.

Welfare: The net gain to consumers, producers, taxpayers, and other affected parties from the production and consumption of some good or service or aggregate of goods and services. The welfare change associated with a policy change or other economic event is the sum of the welfare impacts on each individual or group affected.

White list: A list of species allowed for import that have been proved harmless; also referred to as a clean list. In contrast to a black list of prohibited species that have been proved harmful.

WHO: World Health Organization of the United Nations.

WTO: World Trade Organization. Created by the Uruguay Round of multilateral trade negotiations on January 1, 1995. WTO Administers the GATT, facilitates further trade negotiations, and provides a forum for dispute settlement.

Zoonoses: Pathogens that can be transmitted from animals to humans, giving rise to human disease.

Index

Abamectin, 187
Aceria sp., 230
Administrative Procedure Act, 28
Aesthetics, of trees, cost of preserving, 207-213
Aflatoxins, EU regulations, 46
Agreement on Technical Barriers to Trade (TBT), WTO, 46-47, 52
Agriculture
  losses due to nematodes, 107-117
  red imported fire ants infestation costs, 156-161
Alates, 152
Alfalfa, red imported fire ant damage, 158-159
Alfalfa weevil, 64
Almonds, import prohibitions, 23, 29
Amorbia moth, 185
Anaheim disease, 60-61
Animal and Plant Health Inspection Service (APHIS), 5
  animal import centers, 24
  duties of, 21
  evaluation of Uruguay's FMD-free status, 49
  Karnal bunt regulations, 169
  pest eradication programs, 15
  Plant Protection and Quarantine (PPQ), 21
  regulated pest list, 24
  risk assessment policies, 49-50
  transparency, 50
  Veterinary Services, 21
    reportable diseases, 36-39
Animal health
  APHIS quarantine surveillance centers, 24
  federal regulatory agencies, 21
  international regulation of, 21
  introduced diseases, 20
Animal Health Protection Act, 34
Animal products
  contraband, 26
  import restrictions, 23, 24
Animals
  fallen, 75
  and fire ant stings, 160
  importation of, 21, 23, 24
Animal unit, 236
Antibarberry laws, 20

Antibodies, import permits, 24
Antisera, import permits, 24
Ants. *See* Red imported fire ant
APHIS. *See* Animal and Plant Health Inspection Service
Aphtovirus, 85
Appellate Body, WTO, 41-45
Apples, import prohibitions, 23, 29
Apple trees, 204, 206-207
Appraisal techniques, trees, 207-209
Argentina, foot-and-mouth disease status, 49
Arizona, Karnal bunt in, 169, 174
Arizona Wheat Growers Association, 176
Article XX(b), GATT, 39-40, 52
Ash trees, 204
  appraised value of, 210
  ash whitefly damage to, 205-206
    cost of, 207-213
Ash whitefly, 5
  biology of, 203-204
  cost/benefit analysis of control program, 207-213
  intervention strategies, 204-205
  introduction and spread of, 204
  parties affected by, 205-207
Asian ladybird beetle, 63
Asian longhorn beetle, 179
Asiatic citrus canker, 121, 122
Australia, salmon ban, 42-43, 44, 45
Avocado brown mite, 185
Avocado industry
  California, 185, 191, 192-193
  import restrictions, 29
  welfare effects of trade liberalization and pest infestations, 194-200
Avocado pests
  intervention strategies, 191-192
  parties affected by, 192-193
Avocados
  burrowing nematode infestation, 101
  Hass, 185, 186, 188, 191-192, 196-197
  Mexican, 191-192
  mite-resistant cultivars, 188
  pests of, 185, 186, 191
  subspecies of, 185

251

Avocado seed moth, 186, 191
Avocado thrips, 5, 185. *See also* Avocado pests
  biology of, 186-188
  introduction and spread of, 189-190

Bachelor's button, 230
Backyard operations, livestock, 94, 97
Backyards, citrus canker in, 125
Baits, for red imported fire ants, 153, 154
Bananas, nematode pests of, 100-101
Bangasternus orientalis, 229
Barb goatgrass, 229
Barriers, natural, 12
Beauty, of trees, cost of preserving, 207-213
Beef
  BSE-infected, 82-83
  consumption in U.K., 78-79
  export of and FMD, 89
Beef industry
  by-products, 77
  hormone dispute, 42
  U.S., 82-83
Beekeeping, 160
Bees, 225
Benefits
  consumer, 16
  expected, 161-164, 177-179
  external, 10-11
  marginal, 177
  per acre, 234
  pest control measures, 64-66
  producer, 16
Berkley, Reverend M.J., 58
Bifenthrin, 159
Biodiversity, impact of fire ants on, 161
Biosecurity, and exotic pests, 3
Biotechnology, and trade barriers, 46-47
Bioterrorism
  federal agencies, 21
  and USDA import moratorium, 25
Birds
  fire ant predation of, 161
  and Newcastle disease, 50-51
  seed dispersal by, 228
Bison, foot-and-mouth disease in, 85
Blackfly, 204
Black list, 23
Black scale, 64
Blindness, from fire ant stings, 160
Blister mite, 230
Blood, import permits, 24
Blue-green sharpshooter, 61
Blue oak, 227
Boll weevil, eradication of, 14
Bolting, yellow starthistle, 226

Borders
  exotic pest control at, 11-14, 200
  federal control measures, 14-15
  monitoring, 26
Boundaries, political, 13
Bovine-origin materials, and BSE, 75, 76
Bovine spongiform encephalopathy (BSE), 9
  and bovine-origin materials, 75, 76
  cross-species transmission of, 81
  epidemiology of, 72
  in Europe, 72-76
  intervention strategies, 74-76
  in Israel, 74
  in Japan, 74, 81
  reporting, 26
  risk of U.S. outbreak, 79-81, 82-83
  spontaneous, 80
  spread of, 72-76
  in the U.K., 3, 5, 71-79, 83
  U.S. regulations, 76, 79-81, 83
  and variant Creutzfeldt-Jacob disease, 74
Brains, and BSE transmission, 75
BSE. *See* Bovine spongiform encephalopathy
BSE Inquiry Report, The, 71, 77
Bud weevil, 230
Buffalo, as foot-and-mouth disease carriers, 87
Bulbs, import pretreatment, 25
Bumblebees, 225
Bureau of Land Management, 21
Burning, prescribed, 229, 231
Burrowing nematode, 100-101, 104
  and commodities export, 108-109
  eradication costs, 110

Calf-Processing Aid Program, 78
California
  ash whitefly introduction, 204, 205-206
  avocado industry, 185, 191, 192-193
    avocado thrips, 186, 189-190
  citrus industry, 125
    citrus canker eradication/establishment,
      138-143
    protection of, 122
  crop agriculture, 4
  foot-and-mouth disease outbreak cost model,
    89-97
  grape diseases, 55-61
  history of agriculture in, 55
  Mediterranean fruit fly regulations, 51-52
  nematodes and agricultural losses, 107-117
  pest control
    agencies, 22
    statutes, 35
  quarantines
    legislation, 66

plant, 24
red imported fire ant outbreaks, 151, 152-153, 164
regulatory process, 28
reportable diseases list, 25-26
rice blast disease in, 215-216
trade liberalization and pest infestations, welfare effects of, 194-200
tree crops, diseases of, 61-64
water costs, 110
yellow starthistle in, 225-241
California Academy of Sciences, 65-66
California Action Plan for RIFA, 154
California Administrative Procedure Act, 28
California Airport and Maritime Plant Quarantine, Inspection, and Plant Protection Act, 35
California Burrowing Nematode Exterior Quarantine, 109
California Department of Food and Agriculture (CDFA)
    Animal Health Branch, livestock transport, 26
    exotic pest exclusion programs, 15
    nematode quarantine program, 105-106
    Nematology Laboratory, 104-105
    pest control responsibilities, 22
    Plant Health and Pest Prevention Services, pest ratings, 26
    red imported fire ant control program, 153-154
    reportable diseases and conditions list, 26, 36-39
California Food and Agricultural Code, 35
California Regulatory Notice Register, 28
California Rice Industry Association, 218
California State Agricultural Society, 65
California State Board of Horticultural Commissioners, 63
California State Board of Viticultural Commissioners, 59-61, 66
Camels, foot-and-mouth disease in, 85
Canada, salmon dispute, 42-43, 44, 45
Cancrosis B, 122
Carbamate pesticides, 106
Carcass disposal
    BSE-infected, 72-74
    FMD-infected, 94
Caribbean fruit fly, 22
Carosso, Vincent, 58
Carriers, of foot-and-mouth disease, 86-87
Case law, 27
Cattle
    foot-and-mouth disease in, 85, 87
    importation of, 20
    pleuropneumonia in, 20
    zebu, 21
Cell lines, import permits, 24

Centers for Disease Control and Prevention (CDC), 21
Central America, avocado pests of, 190, 200-201
Central nervous tissue, and BSE transmission, 82
Central Veterinary Laboratory (U.K.), 72
Ceraption bassicorne, 230
Certification
    disease-free, 23
    pest-free, 23
    phytosanitary, 25
    voluntary, 30
Chaetorellia succinea, 229, 230
Change in Aesthetic Value (CAV), 209
Cherries, import prohibitions, 23, 29
Chile, Hass avocados from, 192, 193
Chilies, thrips of, 187
Chlorpyrifos, 159
Chlorsulfuron, 229
Chronic wasting disease, of deer and elk, 81
Chrysanthemum white rust eradication program, 109
Citrus
    as ash whitefly host, 204, 206-207
    cottony cushion scale of, 64
    elasticities of supply/demand, 132, 133
    export of, 109
    import prohibitions, 23, 29
    nematode pests of, 100, 101
    red imported fire ant damage, 157
    spreading decline of, 101, 106
    thrips of, 187
    trees, actuarial value of, 131
    U.S. consumption, 125
Citrus blackspot, 26
Citrus canker, 5
    cost of eradication vs. establishment, 134-135
    eradication programs, 121
        commercial grove, 130, 131-132, 136-137
        cost of, 130-146
        effects of, 125-126
        government urban, 135-136
        urban, 130-131
        welfare effects, 132
    eradication zone, 130, 131
    establishment
        costs of, 132-135
        effects of, 126-128
    intervention strategies, 124
    introduction and spread of, 123-124
    parties affected by, 125
    pathotypes of, 121, 122-123
    quarantines for, 123-124
    symptoms of, 121, 122
Citrus canker-A, 121, 122
Citrus Clonal Protection Program, 30

Citrus market
  equilibrium displacement model, 132, 148-149
  welfare effects of citrus canker eradication
      programs, 137-143
Citrus mealybug, 62
Citrus nematode, 99
Citrus whitefly, 204
CLAMP Project, 22, 26
Clean list, 23
Clitostethus arcuatua, 204-205
Clopyralid, 229
Code of Federal Regulation, 28
Codex Alimentarius Commission (Codex), 22
  growth hormones case, 42
  and SPS requirements, 41
Codling moth, 62
  U.S./Japan dispute, 43
Collective action
  pest control programs, 64-66
  in prevention measures, 93
Colman, Norman, 64
Colonies, red imported fire ants, 151-152
Compensation policies
  for BSE, 77-78
  citrus canker eradication, 128-129
  for FMD losses, 94-95
  foot-and-mouth disease losses, 94-95
  rice blast eradication programs, 222
  tree removal, 130, 131-132
Condition factor, tree value appraisal, 207, 208-209
Conditions, reportable, 36-39
Conformity assessment procedures, TBT, 47
Consumers
  benefits to, 16
  and citrus canker
      impact of quarantine, 179
      welfare effects of eradication programs,
        138-143
  deception of, 47
  effects of trade liberalization and pest
      infestations, 193-200
  surplus losses due to rice blast, 217-223
  willingness to pay, 16
Consumption, nonrivalry in, 10-12, 17
Containment, exotic nematode pests, 106
Contraband, agricultural, 26
Control programs, success of, 93
Corn earworm, 62
Cosmetics, and BSE crisis, 75, 76, 77
Cost/benefit analysis
  ash whitefly control program, 207-213
  citrus canker eradication/establishment, 134-135,
      143-146
  fire ant eradication/establishment, 161-164

foot-and-mouth disease outbreak simulation
      model, 89-97
  Karnal bunt quarantine, 174-179
  nematode eradication programs, 110-111, 114-
      117
  rice blast disease, 219-221
  yellow starthistle control programs, 232-241
Costs
  expected, 161-164, 177-179
  external, 10
  future, 234
  opportunity, 238
  pest control measures, 64-66
Cost shock, welfare effects on avocado industry,
      194-200
Cotton, nematode pests of, 100
Cotton boll weevil, 152, 159
Cottony cushion scale, 62, 63-64
Council of Landscape and Tree Appraisers, 207
Crape myrtle, 204
Creutzfeldt-Jacob disease, variant (vCJD), 5, 71
  and BSE, 74
Crimp, strawberry plants, 102
Crop insurance, federal, 131, 136-137
Crop rotation, for nematode control, 107
Crops
  fire ant damage to, 155
  nematicide treatment, 113-114
  nematode eradication programs, cost/benefit
      analysis, 114-117
Cross-fertilization, yellow starthistle, 225
Currants, 62
Customs duties, 13
Cysts, wind spread of, 103

Dagger nematode, 99
Dairy industry, impact of foot-and-mouth disease
      outbreak, 88, 95-97
Dairy products, import permits, 24
Date palms, reniform nematode infested, 104
Davis, Gray, 60
De Barth Shorb, J., 61
Deer
  foot-and-mouth disease in, 3, 9, 85
  spongiform encephalopathies of, 81
Defoliation, by ash whiteflies, 205, 206, 209
Department of Defense, 21
Department of Health and Human Services, 21
Department of Homeland Security, 15, 17
  and bioterrorism, 21
Department of Interior, regulatory agencies, 21
Depopulation
  and carcass disposal, 94
  foot-and-mouth disease model, 90-92

Destructive Insects Act (England), 20
Diazinon, 159
Dibromochloropropane (Nemagon), 104
Dicamba, 229
2,4-D, 229
1,3-Dichloropropene, 106
Dirty list, 23
Discount rate, 235-236
Disease-free areas, WTO/SPS measures, 45-46
Diseases, reportable, 36-39
Dispersal, yellow starthistle seeds, 227-228
Disposal methods, infected carcasses, 94
Dissemination rates, foot-and-mouth disease model, 90-92
Diversion, of Karnal bunt positive wheat, 177
DNA, import permits, 24
Dominican Republic, Hass avocados from, 192
Dowlen, Ethelbert, 61
Ducharte, Pierre, 58

Economics
    agricultural, 9
    exotic pest policy, 9-17
Ecosystems
    impact of yellow starthistle on, 231 232
    preservation benefits, 234
Ecoterrorism, and foot-and-mouth disease, 85
Eggplant, as golden nematode host, 102
Eggs, and Newcastle disease quarantines, 51
Elasticities, of supply/demand
    citrus, 132, 133
    Hass avocado, 196
    rice blast control simulation, 219-221
Electric equipment, red imported fire ant damage to, 151, 154, 158, 159, 160
Elevators, grain, Karnal bunt and, 173, 182-184
Elk, spongiform encephalopathies of, 81
Embargoes, citrus, 126-128, 129
Embryos, import bans on, 76
Encarsia inaron, 204-205
Encarsia partenopia, 203
Endangered species
    fire ant threat to, 161
    risk from yellow starthistle, 231
Endangered Species Act, 35
England
    beef consumption in, 78-79
    BSE in, 3, 5, 71-79, 83
    foot-and-mouth disease outbreak, 3, 87
    pest legislation, 20
    powdery mildew in, 57
Entry, risk of, WTO/SPS measures, 44
Environment
    impact of exotic pests on, 3, 9

and technical regulations, 47
Enzymes, import permits, 24
Equilibrium displacement model
    citrus market, 132, 148-149
    Mexican Hass Avocado Agreement, 194-200
    rice blast control programs, 216-217
Equipment, agricultural
    fire ant damage to, 151, 154, 158-159, 160
    Karnal bunt and, 172-173, 182-184
    quarantine regulations, 160
Equivalency, of SPS measures, 45, 49
Eradication policies, efficiency of, 91
Eradication programs
    avocado pests, 191
    citrus canker, 121, 124
        boundaries, 128
        cost of, 130-146
    exotic pests, 14, 19
    fire ants, cost/benefit analysis, 161-164
    foot-and-mouth disease, 88-89
    inspection costs, 132
    monitoring costs, 132
    nematodes, 106, 109-111
    red imported fire ants, 153-154, 155-156, 161-164
    rice blast, 221-223
    stem rust, 20
    success of, 93
Eradication zone, citrus canker, 130, 131
Establishment
    fire ants, cost/benefit analysis, 161-164
    nematodes, 106-107, 111-114
    red imported fire ants, 156-164
Estrogens, in meat, 42
Europe, spread of BSE in, 72-76
European shoot moth, 22
European Union (EU)
    aflatoxins regulation, 46
    beef hormones case, 42, 43-44
    biotechnology products labeling, 47
    BSE intervention strategies, 75-76
    public opinion and import policy, 16
Eustenopus villosus, 229
Excise taxes, 13
Excludability, for nonpayers, 12-13
Exclusion
    of avocado pests, 191-192
    of citrus canker, 124
    of exotic nematodes, 105-106
    of pests, 12, 13, 19, 23-24
Exotic pest policy
    definition of, 9
    economics of, 9-17
    evaluation of, 16-17

Exotic pest policy (*continued*)
   funding, 17
   and trade liberalization, 193-200
Exotic pests
   eradication of, 14, 19. *See also* Eradication
      programs
   exclusion of, 13, 19, 23-24
   impact on environment, 3, 9
Exports, beef, and foot-and-mouth disease, 88, 89
Externalities, 10
   and pest control measures, 64-66
Extinguish, 156, 158

Farm machinery, quarantine regulations, 160
Fats, 82
Federal Environmental Pesticide Control Act, 35
Federal Insecticide, Fungicide, Rodenticide Act
     Amendments, 35
Federal Noxious Weed Act, 34
Federal Plant Pest Act, 167
Federal Register, 28, 50
   Karnal bunt quarantine protocol, 167
Federal Seed Act, 35
Feedstuffs
   labeling, 82-83
   mammalian protein in, 81
   use of meat and bone meals, 72-74
Fenoxycarb, 154, 159
Field crops, red imported fire ant damage, 158-159
Field workers, red imported fire ant protection for,
   156, 158
Fipronil, 153
Fire ant. *See* Red imported fire ant
Fish
   federal regulatory agencies, 21
   importation of, 21
Fish and Wildlife Service, 21
Flea beetle, 230
Flooding, continuous, 218
Florida
   avocado production in, 185, 192
   citrus canker eradication program, 121
     boundaries, 128
       compensation policies, 128-129
       quarantine for, 179
     citrus canker outbreak, 124
Florida Fruit Tree Pilot Crop Insurance Provisions,
   131
Flower weevil, 230
FMD. *See* Foot-and-mouth disease
Food and Drug Administration (FDA), 21
Food and Standards Agency (U.K.), 75
Foot-and-mouth disease (FMD), 5
   in Britain, 3, 87
   control and eradication policies, 88-89

disease-free status, standards for, 48-49
early diagnosis of, 93-94
epidemiology of, 85-88
outbreak, cost of, 85, 89-97
reporting, 26
routes of entry, 85
trade and, 95-97
Forage
   reducing weeds in, 228
   and yellow starthistle infestations, 231
Forestry Service, 21
Frankliniella, 200
Free rider, 10
Fruit
   blemished, 121, 122
   premature drop, 121, 122
Fruit flies
   Mediterranean, 51-52
   quarantine for, 191
   tephritid, 186
Fruits
   embargoes, 23
   import prohibitions, 23, 24, 29
   Persea mite infestation, 189
   quarantine acts, 24-25
   red imported fire ant infestation costs, 156-157
Fruits and Vegetable Quarantine, and avocado
     imports, 186
Fruit trees, scales of, 62-64
Fuerte avocado, 188, 192
Fumigation
   before export, 27
   for cottony cushion scale, 64
   soil, 110
Fungicide, for rice blast, 218-219

Galendromus helveous, 189
Gall fly, 230
Garbage
   and foot-and-mouth disease transmission, 85, 86
   foreign, 27
General Agreement on Tariffs and Trade (GATT),
   28
   Article XX(b), 39-40, 52
   history of, 39
   SPS agreement, 4, 39-41
Genetic modifications, and trade barriers, 47
Germany, pest legislation, 20
Germination, yellow starthistle seeds, 226
Germplasm, import of, 29-30
Giant salvinia, 26
Glassy-winged sharpshooter, 25, 61
Globalization
   and spread of pests and disease, 3
   of trade, 29

Glyphosate, 229
Goats, foot-and-mouth disease in, 85, 86
Golden nematode, 100, 102-103, 104
    federal quarantine on, 106, 109
    yield losses due to, 114
Golf courses, sting nematode infestation, 100, 105
Goods, willingness to pay, 16
Government subsidies, rangeland restoration, 233, 237, 239-240
Grafting, grape vines, 59
Grapes
    diseases of, 55-61
    import prohibitions, 23, 29
    nematode pests of, 100
    thrips of, 187
Grasses
    competition with yellow starthistle, 227
    sting nematode infestations, 102
Grazing
    fee, 236
    timed, 228-229
Greenhouses, fire ant treatments, 159
Greenhouse thrips, 185
Grison, A.M., 58
Groves, commercial
    citrus canker eradication programs, 130, 131-132
        federal, 136-137
            welfare effects, 137-143
        investment value of, 132
        red imported fire ant infestation costs, 156
Growers
    citrus canker eradication/establishment, welfare effects of, 138-143
    nematode infestation costs, 108
Growth hormone, EU prohibition, 42
Gwen avocado, 188
Gypsy moths, 167

Habitat, federal regulatory agencies, 21
Hairy weevil, 230
Harmonization, and equivalency agreements, 45
Harvard Center for Risk Analysis, 80
Harvesting, and fire ant mounds, 159
Hass Avocado Agreement, 191-192
Hass avocados, 188, 192
    regulation of, 185, 186
    supply and demand elasticities, 196-197
Hawaii, avocado production in, 185, 192
Hawthorn, 204
Hay
    fire ant infested, 159, 160
    yellow starthistle contaminated, 228
Health certificates, for livestock movement, 25
Heat, for waste treatment, 27
Herbicides, 229

Hessian fly, 62
Hilgard, Eugene, 59, 60-61, 66
Homeowners
    citrus canker effects, 125
    and red imported fire ants, 154
    tree removal compensation, 128, 130
Honeybees, European, 225
Honeydew, whitefly-produced, 203, 206
Hormones, import permits, 24
Horses, yellow starthistle poisoning in, 231
Households, urban, costs due to red imported fire ants, 156
Human health
    and BSE transmission, 73, 74-76
    and exotic pests, 3, 9
    federal regulatory agencies, 21
    and technical regulations, 47
Humans
    as foot-and-mouth disease source, 86
    seed dispersal by, 228
    and spread of citrus canker, 123-124
Husmann, George, 59
Hydramethylnon, 154, 159
Hydrocyanic acid, fumigation with, 64

IMPLAN, 91
Importation
    of animal products, 23
    of animals, 21, 23, 24
    arbitrary/unjustifiable, 46
    of avocados, 185-186
    of birds/poultry, 50-51
    of cattle, 20
    European rendering wastes, 81
    of fish, 21
    fruit, 24
    impact of foot-and-mouth disease outbreak on, 95-97
    meat, 42, 49
    of plants, 15, 23
    of ruminants, 81
Import bans
    embryos, 76
    and risk assessment, 16
Imports
    illegal, 26
    permits for, 24-25
Incineration, for waste treatment, 27
Indemnity payments, 95
India, Karnal bunt in, 168
Infestations, widespread, 19
Insect growth regulators, 154
Insecticides
    contact, 153, 154
    for San Jose scale, 63

Insects, for yellow starthistle control, 229-230
Inspection
    certification, 20
    timely, 46
Inspection stations. *See also* Borders
    agricultural, 66
    and RIFA surveillance, 154
Insurance
    citrus canker, 129
    crop, 131-132
    livestock, 95
Interest, lost, 234
Interest rate, annual, 234
International Animal Health Code, 48
International Convention for the Protection of
    Plants, 21
International Monetary Fund, 39
International organizations, pest control, 22-23
International Plant Protection Convention (IPPC),
    and SPS requirements, 41
International Society of Arboculture, 130
International Trade Organization (ITO), 39
Invasion, by pests, 19
Irrigation, and spread of nematodes, 104
Irrigation lines, red imported fire ants in, 151, 153,
    159
Israel, BSE in, 74
Italy, foot-and-mouth disease epidemic in, 87

Japan
    agricultural import bans, 43, 45, 46
    and American beef imports, 96
    BSE in, 74, 81
    nematode policies, 108
Japonica rice, 215
Java sparrow, 23
Jimsonweed, as golden nematode host, 102

Kaffir lime leaves, 124
Karnal bunt
    quarantines, 169-179
        expected cost/benefits, 177-179
        Federal Register protocol, 167
        welfare effects, 174-179
    regulatory history, 168-172
    risk assessment, 5, 172-174, 182-184
Kocide, 133
Koebele, Albert, 64

Labeling
    biotech products, 47
    European Union rules for, 47
    feedstuffs, 82-83
Lacewings, for avocado thrip control, 187
Lacey Act, 23, 34

Ladybird beetle, 63, 64
Lamb Hass avocado, 188
Land, rental rates, 235-240
Landfills, and carcass disposal, 94
Land restoration
    private cost of, 233, 238-239
    publicly subsidized, 233-241
Lands, public, and yellow starthistle infestations,
    231, 232
Land subsidy values, 233
Land values, impact of yellow starthistle infestation,
    234
Large, E.C., 57
"Latent to infectious", 90
Laws
    antibarberry, 20
    pest abatement, 20
Leafhoppers, 61
Leaves, chlorosis of, 205
Lemon, citrus canker of, 122
Léveillé, J., 58
Levies, 13
Lick, James, 62-63
Likelihood, in risk assessment, 43
Lilacs, 204
Lime, Mexican, 122
Lime-sulfur spray, 64
List A diseases, 23, 26, 48
Livestock
    health certificates for, 25
    interstate movement of, 25
    red imported fire ant damage, 155, 160
Livestock industry
    impact of foot-and-mouth disease outbreak,
        95-97
    insurance, 95
Location factor, tree value appraisal, 207, 208
Logs, import permits, 24
Looper, omnivorous, 185
Los Angeles Vineyard Society, 60
Losses
    consequential, 95
    output, 91-92
    trade, 91-92
    yield, due to nematodes, 114
Lumber, import permits, 24

Mad cow disease. *See* Bovine spongiform
    encephalopathy
Madeira, powdery mildew in, 58
Mango, thrips of, 187
Markets
    beef, 95-96
    effects of trade liberalization and pest
        infestations, 193-200

foot-and-mouth disease-free, 95, 96
impact of citrus canker establishment on, 126-128
private, 11
public, 10-11
Marlatt, C.L., 63
Mealybug, common, 64
Meat
  BSE-infected, 71, 72-74
  imports, 49
  inactivation of foot-and-mouth disease virus, 86
  mechanically recovered (MRM), 73
Meat and bone meals (MBM), 72-74
  EU ban on, 75, 76
  U.K. ban on, 75
Meat products, import permits, 24
Mechanically recovered meat (MRM), 73
Mediterranean fruit fly, 51-52
Medusahead, 229
Melons, red imported fire ant damage, 157 158
Memorandum of Understanding, 27
Metam-sodium (Vapam), 110, 113
Methyl bromide, 106
Mexican Hass Avocado Agreement, 185-186
  welfare effects of, 194 200
Mexico
  avocado imports from, 191-192
  avocado pests of, 200
  Karnal bunt in, 168-169
Michoacán, 191
Microorganisms, import permits, 24
Migratory Bird Act, 34-35
Millfeed, and probability of Karnal bunt outbreak, 172-179, 182-184
Milling rates, rice, 218
Minks, spongiform encephalopathies of, 80
Mites
  of avocados, 185
  Persea, 188-189, 190
  predatory, 189
Mitigation measures, systems approach to pest management, 191
Mold, sooty black, 205
Mongoose, 23
Morrill Act, 65
Morse, F.W., 60
Mounds, red imported fire ants, 154, 159
Mowing, for yellow starthistle control, 228

NAFTA, 4
  SPS agreement, 28-29, 39-40, 46
National Agricultural Society, 65
National Environment Policy Act, 35
National Grape Importation and Clean Stock Program, 30

National Invasive Species Act, 34
National Invasive Species Council, 22
National Marine Fisheries Service, 21
National Plant Board, 22
  quarantine guidelines, 167
National Research Support Project 5, 30
National security
  and border surveillance, 15
  and technical regulations, 47
Nature Conservancy, The, 234
Nemacur, 3, 113
Nemagon, 104
Nematicides, 106
  application costs, 111-114
  efficacy of, 110
Nematodes
  biological control of, 107
  biology of, 100-103
  eradication programs, costs, 109-110, 114-117
  establishment, 106-107
  infestation costs, 115-116
  intervention strategies, 105-108, 109
  introduction and spread of, 103-105
  invasion of, 99
  management costs, 110-117
  potential effects of, 107-109
Neohydatothrips burangae, 200
Neoseiulus californicus, 189
Newcastle disease, 50-51
New Mexico, Karnal bunt in, 169
New York, golden nematode restrictions, 106
New Zealand, Hass avocado exports, 193
Nightshades, as golden nematode host, 102
Nonexcludability, 10, 17
  in eradication programs, 14
Nonindigenous Aquatic Nuisance Species Prevention and Control Act, 34
Nonpayers, and pest control regions, 12-13
Nonrivalry, 10-12, 17
  in eradication programs, 14
North American Free Trade Agreement (NAFTA), 4
  SPS agreement, 28-29, 39-40, 46
North American Plant Protection Organization (NAPPO), 23, 30
Nursery stock
  cost of fire ant treatment, 159-160
  nematodes and, 103, 105
  phytosanitary standards, 29, 30-31
  red imported fire ant damage, 157, 159-160
Nuts, red imported fire ant infestation costs, 156-157

Oat cyst nematode, 103
Objectives, legitimate, 47
Offal, 73, 75

Office International des Épizooties (OIE), 21, 22
  foot-and-mouth disease standards, 48
  free with vaccination, 26
  Newcastle disease regulations, 51
  Reportable Diseases List A & B, 23, 26, 48
  risk assessment standards, 50
  and SPS requirements, 41
Office of Administrative Law, 28
Office of Public Health and Science, 21
Oidium, 57-58
OIE. *See* Office International des Épizooties
Oligonychus persea, 188
Orange, sweet, 122
Orchards
  diseases/pests of, 62-64
  red imported fire ant infestation costs, 156
  thrip control in, 187
Organisms, genetically modified, import permits, 25
Organophosphate pesticides, 106
OTM scheme, 77, 78

Paraguay, foot-and-mouth disease-free status, 49
Parasitoids, for whitefly control, 191, 204-205
Pathogens, plant, import permits, 25
Peaches, import prohibitions, 23, 29
Peacock weevil, 230
Pears, import prohibitions, 3, 29
Pear slug, 62
Pear trees, 204
  appraised value of, 210
  ash whitefly damage to, 204, 205-206
    cost of, 207-213
Period, cost/benefit analysis, 161-164
Permits, import, 24-25
Persea mite. *See also* Avocado pests
  biology of, 188-189
  introduction and spread of, 190
Pest, definition of, 9
Pest control. *See also under individual pest*
  advantages of, 65
  biological
    cost/benefits of, 207-210
    objectives of, 203, 233
    strategies for, 19-20
  border measures, 11-14, 200
  collective action, 64-66. *See also under*
      individual pest
  eradication, 14. *See also* Eradication programs
  federal, 22
  international organizations, 20-23
  interstate, 21-22
  regions, 12-13
  regulatory tools, 23-27
  state, 22
Pest-free regions, 49

WTO/SPS measures, 45-46
Pest-free status, 25
Pesticides
  aerial application, 193
  for fire ants, 159
  for nematode control, 106
  and plant export, 27
  for rice blast control, 223
Pest management, systems approach to, 191
Pests
  actionable, 26
  A-rated, 26, 99
  eradication of, 19
  exclusion of, 19
  exotic. *See* Exotic pests
  invasion by, 19
  Q-rated, 26
Pets, and fire ant stings, 160
Phylloxera, 58-59
Phytosanitary standards, international, 29
Picloram, 229
Pierce, Newton B., 61
Pierce's disease, 60-61
Pigs, foot-and-mouth disease in, 85, 86
Pinkerton avocado, 188
Piperidines, 152
Plantain, nematode pests of, 101
Plant-breeding programs, nematode control,
    106-107
Planting stock, certified nematode-free, 105
Plant pests
  import moratorium, 25
  import permits, 24
Plant Protection Act, 34
Plant Quarantine Act, 20, 167
Plant Quarantine Inspection Act (CA), 35
Plants
  contraband, 26
  import of propagative materials, 25
  import permits, 25
  prohibited/restricted, 23
  quarantine acts, 24-25
Plants, container, fire ant treatment, 159
Plants, ornamental, Persea mite infestation, 189
Plate waste, recycling, 81
Pleuropneumonia, 20
Plum curculio, 22
Plum pox virus, 167, 179
Pomegranates, 204, 206
Pork industry, and BSE crisis, 77
Ports of entry, monitoring, 21, 26
Potato, golden nematode infestation, 102-103
Potato cyst nematode, 100
Poultry
  health certificates for, 25

and Newcastle disease, 50-51
Poultry houses, red imported fire ants in, 160
Poultry industry, and BSE crisis, 77
Powdery mildew, 57-58
Predators, for whitefly control, 204-205
Prevention measures, collective action in, 93
Price, effects of trade liberalization and pest
    infestations on, 194-200
Probability
    break-even, 235
    cost/benefit analysis, 161-164
    Karnal bunt outbreak, 172-174, 182-184
    in risk assessment, 43
    transition, 89, 90
Producers
    benefits to, 16
    citrus canker
        insurance, 129
        welfare effects of eradication programs,
            138-143
    levies, 13
    surplus losses due to rice blast, 217-223
    wheat quarantines and, 175, 179
Product standards, mandatory, TBT, 47
Prohibited articles, 29
Prohibitions, 23
Protection
    arbitrary/unjustified, 44-45
    levels of in SPS measures, 44-46
Protein, mammalian, in feedstuffs, 81
Psilloides nr. *chalcomera,* 230
Public goods, 10-11, 17
    and pest control measures, 64-66
Public opinion, and pest regulation, 16
Public subsidies, rangelands restoration, 233, 237,
    239-240
Puccinia jaceae var. *solstitialis,* 230
Purple needlegrass, 227
Purple scale, 62
Pyricularia grisea, 216
Pyriproxyfen, 154

Quadris, 219
Quaintance, A.L., 63
Quarantine 37, 20, 24-25
    prohibited plants, 23
Quarantine 56, 24
Quarantines
    California legislation, 66
    for citrus canker, 123-124
    of farm machinery, 160
    for fruit flies, 191
    history of, 20
    interior, 25
    interstate, 22

for Karnal bunt, 169-179
of live animals, 23
for Mediterranean fruit fly, 52
nematodes, 105-106, 113-114
for Newcastle disease, 51
penalties for violating, 212
plants, guidelines, 167
for red imported fire ants, 25, 153, 155, 159, 160
U.S., cost of, 167
for vesicular diseases, 88, 93-94
for weevils, 191

Railcars, and spread of Karnal bunt, 172-174,
    177-178
Ranching
    red imported fire ants infestation costs, 160
    and yellow starthistle infestations, 231
Random state-transition model, foot-and-mouth
    disease outbreak, 89-97
Rangeland restoration
    intermountain scenario, 235, 238-239
    private cost of, 233, 238-239
    publicly subsidized, 233-241
Rangelands
    cost of improvements, 236-237
    red imported fire ants infestation costs, 160
    restoration of, 225
    valuation model, 235-236
    yellow starthistle infestations, 225, 231
Red-banded whitefly, 185, 190-191
Red imported fire ant (RIFA), 5
    biology of, 151-152
    cost/benefit analysis of eradication/establishment,
        161-164
    damage to animal industries, 160
    eradication costs, 153-156, 161-164
    establishment costs, 156-164
    intervention strategies, 153-154
    introduction and spread of, 152-153
    mounds, 159
    parties affected by, 154-155
    as predators, 151, 152, 159, 160
    quarantines for, 25, 153, 155, 159, 160
Red scale, 62
Reed avocado, 188
Regionalization, 29, 49, 96
Regulations, federal/state, 27
Regulatory Impact Analysis, USDA, 175, 177
Reinfestation, by red imported fire ants, 153
Rendering, for carcass disposal, 94
Rendering plants, cost of BSE crisis, 77
Rendering wastes, import of, 81
Reniform nematode, 100, 101, 104
    eradication costs, 110, 115
    management costs, 113

Rental rates, land, 235-240
Reptiles, fire ant predation of, 161
Research, as a public good, 10
Resistance, plant, for nematode control, 106-107
Restoration, land
  private costs, 233, 238-239
  publicly subsidized, 233-241
Resurgence, of pests, 189
Rice
  blast-resistant cultivars, 219, 223
  import restrictions, 106
Rice blast disease, 5-6
  in California, 215-216
  control program simulation model, 216-217
  damage due to, 217-218
  eradication programs, 221-223
  impact of control methods, 218-219
Rice foliar nematode, 100, 101-102, 105
  dissemination of, 104
  eradication costs, 115
  yield losses due to, 114
Rice industry, 215
Rice straw, disposal of, 215
RIFA. *See* Red imported fire ant
RIFA Science Advisory Panel, 153
Riley, Charles V., 59, 64
Rinderpest, 21
Ripgut brome, 229
Risk, definition of, 43
Risk assessment
  APHIS policies, 49-50
  BSE, 80
  definition of, 43
  EC beef hormones case, 42
  and import bans, 16
  Karnal bunt, 5, 169, 172-174, 182-184
  OIE standards, 50
  in SPS agreements, 40
  U.S., 50
  WTO/SPS, 41-42, 43, 44-45
Rivalry, in consumption of goods, 10
RNA, import permits, 24
Robison, Solon, 65
Root crops, and nematode quarantines, 113
Root growth, yellow starthistle, 226
Rosettes, yellow starthistle, 226
Row crops, red imported fire ant damage, 158-159
Ruminants
  as foot-and-mouth disease carriers, 86-87
  import of, 81
Rusts, for yellow starthistle control, 230
Rye, Karnal bunt of, 168

Sabadilla, 187
Sacramento Valley

rice production in, 215, 218
  spread of yellow starthistle, 227
Safeguards, systems approach to pest management, 191
Salmon, Australian ban on, 42-43, 44, 45
Sanitary and Phytosanitary (SPS) Agreement
  GATT, 4, 39-41
  NAFTA
    provisions of, 39-40
    transparency, 46
  WTO, 9, 11, 15
    Article 2.3, 40
    Article 4, 45, 49
    Article 5.1, 43
    Article 5.5, 44
    Article 5.6, 45-46
    Article 5.7, 43-44
    Article 7, 46
    Article 8, 46
    compliance requirements, 40-46
    implications of, 47-53
    specifications of, 15
    transparency, 46
San Joaquin Valley
  ash whitefly in, 204, 205-206
  citrus industry, 125
  losses due to nematodes, 107-110
San Jose scale, 62-63
Scales, of fruit trees, 62-64
Scirtothrips abditus, 190
Scirtothrips aceri, 190
Scirtothrips citri, 189
Scirtothrips perseae, 186, 200
Scrapie, 72, 73
Scribner, F.W., 61
Seeding, water, 218
Seeds
  import pretreatment, 25
  nematode treatments, 105
  wheat, Karnal bunt and, 172-174, 182-184
  yellow starthistle, 225, 227-228
Seed weevil, 186
Selective Cull Program, 78
Self-insurance, livestock industry, 95
September 11th, import moratoriums due to, 25
Services, willingness to pay, 16
Sharpshooters, 61
Sheep
  foot-and-mouth disease in, 85, 86, 87
  scrapie in, 80
Shock, anaphylactic, due to red imported fire ants, 152
Simulation models
  foot-and-mouth disease outbreak, 89-97
  rice blast control, 216-217, 219-221

Six spotted mite, 185

Slaughter plants, cost of BSE crisis, 77

Smoke tree sharpshooter, 61

Smuggling, 26
  citrus canker and, 124
  and foot-and-mouth disease, 85
  interdiction, 22

Smut fungus, 26

Sod growers, and fire ant quarantine regulations, 159

Sodium hypochlorite, 133

Sodium o-phenylphenate, 133

Soil
  aeration, 107
  fumigation, 110
  moisture reduction by yellow starthistle, 227, 232
  nematode contaminated, 103, 104
  solarization, 107

SOPP, 133

South, red imported fire ants in, 155

South Korea, and American beef imports, 96

Soybean cyst nematode, 103

Species barrier, BSE transmission, 81

Species Classification and Group Assignment handbook, 208

Species factor, tree value appraisal, 207, 208

Specified bovine offal, 75

Specified risk materials (SRMs), 76

Spillover effects, negative, 124

Spinal cord, and BSE transmission, 75

Spinosad, 187

Spongiform encephalopathies, in U.S., 80-81

Spores, 178

Spreading decline, of citrus, 101, 106

SRMs, 76

Stamping-out
  foot-and-mouth disease model, 89-92
  partial, 88
  total, 88

Standards, TBT, 47

Starthistle, yellow. *See* Yellow starthistle

States, state-transition model, 89

Statutes
  federal, exotic pests and diseases, 34-35
  federal/state, 27

Stem rust fungus, 20

Sting nematode, 100, 102, 105
  eradication costs, 110
  management costs, 113

Stings, red imported fire ants, 152, 160

Strawberry, rice foliar nematode infestation, 102, 105

Strawberry nursery stock, export of, 109

Stunning, and spread of BSE, 82

Sugar-beet nematode, 99

Sugar cane borer, 152, 159

Sulfur, to prevent powdery mildew, 58

Summer dwarf, strawberry plants, 102

Sunflower, 158

Supply curve
  effects of trade liberalization on, 193-200
  rice blast control model, 216-217
  shift due to citrus canker establishment, 126-128

Surplus, losses due to rice blast, 217-223

Sweet potatoes, import prohibitions, 23, 29

Swelling, from fire ant stings, 160

Taiwan, foot-and-mouth disease epidemic in, 87

Tallow, 82

Tariffs, 11, 13

Taxes, excise, 13

Taxpayers
  cost to of yellow starthistle control programs, 232-241
  impact of citrus canker quarantine on, 179

Tea, thrips of, 187

Technical regulations, TBT, 47

Technicians, as foot-and-mouth disease vectors, 86, 87

Teliospores, Karnal bunt, 172, 178

Telone II, 106, 110
  costs of, 114

Temik, costs of, 114, 115

Tenth Amendment, and federal/state regulations, 27

Terrorism, and biosecurity, 3

Texas
  fire ant program, 164
  Karnal bunt in, 169

Thrips
  avocado, 5, 185-190
  predator, 187

Tillage, for yellow starthistle control, 228

Tobacco budworm, 152, 159

Tomatoes, nematodes and, 100, 102

Tortoises, fire ant predation of, 161

Traceability, of imported animals, 81

Trade
  impact of foot-and-mouth disease outbreak on, 95-97
  international, standards for, 48
  technical barriers to, 46-47, 52

Trade agreements, 39

Trade barriers, disguised, 39

Trade liberalization, effects on exotic pest infestations, 193-200

Trade restrictions, Karnal bunt, 168-174

Trade shock, welfare effects on avocado industry, 194-200

Transition probabilities, 89, 90

Transline, 229

Transmissible mink encephalopathy (TME), 80
Transparency
  APHIS, 50
  in international trade, 29
  in SPS measures, 46
Transshipment, Mexican avocados, 192
Tree crops
  diseases of, 61-64
  red imported fire ant infestation costs, 156
Tree removal
  commercial, 131
  residential/urban, 124, 125, 130
Trees
  landscape, appraised value of, 207-209
  ornamental, ash whitefly infestation, 204-205
  replacement costs, 208
  street, density of, 209, 210
  urban, ash whitefly damage to, 205-213
Triclopyr, 229
Triticale, Karnal bunt of, 168
Trunk formula method, 207-209
Tucker, E., 58
Turf grass, sting nematode destruction of, 102
Turkey, rice import regulations, 105, 116
Twig dieback, 121, 122

Uncertainty
  in effects of quarantines, 167
  in eradication programs, 161
United Kingdom (U.K.)
  beef consumption, 78-79
  BSE in, 3, 5, 71-79, 83
  foot-and-mouth disease in, 3, 87
  pest legislation, 20
  powdery mildew in, 57
United Nations, Food and Agricultural Organization
    (FAO), International Plant Protection
    Convention (IPPC), 21, 22
United States
  avocado industry, 192-193, 194
  beef industry, 82-83
  BSE outbreak, risk of, 82-83
  citrus canker outbreak, 121
  citrus consumption, 125
  foot-and-mouth disease-free status, 48
  history of pest control, 20
  poultry imports/exports, 51
  quarantine costs, 167
  rice import restrictions, 106
  risk assessment methodology, 50
  spongiform encephalopathies in, 80-81
U.S. Agricultural Society, 65
U.S. Animal Health Association, 22
U.S. Code Title 7, Agriculture, 34
U.S. Code Title 16, Conservation, 34

U.S. Congress, pest regulations, 27-28
U.S. Customs Service, 21
U.S. Department of Agriculture (USDA)
  APHIS. *See* Animal and Plant Health Inspection
      Service
  Food Safety Inspection Service, 21
  foot-and-mouth disease regulations, 48-49
  Forestry Service, 21
  Hass Avocado Agreement, 191-192
  Karnal bunt quarantine policies, 169-179
  Karnal bunt risk assessment, 5
  Plant Germplasm Quarantine Office, 29-30
  plant quarantines, 167
  quarantine diseases, 23
  Regulatory Impact Analysis, 175, 177
  Reportable National Program Diseases, 26
  veterinary permits, 24
U.S. Department of Defense, 21
U.S. government
  agricultural statutes, 34-35
  ban on live cattle from the U.K., 76
  BSE regulations, 76, 79-81, 83
  foot-and-mouth disease eradication policies,
      88-89
  Mediterranean fruit fly import/export protocols,
      52
  pest control agencies, 21-23. *See also under*
      individual agencies
  quarantine acts, 24-25
  SPS trade restrictions, 43
U.S. Postal Service, 21
U.S. Secretary of Agriculture, 21
University of California
  establishment of, 66
  Exotic Pest Center, 164
  phylloxera research, 59
University of California Riverside, ash whitefly
    control program, 203, 205
Urban areas
  ash whitefly
    control programs, 207-213
    problems associated with, 205-207
Urophora sirunaseva, 229
Uruguay, foot-and-mouth disease-free status, 49
Uruguay Round Agreement, 15, 39-40, 46
  SPS Agreement, 28

Vaccination
  and animal imports, 26-27
  for foot-and-mouth disease, 88, 89
  ring, 97
Value, expected, 161
Vapam, 110, 113
Vedalia, 64
Vegetables

embargoes, 23
import permits, 24
quarantine acts, 24-25
red imported fire ant damage, 157-158
Venom, red imported fire ants, 152
Vesicular diseases
quarantine for, 88
reporting, 93-94
Veterinarians, as foot-and-mouth disease vectors, 86, 87
Veterinary Agreement, U.S.-EU, 49
Veterinary infrastructure factor, APHIS, 50
Viala, Pierre, 61
Vineyards
California, 55-61
red imported fire ant infestation costs, 156

Wasp, parasitic, for whitefly control, 203, 204-205, 209, 212
Waste, plate, 81
Water
cost of, 110
reduction by yellow starthistle, 232
Watersheds, improved, 123
Water spinach seed, 26
Weeds
broadleaf, 229
contraband, 26
as golden nematode host, 102
import moratorium, 25
import permits, 24
Persea mite infestation, 189
reducing, 228
thrips for control of, 201
Weevils
quarantines for, 191
for yellow starthistle control, 230
Western Plant Board, 20
Wheat
Karnal bunt of, 168-172
seed, 172
U.S. exports, 174-175
Whiteflies, control of, 204
Whitefly
ash. *See* Ash whitefly
red-banded, 185, 190-191
wooly, 191, 204
White list, 23

White tip disease, 100, 101-102
Wilder, Marshall P., 65
Wildlife
federal regulatory agencies, 21
foot-and-mouth disease and, 3, 9
impact of fire ants on, 161
impact of pests on, 3
Wooly apple aphid, 62
Wooly whitefly, 191, 204
World Bank, 39
World Trade Organization (WTO), 4
Agreement on Technical Barriers to Trade (TBT), 46-47, 52
SPS agreement, 9, 11, 15
Article 2.3, 40
Article 4, 45, 49
Article 5.1, 43
Article 5.5, 44
Article 5.6, 45-46
Article 5.7, 43-44
Article 7, 46
Article 8, 46
compliance requirements, 40-46
implications of, 47-53
specifications of, 15
transparency, 46
WTO. *See* World Trade Organization

Xanthomonas campestris, 121
Xylella fastidiosa, 61

Yellow starthistle, 6
biological control program, 234-235, 237-238
biology and ecology of, 225-227
in California, 225
cost/benefit analysis of control programs, 232-241
intervention strategies, 229-231
introduction and spread of, 227-228
parties affected by, 231-232
rangeland restoration program, 235-237
Yield losses
due to nematodes, 114
due to rice blast, 218
Yucca plants, reniform nematode infested, 104

Zanardini, Giovanni, 58

Lightning Source UK Ltd.
Milton Keynes UK
16 May 2010

154236UK00001B/4/P

9 780813 819662